Computer-Aided
Facilities Planning

INDUSTRIAL ENGINEERING

A Series of Reference Books and Textbooks

Editor

WILBUR MEIER, JR.
Dean, College of Engineering
The Pennsylvania State University
University Park, Pennsylvania

Additional Volumes in Preparation

Computer-Aided Facilities Planning

H. LEE HALES

Management Consultant
Houston, Texas

Marcel Dekker, Inc.   New York and Basel

Library of Congress Cataloging in Publication Data

Hales, H. Lee
    Computer-aided facilities planning.

    (Industrial engineering ; v. 9)
    Includes index.
    1. Factories--Design and construction--Data processing.
2. Plant layout--Data processing. 3. Facility manage-
ment--Data processing. I. Title. II. Series.
TS177.H35 1984   725'.4'02854   84-14249
ISBN 0-8247-7240-7

MARCEL DEKKER, INC.
270 Madison Avenue, New York, New York 10016

Current printing (last digit):
10 9 8 7 6 5 4 3 2 1

PRINTED IN THE UNITED STATES OF AMERICA

Preface

This book is for facilities planners and managers. It is a survey of current practice in both planning and computer aids. The ideas and computer applications contained in this book can help you "work smarter," instead of harder, on future planning projects.

I have assumed that you have some familiarity with computers and that you have seen a personal computer, a large mainframe, and probably a large minicomputer. I have also assumed that you know what a computer program is and does, that you have heard about such things as service bureaus and time-sharing networks, that you have read about or seen a demonstration of computer-aided design.

We will use a broad definition of "facilities" to include: land and its improvements; buildings and other structures; space within buildings; furniture, equipment, and machinery. In discussing industrial facilities, we will include material handling equipment.

We will view facilities planning as one stage in the facilities management cycle. This cycle, depicted in the figure, encompasses the provision or construction of facilities, as well as their ongoing review and management.

Facilities planning is a broad endeavor that cuts across other disciplines of planning, engineering, and design. To keep this book manageable, it has been necessary to limit its scope where these allied disciplines are concerned. We will not cover the use of computers for such engineering tasks as structural analysis, electrical schematics, piping design, civil engineering, and energy simulation. The planning of industrial facilities is sometimes inseparable from process and methods planning. Computer-aided process planning, however, is a subject unto itself, and will not be covered here. I have taken the same position with respect to computer simulation of materials handling and production systems. While simulation is an indispensible tool in certain facilities planning tasks, it is also a field of its own and is already the subject of many books.

The scope of this book, then, is largely confined to space projections, block and detailed layout planning, material flow analysis, plan and elevation drawings--the core activities of most facilities planners. The management of the planning project and of the facility itself are also covered.

You will be a much better planner if you systematize first, and computerize second. For this reason, Part I is about facilities planning. In it I have

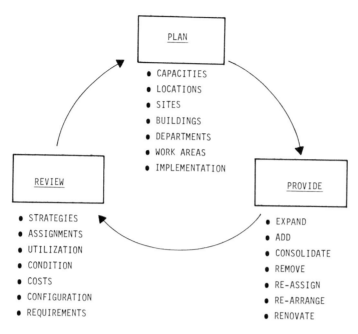

PLAN

- CAPACITIES
- LOCATIONS
- SITES
- BUILDINGS
- DEPARTMENTS
- WORK AREAS
- IMPLEMENTATION

REVIEW

- STRATEGIES
- ASSIGNMENTS
- UTILIZATION
- CONDITION
- COSTS
- CONFIGURATION
- REQUIREMENTS

PROVIDE

- EXPAND
- ADD
- CONSOLIDATE
- REMOVE
- RE-ASSIGN
- RE-ARRANGE
- RENOVATE

The facilities management cycle.

outlined the kinds of systematic planning approaches and techniques that lend themselves well to computer support. My experience suggests that the computer will not make you systematic although it will force you in that direction. The computer will make you better and rapidly more effective if you already have good manual approaches. For this reason, I urge you to read Part I carefully and compare your present practices to the methods it describes.

Part II is about computers--describing what is available, and providing typical examples of computer-aided facilities planning. Part II is a survey. It covers a wide range of applications in layout planning, space estimating, design and drafting, inventories, etc. We will discuss the purpose and benefits of these applications. Inner workings will be covered only as required to understand the strengths and weaknesses of a particular computer aid. We will not look at code, or programming, or the inner workings of computer equipment.

I have limited Part II almost exclusively to commercially available software and systems. With dozens of proven products available there is little need today to develop one's own. Internal development efforts should generally be limited to the construction of large data bases, specialized facilities management systems, or the development of links between separately sourced systems and programs.

Part III is about implementation. Finding computer aids is easy; there are a great many around. Others are easily developed. But getting them installed and achieving good results is often another story. In Part III, I have compiled checklists and pointers from past installation efforts. If observed, you have a good chance of avoiding others' mistakes.

Finally, in Part IV, I have included two appendixes--one on information sources and a second on consultants--and a reference section. I do not claim

completeness for these materials. They have been compiled from my personal
experience and my awareness of what others are doing. Because this book is
a survey of a rapidly changing field, Part IV will help you keep current and
gain additional depth in the areas of your primary interest.

I first outlined this book in December 1980. Had I written it then, it would
be hopelessly obsolete today. Such is the pace of change in computer-aided
facilities planning. The past three years have brought the amazing spread of
personal computers; the electronic spreadsheet, making it easy to "program"
many routine tasks; integrated software combining spreadsheets, graphics,
and word processing; and the 32-bit microprocessor, dropping the price of
computer-aided design systems from $100,000 to as low as $25,000.

During this time we have also seen the formation and growth of new professional groups devoted to facilities planning: the International Facility Management Association, the Society of Property Administrators, the Organization
of Facilities Managers and Planners. Each of these groups has provided a forum
for the discussion and spread of computer aids. New journals have also appeared such as *Computers in Design and Construction*, *Facilities Design and
Management*, and *Corporate Design*, each devoting space to computer aids and
case studies.

I am convinced that facilities planners are just beginning to make effective
use of computers. The price of dedicated devices is now within our reach.
Computers are also much easier to use. And, with these new forums available, the spread of computer aids should be very rapid indeed. While this
book includes current applications, I am sure some exciting new ones will
come along in the future. Appendix A has been included to help you keep up
with these developments yet to come.

H. Lee Hales

Acknowledgments

I had the good fortune to learn facilities planning from Richard Muther, working in his consulting firm for eight years. We also coauthored two books. The first was a brief survey of office space planning. The second was a two-volume, 600-page book, *Systematic Planning of Industrial Facilities*. One does not write a 600-page technical book without drawing on it heavily in subsequent works. So you will find that much of Part I here is written around the Richard Muther techniques as described in our joint efforts, and in his earlier well-known works on facilities planning.

In 1980, I went to the Massachusetts Institute of Technology, spending much of the year studying computers. During this time I met and learned from some true pioneers in the field of computer-aided facilities planning.

From Kreon Cyros, M.I.T.'s director of facilities management systems, I learned about large-scale information systems and data base design. Software developed by Kreon Cyros and his staff is currently used to manage millions of square feet in universities, hospitals, offices, and plants. While living in Boston, I was able to spend time with John Nilsson, an M.I.T. graduate and early pioneer in computer-aided architectural design. As president of Decision Graphics in Southboro, Massachusetts, John Nilsson shared with me his great knowledge of CAD and its evolution in architecture and facilities planning.

My research at M.I.T. led me to Howard Berger, founder of Micro-Vector, Inc., in Armonk, New York. From him I learned about the surprising power of personal and microcomputers when applied to facilities planning. Since the early 1970s, Howard Berger has written dozens of programs. He is one of the most creative and prolific software writers I know. His programs have been used by a variety of professionals to plan millions of square feet.

In New York, I also met and learned from Marshall Graham, senior vice president of Environetics International, one of the first firms to apply computer-aided design to interior layout planning. Marshall Graham and his staff have been practicing computer-aided facilities planning for more than 10 years.

Still another New Yorker, Allan Cytryn, amazed me with his algorithms for multistory and block layout planning. Currently running on an Apple IIe, they are the latest in a series of programs that originally required large mainframe computers.

I have learned the most about algorithms from James M. Moore, president of Moore Productivity Software in Blacksburg, Virginia. As a professor and consultant, Jim Moore has developed encyclopedic knowledge of algorithms, their origins, and inner workings. He is the developer of CORELAP, one of the most widely used algorithms for plant layout. Fortunately for us all, Jim Moore shares his knowledge through prolific writings on computer-aided facilities planning.

Recently I had the good fortune to work with David Arrigoni and David Albert at Arrigoni Computer Graphics. Since the mid 1960s, David Albert has contributed to the development of several popular CAD systems. He is currently president of Vulcan Software, Inc., in Campbell, California. David Arrigoni pioneered the development of low-cost, easily used CAD systems for architectural drafting. Both of these men shared their knowledge and experience with me.

My recent consulting work led me to Dennis and Rose Erickson, president and vice president respectively of BASICOMP, Inc., in Mesa, Arizona. The Ericksons and their staff are a remarkable team. They have spent years perfecting a variety of low-cost aids to interior design and facilities planning.

Two others deserve special mention. First, Harvey Jones, Jr., friend and fellow student at M.I.T., who suffered my ignorance of bits and bytes while we wrote a thesis together on computers. Harvey brought me up to speed with patience and much repetition. He is now senior vice president, marketing, for Daisy Systems Corporation. And finally, my wife, Pamela--the real force behind this book. After two years of hearing about it, she said "Publish or perish!" And she still loved me enough to type the manuscript. No, we don't have a home computer . . . yet.

Contents

Part I

HOW FACILITIES ARE PLANNED

i

1

An Organized Approach to Planning

Levels of Planning Decisions

Facilities plans and decisions are made at several levels. The highest, most abstract decision level is that of capacity planning--providing enough productive capacity of all kinds to meet the needs of the organization. At some point, this capacity must be related to specific quantities and conditions of floorspace, land, buildings and equipment.

From overall capacity decisions, the planner moves "down" to locations--the geographic placement of capacity. At this level, site planning or land use decisions are made, followed by building decisions covering both the interior and the structure or shell.

Next comes the "department level," where decisions involve groups of people and equipment and their day-to-day activities. Once group or department level decisions are set, the planner looks at workplace design. Finally, plans are made to implement the foregoing decisions.

In practice, of course, there is considerable overlap. It is desirable, but not necessary, to proceed in rigorous "top-down" sequence, but the decisions made at any level must be compatible with those at other levels. Figure 1.1 illustrates the levels of planning. Note the slight variation in issues among the different types of facilities.

Facilities Planning Phases

In addition to levels, let us also introduce the concept of facilities planning phases. Each planning project can be said to move from some existing facilities or conditions to some future stage of development or desired conditions. The project is accomplished by proceeding logically through a sequence of planning phases. Each *phase* addresses a different *level* of planning and decision.

The phases overlap in recognition of the need to integrate the decisions made at different levels of planning. The concept of phases is illustrated in Figure 1.2 and defined further in Figure 1.3.

The difference between overall (Phase II) and detailed (Phase III) planning needs to be stressed. Overall planning focuses primarily on form and

LEVEL OF PLANNING	GENERAL ISSUES	SPECIFIC ISSUES		
		INDUSTRIAL SETTINGS	ADMINISTRATIVE SETTINGS	INSTITUTIONAL SETTINGS
CAPACITY	Rate of output. Amount of space. Make or buy. Subcontracting. Own or lease.	Production rates. Number of shifts. Degree of auto-mation. Methods.	Space standards. Types of furnish-ings. Interior concepts.	Number, size and duration of programs and projects. Limits to growth.
LOCATIONS	Number of sites. Geographic locations. Utilities.	Transportation. Labor costs. Zoning. Taxes.	Rental rates. Convenience. Labor costs.	Acquisition of adjoining property for growth. Zoning.
SITES	Access/egress. Traffic patterns. Space allocation. Future growth.	Topography. Ratio of yard to underroof space.	Aesthetics. Solar energy.	Density. Aesthetics.
BUILDINGS	Size, orientat-ion and placement on site. Type of structure. Possible future uses.	Bay size. Clearances. Floor strengths. Utilities. Flow patterns.	Building module. Number of floors. Size of floors. Exterior finish. Aesthetics.	Degree of specialization. Recycling and reuse of existing facilities. Aesthetics.
DEPARTMENTS	Effective use of space. Proper placement of people and activities.	Work flow. Production methods and equipment. Materials hand-ling.	Communications patterns. Report-ing relationships	Communications patterns. Report-ing relationships Location of specialized equipment.
WORK AREAS	Individual workplace productivity.	Methods and standards. Materials handling. Safety. Workplace layout.	Systems and procedures. Furniture specifications. Status. Appearance.	Systems and procedures. Scheduling, placement and control of specialized equipment.
IMPLEMENTATION	Budgets. Sequencing. Timing.	Schedules. Lost production.	Schedules. Disruption.	Schedules. Disruption.

Figure 1.1. Levels of facilities planning.

the arrangement of facilities. It addresses major operating activities and key economic issues. Detailed planning focuses primarily on the position and place-ment of dimensioned objects to realize an overall concept or plan.

Consider the planning of a corporate office building. Deciding to use a precast concrete, multistory structure is a Phase II decision. So too is the de-cision to place Finance next to the Executive Offices on the top floor. Position-ing the receptionist's desk to face the elevator is a detailed (Phase III) decision. So too is the placement of lighting, electrical, and telephone outlets to serve the receptionist's desk.

The Process of Planning Facilities

Putting together our levels and phases, we arrive at the process of planning shown in Figure 1.4. This basic process can be followed for any facility,

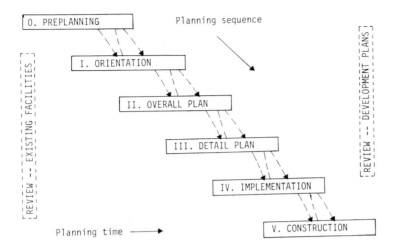

Figure 1.2. Facilities planning phases defined by Richard Muther. Phases I through IV represent pure, facilities planning activities. Phase O represents the on-going interface with business or corporate planning. Phase V is actual construction, renovation or installation. These phases are "framed" by the on going review of existing facilities, future requirements and standing facilities plans.

PHASE	SPIF NAME	ACTION	RESULT/OUTPUT	GENERAL TERMINOLOGY
0	Preplanning	Preplan	Stated Needs of the Business	Business Planning
I	Orientation	Localize	Determined Location and External Considerations	Physical Facilities Planning
II	Overall Plan	Plan the Whole	Overall Plan (Solution in Principle)	
III	Detail Plans	Plan the Parts	Detail Plans (Solution in Detail)	
IV	Implementation Planning	Plan the Action	The Plan for Implementation	
V	Execution of Plans	Act/Do	The Facility Completed, Ready to Operate	Constructing, Rehabilitating, &/or Installing

Figure 1.3. Extended framework of phases summary with planning phases defined. (From Ref. 8, Vol. 1. Copyright 1979, Richard Muther.)

regardless of type. Only the specific techniques and decisions within each phase will vary, depending upon the type of facility--industrial, administrative, institutional, or other.

Phase 0 - PREPLANNING

- Compile Basic Needs--(Company policy, business plans, and general goals)--and forecast non-physical requirements (What do we want?)

- Evaluate Existing Facilities (What do we have now?)

- Forecast Plan for Capacity Requirements--sizing, dimensioning, conceptual plans...(What do we need?)

- Check for Feasibility (Is it economical to go after the need?)

- Develop a Project Plan--Breakdown of phases, steps, responsibilities, and schedule--(How will we plan to go after the need?)

Phase I - ORIENTATION

- Convert the non-facility objectives and existing conditions to physical facility requirements.

- Locate the site, the facility(ies) on the site, or the department in the plant, and identify its external opportunities and constraints.

Phase II - OVERALL PLAN

- Convert the physical requirements into overall plan of physical facilities--that is, plan the overall facility.

- Establish a solution in principle.

Phase III- DETAIL PLANS

- Convert the physical requirements and physical constraints for subdivided areas or components of the overall plan into more detailed facilities.

- Establish solutions in detail--details of major features.

Phase IV - IMPLEMENTATION

- Convert the plans of physical facilities into a program of action; planning the "Do".

- Plan the construction, renovation, and/or installation.

Phase V - CONSTRUCTION, RENOVATION AND/OR INSTALLATION

Follow-up or carrying out of the planning

Figure 1.4. The process of planning. (Courtesy of Richard Muther & Associates, Inc., Kansas City, Missouri.)

The Five Components of a Facility

One final concept is necessary to completely organize our approach. Within each planning phase, it is very useful to distinguish the different parts of the facility that must be planned or designed.

We can do this very easily with Richard Muther's five components. These are:

1. *Layout*: the arrangement of activities, features, and spaces around the relationships that exist between them
2. *Handling*: the methods of moving products, materials, people and equipment between various points in the facility
3. *Communications*: the means of transmitting information between various points in the facility
4. *Utilities*: the conductors and distribution of substances like water, waste, gas, air, and power
5. *Building*: the form, materials, and design of the structure itself.

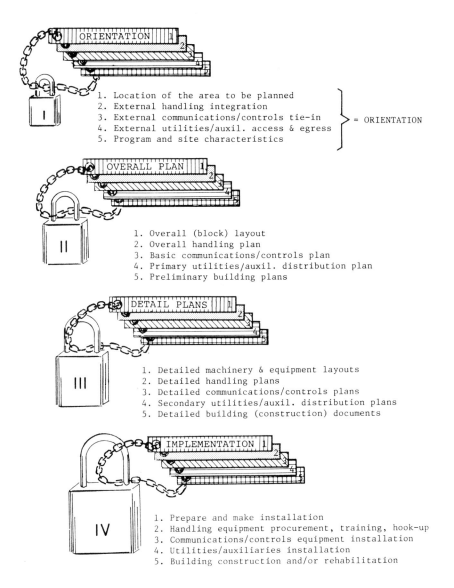

ORIENTATION

1. Location of the area to be planned
2. External handling integration
3. External communications/controls tie-in   } = ORIENTATION
4. External utilities/auxil. access & egress
5. Program and site characteristics

OVERALL PLAN

1. Overall (block) layout
2. Overall handling plan
3. Basic communications/controls plan
4. Primary utilities/auxil. distribution plan
5. Preliminary building plans

DETAIL PLANS

1. Detailed machinery & equipment layouts
2. Detailed handling plans
3. Detailed communications/controls plans
4. Secondary utilities/auxil. distribution plans
5. Detailed building (construction) documents

IMPLEMENTATION

1. Prepare and make installation
2. Handling equipment procurement, training, hook-up
3. Communications/controls equipment installation
4. Utilities/auxiliaries installation
5. Building construction and/or rehabilitation

Figure 1.5. All five components are coordinated and approved (locked) at each planning phase. (From Ref. 8, Vol. 1. Copyright 1979, Richard Muther.)

These components are often planned by different parties in a project team, representing the different professional points of view. By recognizing the need to make "component decisions," we can "lock together" our planning efforts by phase and level, as shown in Figure 1.5.

The Economic Consequences of Facilities Planning

Facilities decisions can have a direct and lasting impact on financial resources and operating efficiency. An expensive plan or design will consume extra cash or incur extra debt during its construction. An inefficient plan or design will consume extra cash over its entire life. A shortsighted plan will often be short-lived. This may require expensive adjustments later that should have been avoided.

It turns out that the economic consequences of facilities planning can be related directly to the phases we introduced earlier. In Figure 1.6, we see two curves representing the consequences of planning. On the one hand, we have *resources invested* which rise rapidly during detailed planning and construction. On the other hand we have *influence on profit and operating efficiency* which is greatest during preplanning and falls off as decisions and designs become firm. The influence curve declines because the level of decision declines in each successive phase.

Capacity decisions (preplanning) have the greatest influence on profit. Too much can be a fatal drag, too little and the market may get away.

Location decisions (orientation) are the next greatest influence. The best site plan, the best building design, can be of little value if the facility is poorly located. On the other hand, the best location may not overcome inadequate or excessive capacity.

Similarly, the best workplace designs may not overcome the inefficiencies of poor departmental placements in the overall plan.

The Key Issues in Facilities Planning

We can summarize our discussion of economic consequences by listing the key issues in facilities planning.

Size of facilities: maximum and minimum practical sizes in terms of operating cost, organizational effectiveness, exposure to risk, etc.

Location of facilities: stay or move; expand existing locations, or start new ones. Where?

Site missions: assignments and allocation of activities among sites. Roles of each site. Focus of each site or facility.

Financial position: lease or buy. Modernize or build new. Over-build or under-build. When?

Operating efficiency and profit: flexibility, adaptability. Proper configuration and arrangement.

Clearly, the issues above are economic in nature and impact. The manner in which they are resolved is important and requires a sound approach.

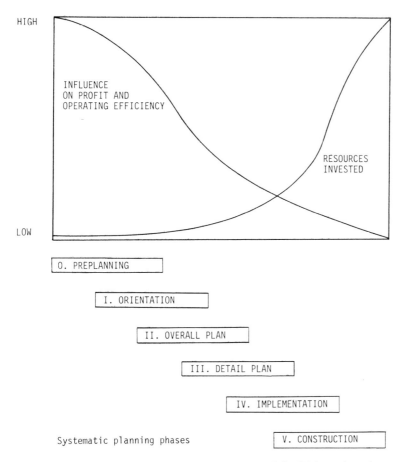

Figure 1.6. The economic consequences of facilities planning.

Organizational Settings

Who resolves key issues? Typically it is a team effort. Members of the team
vary, depending upon the project and type of facility being planned. Figure
1.7 shows some of the many different parties that may be on the team.

In general, we may view facilities planning and management as being internally and externally provided. Most organizations use both sources. Internal
efforts are provided by central (headquarters) and "on-site" (satellite) staffs.
External efforts are provided by specialized consultants, design and engineering firms, and contractors and developers.

Facilities Planning in Large Corporations

Large corporations have rather complex organizational approaches to facilities
planning. These merit special attention.

Corporations involve both line managers and staff specialists in reaching
facilities decisions. The roles played by line and staff are outlined in Figure 1.8,

TYPES OF PLANNERS AND MANAGERS / TYPES OF FACILITIES	INTERNAL GROUPS		OUTSIDE SERVICES		
	ON-SITE STAFF	CENTRAL STAFF	PLANNING CONSULTANTS	ARCHITECT-ENGINEERS	DESIGN-BUILDERS
INDUSTRIAL Plants and Warehouses	Plant engineers. Division or local staff-- Architects, planners.	Construction and real estate professionals. Corporate consultants.	Independent specialists. Large, diversified firms.	Small, local partnerships. International giants.	Industry specialists. Diversified firms.
ADMINISTRATIVE Offices, laboratories, and institutions	Field office managers. Plant engineers. Architects.	Corporate administrative staff. Architects. Planners.	Interior designers. Furniture suppliers.	As above.	Developers.

Figure 1.7. Facilities management: organizational settings.

FACILITIES PLANNING PHASES	CHANGING ROLES OF THE FACILITIES STAFF	LINE MANAGEMENT				
		NEW SITE	EXISTING SITE			
		NEW CONSTRUCTION	MAJOR NEW CONSTRUCTION	MINOR NEW CONSTRUCTION	MODEST CHANGES	MINOR CHANGES
0 PREPLANNING	Business planner	Top management leads or is actively involved.				
I ORIENTATION	Business planner		Significant interest by top management. Operating management leads or is actively involved.			
II OVERALL PLAN	Mfg. planner Process/opns. analyst			Modest interest by top management.		
III DETAIL PLAN	Designer Supervisor of arch./engr.			Significant attention by operating management. First-level supervisors actively involved.		
IV IMPLEMENTATION	Project mgr. Project planner			Minor interest by top management. Moderate interest by operating management. Significant attention by first-level supervisors.		
V CONSTRUCTION	Cnstr. mgr. Project mgr. Project coord.			Chiefly of interest to operating management and first-level supervisors.		

Figure 1.8. The roles of line and staff. (Adapted from Ref. 8, Vol. 2. Copyright 1980, Richard Muther.)

by planning phase and by scope of project. While facilities planning is generally a staff activity, the decisions to adopt or change a plan are typically made
by various levels of line management. This is almost always the case with industrial facilities. The distinction is less clear in administrative settings.

The interest and involvement of top management is naturally greatest when
new sites and major investments are involved. A top manager may even lead
the effort. As a rule, the interest and involvement of top management drops
off rapidly after the early phases of a project.

The interests of middle-level, operating managers and first-level supervisors are similarly, but less significantly, correlated with phases and scope
of project.

Note the different roles played by the "facilities staff" in Figure 1.8, as
a project moves from Preplanning through Construction. Some of these roles
may be externally provided by consultants, architects and contractors, as
suggested earlier in Figure 1.7.

Virtually every large corporation has at least one top staffer in the corporate offices to advise top management on key issues, and to lead or participate with staff specialists in reaching downstream, field-level decisions.

Typical staff rolls for industrial and administrative settings are shown, by
phase, in Figures 1.9 and 1.10. The nature of manufacturing organizations--
geographically decentralized along product lines--often leads to more levels of
staff and decision-making in facilities planning projects.

The typical roles illustrated here presuppose that an outside or external
architect-engineer will be retained for Phase II and III. This work may be
preformed by staff architects in large corporations. It is also pre-supposed
that local, on-site staff ("Plant Engineer" or "Facilities Manager") will plan
and manage the implementation and construction phases of the project.

Shared-Site or Area Staff Groups

It is quite common in large corporations to find two or more profit centers or
business units occupying a single site, or even a single building. Where this
occurs, there is a real need to coordinate and integrate the "claims" of each
upon the facility.

This need is addressed in one of two ways. Most often, the largest business in a facility acts as a landlord for the smaller ones. The facilities decisions of all units are interrelated, so the largest unit ends up doing the longer-
range planning for the smaller ones, insofar as the facilities are concerned.

The other approach is to create a special facilities organization acting as
landlord for all the businesses present--both large and small. This facilities
group then reports to a line manager well up in the organization--high enough
to resolve the inevitable conflicts that occur.

Property Management Divisions

Some of the largest U.S. corporations, especially financial and service firms,
have a subsidiary property management or real estate division which owns the
company facilities and provides planning services to its "tenants." This is
done for a variety of financial and operating reasons, especially with office
buildings, where portions may be leased to outsiders.

CORPORATE STAFF LEVELS	FACILITIES PLANNING PHASES						
	0 PREPLANNING	I ORIENTATION	II OVERALL PLAN	III DETAIL PLAN	IV IMPLEMENTAT'N	CONSTRUCTION	ON-GOING REVIEW
Corporate staff and consultants	May provide expert advice.	May lead site selection project. Review existing corporate locations.	Review group or division plans. May provide expert advice to lower staff levels.	May provide expert advice to lower staff levels. May provide A&E services to lower levels	May provide planning service to lower staff levels.	May provide project management service to lower staff levels.	5 - 10-year needs -- all business units.
Group or division	As above.	Review existing group or division locations.	As above.	As above.	Review.	Review.	3 - 5-year needs -- all units in division.
Business unit, profit center or division. (HQ location)	Define capacity requirements	Review existing business unit locations.	Review and supervise outside A&E services.	Review and supervise outside A&E services.	Review and supervise progress.	Review and supervise progress.	2 - 3-year needs -- all unit locations.
Satellite locations	None.	Assist in site selection.	Develop plans.	Develop plans.	Plan project.	Manage project.	immediate needs -- local site.
Shared-site or area staff	None.	As above.	Integrate unit plans into master plan.	Coordinate. May provide A&E service.	May provide planning service to "tenant" business units.	May provide management service to "tenant" business units.	2 - 3-year needs -- all "tenant" business units.

Figure 1.9. Typical staff roles in planning industrial facilities.

CORPORATE STAFF LEVELS	FACILITIES PLANNING PHASES						
	0 PREPLANNING	I ORIENTATION	II OVERALL PLAN	III DETAIL PLAN	IV IMPLEMENTAT'N	V CONSTRUCTION	ON-GOING REVIEW
Corporate or central staff*	Define capacity requirements	Review existing corporate locations. May lead site selection project.	Review local plans. May provide or supervise A&E services.	May provide or supervise A&E services May provide expert advice.	May provide project planning service to local staff.	May provide project management service to local staff.	3 - 5-year needs -- all sites and activities
Local or on-site staff	None.	Assist in site selection.	Develop plans. Retain and supervise outside design services.	Develop plans. Retain and supervise outside design services.	Plan and schedule installation and con-struction.	Manage installation and con-struction projects.	1 - 3-year needs -- local site and activities

* When major, new facilities are provided, the central staff typically performs all functions of the local or on-site staff.

Figure 1.10. Typical staff roles in planning administrative facilities.

Implications for Computer Aids

Four basic questions emerge from the foregoing discussions of facilities planning. These are:

1. Which decision levels or phases do we want to support with computer aids?
2. Which components of the facility will be of greatest and most frequent concern: layout, handling, communications, utilities, buildings?
3. Which members of the planning team need computer support? Line management, staff, internal planners, external parties, headquarters, field locations--all are potential users but each may have a different need.
4. How do we provide access to potential users, recognizing that "the team" may be geographically dispersed and even international in nature?

We will return to these questions in Part II when we examine specific forms of computer support.

2

Data Collection and Survey Techniques

Key Input Data

Before we can plan, we need information. The following key input data defined by Richard Muther represent a good summary of what to collect.

Products/Personnel: To plan industrial facilities, we must consider the present and likely future characteristics of products and materials-- such things as size, weight, composition, finish, fragility, value, etc. To plan administrative facilities and institutions, we must consider the various personnel or positions involved and their characteristics--such things as skills, attitudes, working hours, privacy requirements, physical needs, etc.

Quantities: We must know present and likely future quantities for each product or material, in the case of industrial plants, and for each position, in the case of offices and institutions. How much will be produced? How many personnel will be involved?

Routings/Process Sequence: To plan industrial facilities, we must know the present and proposed routings or manufacturing process for each product and material. The same type of information is required to plan administrative or institutional facilities, but in this context we generally refer to "systems and procedures." The planner must know how work will be performed.

Supporting Services: Must be considered in every facilities plan. There are two kinds--building support and personnel support. Under the heading of building support, we need input on mechanical and electrical systems, plumbing, ventilation and air conditioning, lighting, auxiliary power, maintenance, waste disposal, pollution control, fire control, etc. Under the heading of personnel support, we need input on food services, break and recreational facilities, parking, credit union, first aid, etc.

Timing/Time-Related Factors: Are often overlooked. In particular, sound planning requires input on working hours and activity cycles. The planner needs to know the extent to which overtime and extra shifts can be used to overcome limited capacity and space. The planner also

needs to know the extent of peaks (and valleys) in the activities to be
housed. Do peaks occur in a regular cycle: weekly, monthly, quarterly, annually, every five years, etc.? The interrelationship of extra
working hours and activity peaks is an important factor when sizing
production equipment and floor areas.

Key input data are usually collected with surveys and interviews and summarized in outline or brief report form. It is essential that data be cataloged,
and retrievable since it forms the basis for many planning assumptions that will
be made later on.

Activity-Areas

Every organization has its departments, cost centers, work centers, or other
similar reporting units. However, it is often impractical to build facilities plans
solely or completely around them. Existing departments or cost centers are
often too large or too small to be used without splitting or combining. Moreover, reporting units do not generally reveal the support areas that must be
considered in planning.

In your author's experience, it is necessary to define a set of activity-
areas at the start of each project, tailored to the needs of the project. These
will be cross-referenced to, but are not necessarily the same as, existing reporting units.

On large projects, there will be one set of activity-areas used to arrive at
and illustrate the overall or conceptual facilities plan (developed in Phase II
of the planning process). Each of these activity-areas will be broken into sub-
areas for detailed planning and design (Phase III).

The planner should never define more than 40-50 activity-areas. Above
this range, finished plans become very difficult to illustrate, interpret and
compare. On small projects, 10-15 activity-areas are usually sufficient to produce a sound plan.

Two practical ways to define activity-areas are illustrated in Figure 2.1.
Once defined, all subsequent project data on flow, communications, space,
equipment, personnel, etc. will be collected and reported by activity-area.

Material Flow

Once activity-areas are defined and routings known, the next step for industrial plants is the collection of quantitative material flow data. The planner
needs to know how much of a given product or material is flowing between each
activity-area. This data can be obtained in several ways:

1. *By relative, judgmental estimates*: in this quick and dirty approach,
   a knowledgeable planner or team sets a value of one (1) or 100 to a
   specific product flow over a specific route. Other route-product flows
   are estimated in relative terms to the base value.
2. *By work sampling or other formal survey*: The planning team, or temporary analysts, or the material handlers themselves, fill out sampling

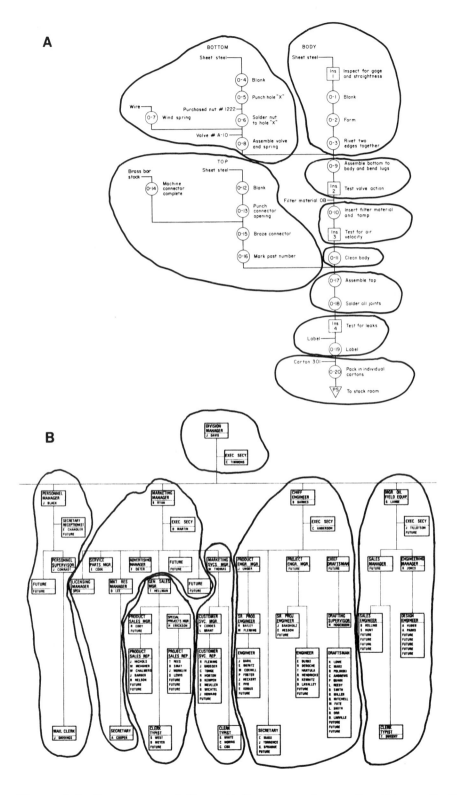

Figure 2.1. Two ways to define activity-areas. A uses the process chart to define related operations. B uses the organizational chart to define related groups. (Courtesy of Richard Muther & Associates, Inc., Kansas City, Missouri.)

forms or structured reports for a set period of time. Several time periods may be used and the results averaged to give a better result.

3. *By extraction from production control reports*: The routings and production quantities tracked in a computer-based production control system can sometimes be used to give a partial picture of flow between activity-areas.

4. *By automatic scanning*: In some cases, flow can be determined by attaching scannable or bar-coded labels to products, materials, containers, or vehicles. Scanners along each route and at the entrance and exit of each activity-area can be used to record material flow. The results can be interpreted and reported by computer.

The choice of survey technique depends upon the level of detail required and the nature of the planning project. Once collected, flow data is usually summarized in one of the two formats illustrated in Figure 2.2.

Communications

Administrative and institutional facilities are planned primarily around communications patterns. A common starting point is the communications survey. In its extreme form, an open-ended questionnaire is given to each employee--or possibly to a large sample of employees.

Each respondent is asked to name the fellow workers with whom communications most frequently occur. For each person named, additional questions must be answered about the nature, purpose and frequency of communications: face-to-face, written, phone? supervision, research, assistance ? hourly, daily, weekly?

Each response is weighted and fed into a computer program for scoring. Most programs use mathematical scoring techniques based on cluster analysis, factor analysis, or hierarchial decomposition. The result is a series of lists of closely related workers who should be placed together in the plan. A numerical score shows in relative terms how closely specific individuals and groups are related to one another. The Herman Miller company has published the very refined communications survey shown in Figure 2.3. Many consultants, architects, and interior designers use similar questionnaires.

Communications surveys are sometimes used to define activity-areas. This is an alternative to the organizational charting approach illustrated earlier in Figure 2.1. The communications survey approach emphasizes the "lateral or horizontal" links in the organization--those which cut across departmental boundaries. While this gives useful insights, it may lead to impractical results. The clusters defined can cut across several reporting units, making it difficult for the planner to correlate them to space and other information. Further, most line managers will resist a plan that fragments their reporting units--no matter how well such a plan may support day-to-day communications. Other problems that may arise include:

1. *Biased responses*: Some employees will use the survey simply to state their seating request in the proposed plan. Their responses may reflect social patterns and cliques as much as bona fide needs.

FROM - TO CHART

Item(s) Charted __ALL ITEMS__   Basis of Values __UNIT LOADS/DAY__

Activity or Operation TO / Activity or Operation FROM	PRESS 1	WELD 2	MACHINE 3	ASSEMBLY 4	PAINT 5	STEEL STOR. 6	PARTS & SUPPLY 7	FINISHED STOR 8	DRIVEWAY 9	10	11	12	13	14	15	16	17	18	19	20	TOTALS
PRESS 1	=	15	12	–	–	–	–	–	–												27
WELD 2	–	=	–	–	11	–	–	2	–												13
MACHINE 3	–	3	=	15	–	–	2	12	–												32
ASSEMBLY 4	–	–	–	=	3	–	–	18	–												21
PAINT 5	–	–	7	13	=	–	–	–	–												20
STEEL STOR. 6	12	5	1	–	–	=	–	–	–												18
PARTS & SUPPLY 7	–	1	4	3	1	–	=	–	–												9
FINISHED STOR 8	–	–	2	–	–	–	–	=	20												22
DRIVEWAY 9	–	–	–	–	–	8	11	–	=												19
10																					
20																					
TOTALS	12	24	26	31	15	8	13	32	20												181

ROUTE-PRODUCT MOVEMENT SUMMARY

Unit of Flow in __UNIT LOADS__ per __DAY__

ROUTE ☒ FROM - TO ☐ BOTH DIRECTIONS	PURCH. PARTS	RAW STEEL	MACH'D PARTS	W.I.P.	FIN. GOODS	PRODUCT-MATERIAL									TOTAL
1 PRESS - WELD	–	–	–	15	–										15
2 PRESS - MACHINE	–	–	–	12	–										12
3 WELD - PAINT	–	–	–	11	–										11
4 WELD - FIN. STOR.	–	–	–	–	2										2
5 MACHINE - WELD	–	–	3	–	–										3
6 MACHINE - ASS'Y.	–	–	15	–	–										15
7 MACHINE - PARTS	–	–	2	–	–										2
8 MACHINE - FIN.STOR.	–	–	–	–	12										15
9 ASSEMBLY - PAINT	–	–	–	3	–										3
10 ASS'Y - FIN. STOR.	–	–	–	–	18										18
11 PAINT - MACHINE	–	–	–	7	–										7
30															
TOTAL	20	26	20	61	54										181

Figure 2.2. Two ways to summarize material flow data. The movement summary is the most effective device, although it takes longer to construct. (Courtesy of Richard Muther & Associates, Inc., Kansas City, Missouri.)

Responses will be biased to current personalities, methods, procedures, organization, and job knowledge--in short, to the status quo. A new worker, relying heavily on a nearby co-worker for instructions, will project this reliance into the survey, even though it may be a temporary condition. Respondents are typically uninformed about

	Co-worker
1 Do you receive new information from this individual which may eventually be applied to your area of responsibility?	Very Frequently 1 / Frequently 2 / Sometimes 3 / Infrequently 4 / Almost Never 5
2 Does this individual direct the course of your task-related activity?	Very Frequently 1 / Frequently 2 / Sometimes 3 / Infrequently 4 / Almost Never 5
3 Does the information received from this individual provide you with alternative approaches and/or solutions to your particular task?	Very Frequently 1 / Frequently 2 / Sometimes 3 / Infrequently 4 / Almost Never 5
4 Do you retain the information received from this individual for future reference?	Very Frequently 1 / Frequently 2 / Sometimes 3 / Infrequently 4 / Almost Never 5
5 Does the information supplied by this individual outline the methods or procedures which you follow when performing your task?	Very Frequently 1 / Frequently 2 / Sometimes 3 / Infrequently 4 / Almost Never 5
6 Would you seek an opinion from this individual even though it may challenge your decision regarding a specific subject?	Very Frequently 1 / Frequently 2 / Sometimes 3 / Infrequently 4 / Almost Never 5
7 Do you discuss general ideas or concepts with this individual which pertain to your work activity?	Very Frequently 1 / Frequently 2 / Sometimes 3 / Infrequently 4 / Almost Never 5
8 Do you receive announcements or directives from this individual which concern the total organizational activity?	Very Frequently 1 / Frequently 2 / Sometimes 3 / Infrequently 4 / Almost Never 5
9 Do you confer with this individual on topics relative to your particular task prior to taking action on them?	Very Frequently 1 / Frequently 2 / Sometimes 3 / Infrequently 4 / Almost Never 5
10 Do you supply this individual with general information which applies to his/her area of responsibility?	Very Frequently 1 / Frequently 2 / Sometimes 3 / Infrequently 4 / Almost Never 5
11 Do you direct the course of this individual's work activity?	Very Frequently 1 / Frequently 2 / Sometimes 3 / Infrequently 4 / Almost Never 5
12 Do you advise this individual concerning decisions which he/she must make?	Very Frequently 1 / Frequently 2 / Sometimes 3 / Infrequently 4 / Almost Never 5
13 How would you rate the importance of communication contact with this individual in terms of your activity?	Very Great 1 / Great 2 / Some 3 / Little 4 / Very Little 5
14 How frequently do you communicate with this individual when performing your task activity?	Continually 1 / Several times per day 2 / Several times per week 3 / Several times per month 4 / Several times per year 5

Figure 2.3. Communications Interaction Analysis Questionnaire. (Courtesy of CORE, Herman Miller, Inc., Grandville, Michigan.)

impending systems or procedural changes which can seriously affect communications.

2. *Self-fulfilling results*: People and groups who already communicate a great deal will be favored by close positioning. Those who do not will be relatively further apart. This may not reflect the true needs of the organization.

   For example, a currently low level of communications between two groups may be caused by personality conflicts among the supervisors and employees. If reflected in survey responses, and not corrected by the planner, this condition--right or wrong--can be sustained in the new plan.

3. *Rapid decay*: Normal personnel turnover between survey and installation can be 10-30%. Over the life of a plan it may be 50-100%. We cannot place too much weight on individual responses with turnover this high. The systems and procedures changes mentioned above also conspire to make survey responses obsolete.

In view of these limitations, communications surveys should be used with a degree of judgment. They may yield a great amount of short-lived detail. For best results, the planner should look at the overall pattern of communications and the aggregate relationships between groups of workers.

Activity Relationships

We have discussed the problems of over-relying on communications in administrative facilities and institutions. In much the same way, industrial planners often over-rely on studies of material flow. While communications and flow are the chief bases of relationships between people and departments, there are other issues to consider. These include (in no particular order):

1. Shared supervision
2. Shared equipment
3. Shared records
4. Shared utilities
5. Reception of visitors
6. Appearance
7. Convenience
8. Safety
9. Noise
10. Contamination
11. Code compliance

In some projects, these issues are of considerable importance. We need a convenient, practical way to survey them along with flow and communications.

One way is to use the relationship survey form shown in Figure 2.4. The Richard Muther vowel-letter rating system is used to rate closeness. This scheme is popular among industrial engineers. Numerical values can be substituted for the vowels but, in your author's experience, this use of numbers implies more precision than is really present in such a subjective effort.

RELATIONSHIP SURVEY

ACTIVITY-AREA _CLERK TYPISTS_

Sub-Area or Individual _____

Date __MAY 19__ Sheet __3__ of __12__
By __LH__ With __DK__
Period Covered __NOW__ to __+2 YRS.__
Space Assigned __760 SQ. FT. NET__
Location __2ⁿᵈ FLOOR – SOUTH__

> Note the activity-area, sub-area or individual given above. Use the vowel-letter ratings below to rate the degree of closeness desired between this activity-area and the others listed at the bottom of this page. Ratings should represent closeness desired in an ideal situation. For each rating other than U (Unimportant) give one or more reasons using the code numbers below. Most ratings should be U's and O's. There should be very few E's and perhaps only one or even no A's. Make ratings for the period above.

CLOSENESS DESIRED RATINGS	REASONS FOR CLOSENESS
CUSTOMER SERVICE EXAMPLES	1. Flow of material/paperwork
A – Absolutely Necessary _COURIER CLERK–COPY MACHINE_	2. Personal contact required 3. Ease of supervision
E – Especially Important _FILES–FILE STORAGE_	4. Use same records 5. Use same personnel
I – Important _COORDINATORS–BAD DEBT FILE_	6. Use same equipment 7. Use same utilities
O – Ordinary Closeness OK _SHIRLEY–BAD DEBT FILE_	8. Reception of visitors 9. Visitor convenience
U – Unimportant _RETAIL GROUPS–DORA_	10. Employee convenience 11. Light, natural conditions
X – Not Desireable _TELEPHONE CLERKS–COPY MACHINE_	12. Noise, hazard, dirt 13. _CONTROL OF WORK_ 14. _____

NO.	ACTIVITY-AREA	Rating (Letter)	Reason (Number)
1	COORDINATORS	O	10
2	TELEPHONE CLERKS	U	—
~~3~~	~~CLERK TYPISTS~~	—	—
4	MCST–MEMORY	E	13
5	COURIER CLERK	U	—
6	COPY MACHINE	U	—
7	BAD DEBT & OVERPAYMENT FILE	I	13
8	INQUIRY STORAGE	A	13
9	FILES	A	13,10
10	FILE STORAGE	U	—
11	SHIRLEY	U	—
12	DORA	O	2,13
13	DESIGN TREASURY GROUP	U	—
14	RETAIL GROUPS	U	—

Notes: @ CODE NUMBERS WILL BE CHANGED WHEN POSTING TO RELATIONSHIP CHART: 10 SHOWN AS 2; 13 AS 5; 2 AS 1.

RICHARD MUTHER & ASSOC. INC. – 129 (Filled-in: RMA-C-2104)

Figure 2.4. Relationship survey. (From R. Muther and L. Hales, *Office Layout Planning*. Kansas City, Missouri: Management and Industrial Research Publications, Inc., 1977.)

The survey form can be used to guide interviews or as a questionnaire. If used as a questionnaire, it should be sent under a cover letter from top management, explaining its purpose. There should be one survey form for each activity-area. Typically these are distributed to department heads or first-line supervisors, each of whom will "speak" for one or more activity-areas.

When the forms are completed, there will be two ratings for each relationship or pair of related areas. Each rating will be supported with its reason(s). The planner makes a list of any discrepancies in the ratings. Do the file clerks and clerk typists, for example, both rate themselves "A" to each other, or does one group say the rating is "E" or perhaps even "O"? If there is a disagreement it must be resolved.

There is often a tendency to over-assign "A" ratings. If done properly, "A's" will amount to about 2% to 5% of the total potential number of relationships. "E" ratings will constitute 3% to 10%, "I's" between 5% and 10%, and "O's" from 10% to 25%. "U's" will make up over half the ratings. The number of "X" ratings will usually be very small. These percentages are based on 25 years of experience using this rating system. They will apply to any similar numerical rating scheme, if properly used.

For best results, the planner should establish some typical relationships before making the survey--examples that are obvious to all respondents as "A's," "E's," "I's," and so forth. These can be entered on the survey form or explained in the cover letter. The planner may also define a set of vowel-letter benchmarks like those shown in Figure 2.5.

Once all the survey forms have been reviewed and the discrepancies resolved, the ratings and reasons are posted to a triangular relationship chart like that shown in Figure 2.6. This chart is an excellent way to summarize the closeness desired between activity-areas. Note that every vowel-letter rating is supported by one or more reasons. Too many planners focus on the relationship and forget to record the reasons. This can be embarrassing and annoying later in the project, if the reasoning is questioned and the planner cannot recall it.

The organization of the chart requires the planner to consider every pair of activities and the degree of relationship between them. In this way, a comprehensive analysis is achieved and summarized on a single page.

If the survey approach has been used, the chart entries will come from the reconciled survey responses. If not, the ratings may come from interviews with one or more key people, the planner's informed judgment, or a combination of the two. Charts like the one shown in Figure 2.6, including the vowel-letter ratings, have become standard practice among industrial engineers, and serve as input forms for several computer-aided layout planning routines. We will examine the uses of the Relationship Chart further in the next chapter.

Space Utilization

In addition to key inputs and relationship data, we need information on space--its configuration, condition, and assignment. Space data is often kept on file by physical entity--site, tract, building, floor, room. Indeed, when making a survey, this is the most effective way to collect the required information. However, the planner should always be able to relate physical entities, or portions thereof, to the activity-areas used in flow studies, communications and relationship surveys.

VOWEL-LETTER CLOSENESS RATINGS	OFFICE PLANNING BENCHMARK RELATIONSHIPS			
	BLOCK LAYOUT		DETAILED LAYOUT	
	PAIRS OF ACTIVITY-AREAS	REASONS FOR CLOSENESS OR SEPARATION	PAIRS OF ACTIVITY-AREAS	REASONS FOR CLOSENESS OR SEPARATION
A	President - Exec. V.P. Restrooms - Lunchroom Purchasing - Lobby	Personal contact Movement of people, utilities, convenience Visitors	President - Exec. Secy. Copy clerk - Copy machine	Personal contact Use of equipment
E	Sales V.P. - General sales office Accounting - Data processing Central stores - Truck dock	Personal contact Personal contact, shared supervision, convenience Movement of supplies and equipment	Clerk typists - Automatic typewriters Secretary - Entrance	Use of equipment Visitors
I	President - Sales V.P. President - Lobby Operations V.P. - Plant	Personal contact Convenience Personal contact	Sales Manager- Sales Representative Copy machine - Storage	Personal contact Movement of supplies, convenience
O	Accounting - Purchasing Operations V.P. - Lobby Production Planning - Restrooms	Paperwork flow, convenience Convenience Convenience	Sales Manager - Entrance Secretary - Storage	Convenience Convenience
U	President - Mail Room Central stores - Restrooms Purchasing - General sales office	No contact No contact Little contact	Files - Storage Loan Manager - Security guard	No contact No contact
X	President - Word processing Data processing - Lobby Lunchroom - Boardroom	Noise Security Appearance	Telephone clerk - Automatic typewriters Copy machine - Water fountain	Noise Unnecessary visiting

Figure 2.5. Benchmarks for use in vowel-letter closeness rating. (Courtesy of Richard Muther & Associates, Inc., Kansas City, Missouri.)

As a practical matter, this means that activity-areas must be defined *before*, and kept in mind during, the survey of space utilization. Typically the planner will collect maps or plans of the spaces under study and mark the activity-areas on them. These marked-up documents are then taken along on the survey for verification or adjustment.

If no drawings or plans can be found, then the planner must sketch the space during the survey, noting the boundaries of the activity-areas along with the physical features of interest. These will generally include:

1. Permanent building walls and thickness
2. Partitions
3. Doors and direction of swing
4. Windows
5. Furniture, machinery, and equipment
6. Columns
7. Electrical, telephone, and other utility outlets
8. Aisles, halls, and circulation spaces

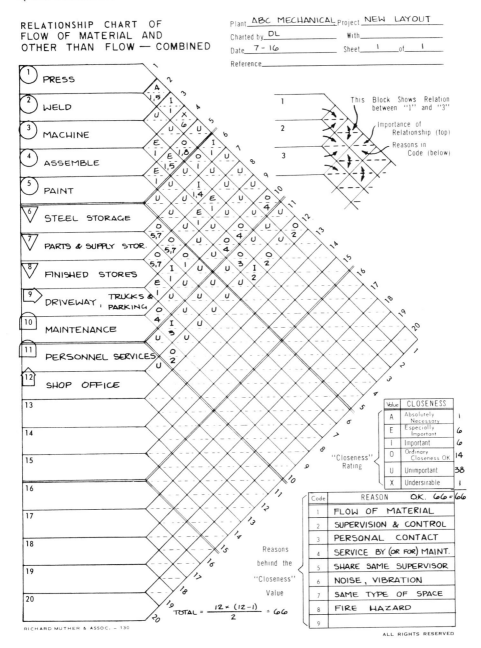

Figure 2.6. Relationship chart showing closeness desired between 12 activity-areas in a manufacturing plant. Both flow and other-than-flow reasons have been given for closeness. A technique for combining flow and other-than-flow factors is presented in Chapter 3. (Courtesy of Richard Muther & Associates, Inc., Kansas City, Missouri.)

Once the existing spaces have been visualized, the planner records information about them. The specific data collected depends upon the level of planning decision. But in general, we are interested in the following attributes:

1. The amounts of space by physical entity (sites, tracts, buildings, floors, rooms, bays, etc.)
2. The occupancy of space by organizational entity--company, division, department, cost center, work center, individual
3. The occupancy of space by project-specific activity-area, cross-referenced to physical and organizational entities just described
4. The type of space (definition varies by type of facility and project)
5. The shape, configuration, or dimensions of major structures and spaces
6. The age and condition of space (good, fair, poor, etc.)
7. The capacity of space--both in physical terms (floor load, suspended load, seating, etc.) and in throughput if applicable
8. The suitabilities of space--for existing and possible, future uses

Collection of space data is made easier by using standard forms, coding and classification schemes. Figure 2.7 shows a building condition report covering a single structure. Figure 2.8 shows a condensed summary of multiple structures. Both are examples of industrial space surveys. Similar forms can be easily developed for administrative and institutional facilities. Properly designed, such forms can record information down to the room level.

The space survey form in Figure 2.9 is a useful way to record the occupancy of space in administrative settings. With minor adaptation it can also be used in industrial and institutional facilities. The form uses three convenient categories or types of space to comprise activity-areas: (1) work stations (2) support and (3) main circulation. These categories can be further defined by the type of space code--Private, Semi-Private, Open, General.

Work station space includes private offices, desk areas, equipment areas, or other places assigned to specific individuals.

There are two types of *support* space in most settings. The first is special, work station support typically dedicated to specific work groups. In offices, such space includes: file areas, data processing equipment areas, copying, reference stations and the like. The other type of support space is general in nature, including rest rooms, vending areas, lobbies, mechanical equipment rooms, and the like.

Main circulation covers hallways, stairs, elevator shafts, and major aisles between activity-areas. It is also convenient to include *building losses* in this category--to walls, columns, partitions, and the like.

It is best to conduct space surveys during business hours. That way the planner can ask the occupants of the space to resolve unclear issues and insure the accuracy of the data.

The collection of space data is not complete without some indication of trends--past and future. One way to identify trends is to interview managers, supervisors, or other key individuals, using the worksheet shown in Figure 2.10. This form or one like it records the underlying reasons for the growth, stability, or decline of each activity-area. Such information is necessary input to space projections later in the planning project.

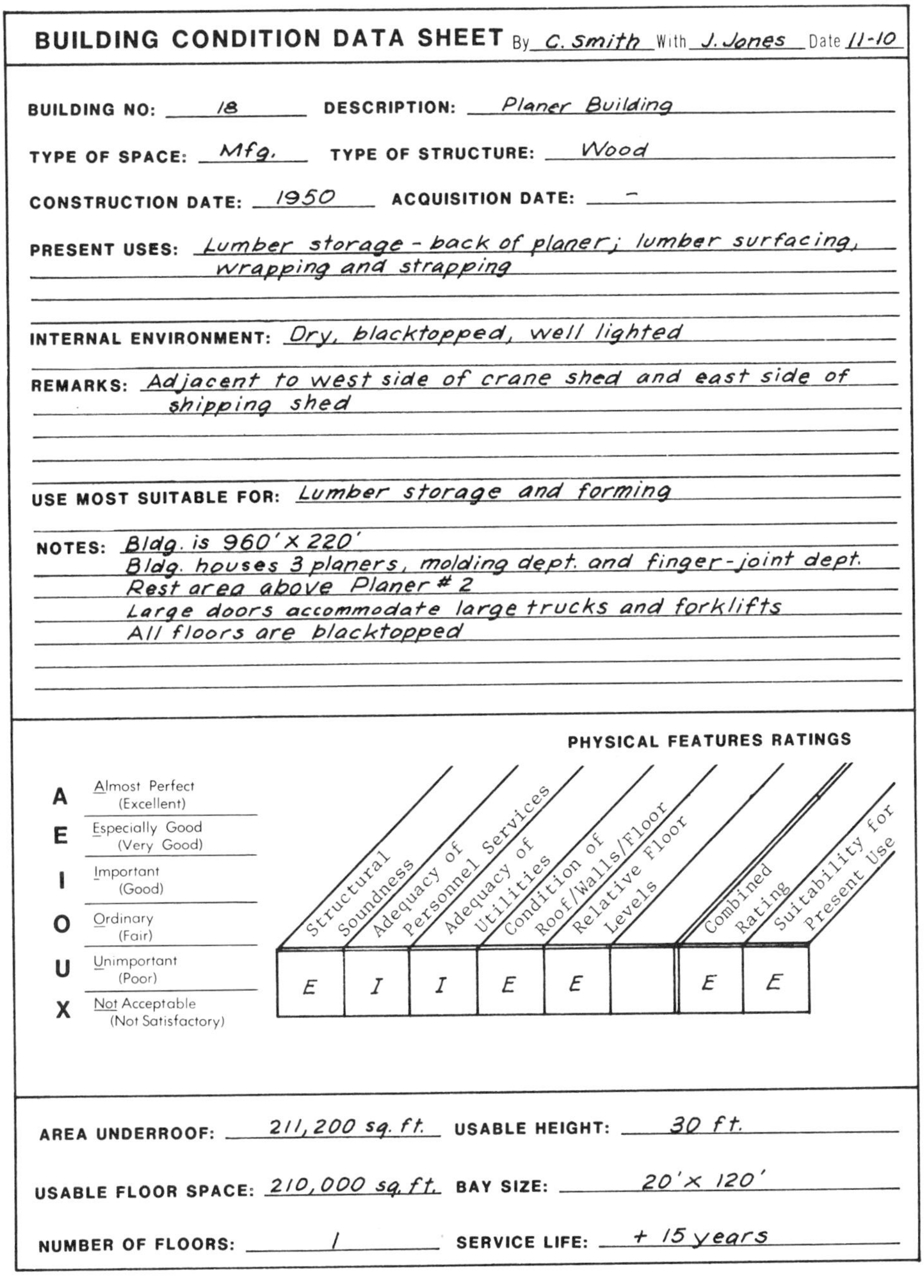

BUILDING CONDITION DATA SHEET By *C. Smith* With *J. Jones* Date *11-10*

BUILDING NO: _____*18*_____ DESCRIPTION: _____*Planer Building*_____

TYPE OF SPACE: _____*Mfg.*_____ TYPE OF STRUCTURE: _____*Wood*_____

CONSTRUCTION DATE: _____*1950*_____ ACQUISITION DATE: _____*—*_____

PRESENT USES: _*Lumber storage – back of planer; lumber surfacing, wrapping and strapping*_

INTERNAL ENVIRONMENT: _*Dry, blacktopped, well lighted*_

REMARKS: _*Adjacent to west side of crane shed and east side of shipping shed*_

USE MOST SUITABLE FOR: _*Lumber storage and forming*_

NOTES: *Bldg. is 960' × 220'*
Bldg. houses 3 planers, molding dept. and finger-joint dept.
Rest area above Planer # 2
Large doors accommodate large trucks and forklifts
All floors are blacktopped

PHYSICAL FEATURES RATINGS

A — Almost Perfect (Excellent)
E — Especially Good (Very Good)
I — Important (Good)
O — Ordinary (Fair)
U — Unimportant (Poor)
X — Not Acceptable (Not Satisfactory)

Structural Soundness	Adequacy of Personnel Services	Adequacy of Utilities	Condition of Roof/Walls/Floor	Relative Floor Levels	Combined Rating	Suitability for Present Use
E	I	I	E	E	E	E

AREA UNDERROOF: _*211,200 sq. ft.*_ USABLE HEIGHT: _*30 ft.*_

USABLE FLOOR SPACE: _*210,000 sq. ft.*_ BAY SIZE: _*20' × 120'*_

NUMBER OF FLOORS: _*1*_ SERVICE LIFE: _*+ 15 years*_

Figure 2.7. Survey form for a single building. (From Ref. 8, Vol. 2. Copyright 1980, Richard Muther.)

| PHYSICAL CONDITION OF BUILDINGS | | Plant **Texas** Project **1311** By **J.D. West** With **G.R. Wilson** Date **25 Mar. 80** Sheet **1** of **1** |

Bldg. No.	Type of Space	Building Description	Construction Date (a)	Acquisition Date (d)	Area U–Underroof F–Floor Space in sq.ft.	Ceiling Height	Age	Structural Soundness	Floor Condi- tion & Levels	Utilities Avail. & Cond.	Condition of Walls, Roof	Flexibility & Adaptability	Combined Rating	Suitability for Present Occupancy	Remarks
A₁	○	Die casting – north bay (e)	1937	–	6206	14'	43	I	U (g)	I	I	I-	I	I-	Needs 10-15% more space. Cols. limit flexibility
A₂	○	Die casting – center bay	'58	–	3253	12'3"	22	E	U (g)	I	I	O-	I-	I-	Low ceiling
A₃	○	Die casting – south bay	'66	–	3400	18'	14	A	I (g)	I	E	I	E-	E-	Needs 15-20% more space
A₄	○	Melting room (f)	39	–	1857	28'	41	E	I	I	E	U	E-	I	Arrangement and expansion limited
A₅	⌂	N.W. service area	'55 & '77	–	1360	12'	24/3	I	I (g)	I	O	U	I-	O	
A₆	○	Old machine shop	'55	–	1365	~16'	25	O	O	I	U	U	O	O-	
A₇	○	New machine shop (b)	'70	–	7811	12'6"	10	E	I	O	I	I-	I	I-	Needs 15% more space
E	▽	Storage – WIP (c)	'61	'70	4960	14'	19	O	U (g)	U	U	U	U	U	
F	▽	Storage – finished goods	'48	'69	5874	11'6" 13'6"	32	O	O-	U	U	U	O-	U	Temperature control, wiring & column problems

Notes
(a) Dates on older buildings may be year or two off
(b) Includes annex (N.E.) & process water disposal problem
(c) Includes 448# machining, 832# casting (trim)
(d) Includes service areas in west end of bay
(e) Roof raised ~1956
(f) Roof raised ~1946
(g) Floor on unstable subsoil

Figure 2.8. Survey form for several buildings or additions. This is a summary. More detailed information may be recorded for each building using a form like the one shown in Figure 2.7. (From Ref. 8, Vol. 2. Copyright 1980, Richard Muther.)

Furniture and Equipment

If the planning project may lead to a rearrangement, the planner will need complete information on the machinery, furniture, or equipment housed in each space or activity-area.

Such information is time-consuming to collect and is best accomplished with the use of standard forms. Figure 2.11 illustrates the detail typically required for one machine in an industrial facility. The information on this form should be summarized for all machines in an activity-area or building, as shown in Figure 2.12.

When planning for office rearrangements, it may be necessary to collect the type of individual workplace detail shown in Figure 2.13. These survey forms are designed as companions to the space survey described earlier.

Institutional settings are extremely varied. Data collection for laboratories, hospitals, schools, and the like is usually done with a blend of the forms shown here for industrial and administrative facilities.

If detailed layouts of machinery or furniture are going to be required, the planner will need drawings of the equipment involved. Hopefully these can be obtained from the manufacturer's catalog or possibly from a template

SPACE SURVEY

Company LOOART PRESS Date MAY 19 Sheet 3 of 14 By LH With ___ [X] Present [ ] Plan for year

ACTIVITY-AREA CLERK TYPISTS

Sub-Area or Individual SUMMARY Location 2nd FLOOR-SOUTH Space Assigned 586 SQ.FT. NET

WORKSTATIONS List by type or activity (a)	Total Area Incl. Internal Aisles (b)	Min. Width of Area (c)	No. of Sta's in Area (d)	Space per Station (e)	Type of Space Code* (f)
CLERK TYPISTS (a) — ROUGH MEASUREMENT					
ELMA	5x7=35 SQ.FT.				
MARY K.	=35 SQ.FT.				O
JOAN	=35				
CAROL	5x5=25				
JANET	" =25				
MARY S.	" =25				
MARGIE	5x7=35				
BOBBIE	5x9=45				
DOLLY	6x10=60				
BETTY	" =60				
MAXINE	5x7=55				
SUB-TOTAL	425				
SENIOR COORDINATOR (b)					
BEVERLY	7x9=63				E
TOTAL	488				

SUPPORT List by type or activity (g)	Total Area Incl. Internal Aisles (h)	Min. Width of Area (j)	Type of Space Code* (k)
SORTING TABLE	7x6=42 SQ.FT.	6'	O
FILES (2)	6x6=36	6'	O
STORAGE CABINET	4x5=20	5'	O
TOTAL	98		

CIRCULATION Main Aisles & Building Losses	Total Area Used or Req'd. (m)	TOTAL SPACE Workstation, Support, and Circulation (b+h+m)
ACCESS IS FROM 2nd FLOOR CENTER AISLE; ALLOCATE 1/2 OF 3' WIDE AISLE FOR DISTANCE OF 40'		
SAY 60 SQ.FT. (c)		
TOTAL	60	646 SQ.FT.

Notes: (a) THIS AREA MUCH TOO CROWDED. DOLLY & BETTY OKAY AT 60 SQ.FT. OTHERS UNACCEPTABLE; CAN'T GET IN AND OUT OF WORK SPACES.
(b) BEV NEEDS ABOUT 90 SQ.FT.
(c) CENTER AISLE SHOULD BE 5 FT. WIDE.

*Type of Space Code: P-Private; E-Semi-Private Enclosure; O-Open or Semi-Open; S-Service or Special Areas (attach sketch); G-General areas not assigned

RICHARD MUTHER & ASSOC. INC. – 165 [Filled-in: RMA-C-2156]

Figure 2.9. Space survey. (From R. Muther and L. Hales, *Office Layout Planning.* Kansas City, Missouri: Management and Industrial Research Publications, Inc., 1977.)

SPACE ESTIMATING WORK SHEET - A   DATE: 3 Jan 79

Dept: 34
Act.: 13, 14, 15, 16   BY: Lee Hales

ACTIVITY-AREA: Computer Services   PAGE: 13 OF

TIME PERIOD	FACTORS INCREASING SPACE REQUIRED	FACTORS DECREASING SPACE REQUIRED
LAST 10 YEARS '60's & '70's	. Development of claims processing applications.	. Installation of modular furniture
NEXT 2 YEARS THRU 1980	. Statewide Medicare A . Upgrading computer equipment . Continued development of new applications.	. Tape-to-tape data entry
3-5 YEARS EARLY 1980's*	. Continued development of new applications - especially information systems for management. . Statewide Medicare A & B	. None
5-15 YEARS '80's & '90's*	. As above . Regional Medicare	. Distributed processing in user departments. . Reduced need for programmers. . Systems in place, development completed.

Figure 2.10. Worksheet for space projections.

supplier. If not, drawings will have to be made--ideally at the time when the other furniture and equipment data are being collected. This is a time-consuming task, and the level of detail in the drawings should be kept to a bare minimum.

Dominant Considerations

There is one type of project-specific data often overlooked. This category of information can best be described as "dominant considerations." Virtually

MACHINERY & EQUIPMENT LAYOUT DATA

Plant _R.M.A. Printing Co._ Project _Plant Relocation_
Prepared by _D.B._ With _L.H._ Date _19 Mar._

Water	—	Steam	—	Drains	—
Comp. Air	—	Gas	—		
Foundation	—	Pi:	—	Level/Lag _Level_	
Exhaust	—			Sp'l. Elect. —	

Electrical	H.P.	Volts	Cycle	Phase	Amps.
Drive Motor	15	220	60	3	100
Auxiliary Motor	1	110	Internal		
Auxiliary Motor	1	110	Internal		

Max. Height _54 in._
Weight (less attach.) _4060 lbs._

Name/Type _Tandemer 8¼_ File _R.O. Press Dept._

Manufacturer _Didde-Glaser_	Size/Model _102-8¼_
Speed/Capacity _354 Ft./Min. @_	Signif. Ident.
Left-Right _13 Ft. 8 In._	Co. Mach./Eqpt. Ident. Numbers
Front-Back (Max.) _4 Ft. 6 In._	Covered by this Sheet
Net Floor Area _41 Sq. Ft._	
Worker/Maint. Area _53 Sq. Ft._	_1028_
Material Set-down _36 Sq. Ft._	_1029_
Area for Aisles _Not Incl._	_1030_
Service/Other _43 Sq. Ft._	
Gross Area (plan for) _175 Sq. Ft._	

PLAN VIEW SCALE _¼"=1'0"_

(Plan view labels: Drive Module; Printing & Sheeting)

ELEVATION SKETCH OR PHOTO SOURCE _Supplier Catalog_
Reference Notations/changes _@ 62,000 I.P.H._

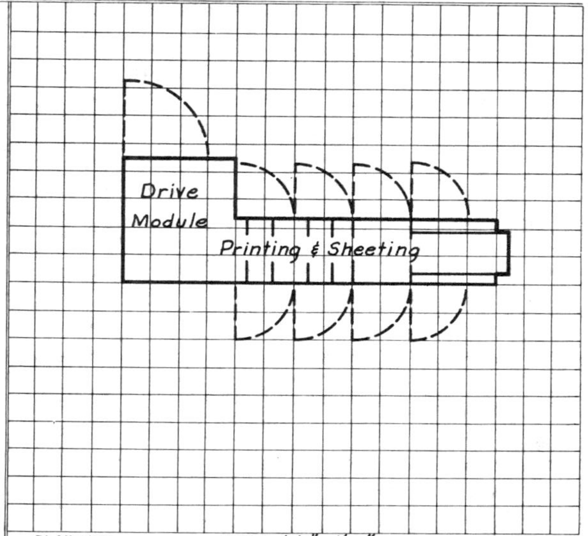

RICHARD MUTHER & ASSOCIATES – 154

Figure 2.11. Equipment information. (From Ref. 8, Vol. 2. Copyright 1980, Richard Muther.)

MACHINERY & EQUIPMENT AREA & FEATURES SHEET

Company/Plant _Spot Manufacturing_ Bldg/Dept/Area _Production_

Project _Plant Survey_
By _D. Johnson_ With _L. May_
Date _11-5_ Sheet _1_ of _1_

Machine or Equipment Identification Number	Name and/or Description	Left – Right ft.	Front – Back ft.	Height ft.	Area for Machinery or Eqpt. in sq. ft.	Operator(s) Work & Maintenance Area	Material Set-down Area	Total Area each Machine or piece of Eqpt.	No. of Mchs Eqpt.	Total Net Area* (in sq. ft.)	110-A.C.	220-A.C.	Other Power	Ampere Rating	Water	Steam	Drains	Compressed Air	Other Piping	Foundation/Pit	Exhaust Hood	Dust Collector	Utilization	Condition	Suitability
P-1059	Webtron 650	17	5	6	85	185	100	370	1	370	✓	✓	–	–	–	–	–	✓		–	✓	–	A	A	A
P-1058	Mark Andy No. 9	12	7	6	85	160	115	360	1	360	✓	✓	–	–	–	–	–	✓		–	✓	–	A	A	A
P-1056	Mark Andy No. 5	12	7	6	85	160	115	360	1	360	✓	✓	–	–	–	–	–	✓		–	✓	–	E	A	A
P-1057	Mark Andy No. 7	12	7	6	85	160	115	360	1	360	✓	✓	–	–	–	–	–	✓		–	✓	–	I	E	E
P-1054	C-Model Press	12	4	5	50	95	35	180	1	180	–	✓	–	–	–	–	–	–		✓	–	–	A	I	I
P-1048-50	Grebex Press	5	3	5	15	75	(a)	90	3	270	–	✓	–	–	–	–	–	–		–	(b)	–	I	I	A
P-1051	Allied Press	5	3	5	15	75	(a)	90	1	90	–	✓	–	–	–	–	–	–		–	(b)	–	O	U	U
P-2001	Die-cut Press	5	3	5	15	25	(a)	40	1	40	–	✓	–	–	–	–	–	–		–	–	–	E	O	E
P-1055	Doboy with tunnel	9	5	7	45	70	85	200	1	200	–	✓	–	–	–	–	–	✓		–	✓	–	E	I	A
E-1052-53	Reroll Table	5	5	3	25	(c)	20	45	2	90	–	✓	–	–	–	–	–	✓		–	–	–	A	I	O
E-1047	Slitting Table	3	3	3	10	45	(a)	55	1	55	✓	✓	–	–	–	–	–	–		–	–	–	A	I	E

* Required space for main or delivery aisles and service areas not included.

Total Net Area Required* (in _sq. ft._) 2375

Aisles .... N.A.
Services .... N.A.
Other ... (d) ... 250
Total Area Rq'd. 2625

Reference Notes:
a. _Included in operator area_
b. _Exhaust in general area_
c. _Included in machine area_
d. _Ink room_
e.
f.

Figure 2.12. Equipment summary. Note the vowel ratings of space utilization, condition and suitability. In this context, "A" is almost perfect; "E" especially good; "I" important results; "O" ordinary results; and "U" unimportant results. An "X" would imply "not acceptable." (From Ref. 8, Vol. 2. Copyright 1980, Richard Muther.)

every facilities planning project is subject to certain constraints such as a budget, or a zoning restriction; or to certain environmental influences such as topography, or soil strength, or prevailing winds; or even to the attitudes of key personnel who will use or work in the facility.

Making a list of these constraints, special considerations, and influences of all kinds is an important step toward good facilities plans. The information involved may not be picked up in surveys of relationships, space, and equipment, or even in the collection of the key input data, though this is a good source of the information we are seeking. Figure 2.14 is an example of the kind of lists that should be developed.

Phasing of Data Collection

Major facilities projects can span periods of one to several years. Even small projects can take months to plan and install. Under these conditions, it is a mistake to collect detailed data too soon. It may have decayed by the time detailed planning comes about.

The planner must continually cope with reassignments of people, space, and equipment, methods changes, and even changes in the scope and purpose

FURNITURE SURVEY

Company __LOOART PRESS__ Sheet __36__ of __45__
By __LH__ With __MK__ Date __MAY 19__

ACTIVITY-AREA __CLERK TYPISTS__

[X] Present [ ] Plan for year _____
Space Assigned __35 SQ. FT.__

Sub-Area or Individual __MARY KENT__ Location __2nd FLOOR - SOUTH (NW CORNER)__

ACTIVITY/JOB DESCRIPTION __TYPE LETTERS AND FORMS RELATED TO LARGE, SPECIAL ORDERS, LOST ORDERS, CUSTOMER COMPLAINTS, ETC. REFER TO HISTORICAL INFORMATION ON MICROFICHE.__

Male(s)	L. Handed	R. Handed	Man-days per mo. out of office	NONE
Female(s) ✓	L. Handed	R. Handed ✓	Working/operating hours or period of use	8AM - 4:30 PM.

Enter checks, numbers or descriptions as appropriate. Make notes in unused part of form or on back DESKS / TABLES	Quantity	Size in INCHES Left - Right	Front - Back	Height	Model No., Asset No., Move No. Condition, Material, or other info.	Number of drawers Pencil	Box-Type	File	Materials normally in use on work surface. EDP Reports, Letters, Books, Order Forms, Ledgers, Etc. (Also see Equipment Survey form)
Primary Desk/Table	1	48	30	29	1603-D-4	1	–	1	LETTERS, FORMS,
Secondary Desk/Table									FILE FOLDERS
Typing Return/Stand	1	30	18	26	1603-D-5	–	–	–	
Executive Return									
Drafting Table									
CRT STAND	1	30	24	26	1603-D-6	–	–	–	(SHARED WITH JOAN)

PANELS / SCREENS						No. Drawers	No. Shelves	% Used per week	Reach while telephoning	Shared with (Individual or Group)	Primary Contents: Letters, Reports. Supplies. Equipment, Forms, Etc.
TAN METAL (SUPPORTS PRIMARY SURFACE AND SHELVES ABOVE)	1	48	–	60	1603-P-2						
TAN METAL	1	24	–	60	1603-P-3						

STORAGE / FILES											
Conventional Letter											
Conventional Legal											
Lateral Letter											
Lateral Legal											
Fireproof											
Check/EDP											
Flat/Map/Plan											
Cabinet											
Shelving (HANGS ON 48" PANEL)	1	46	13	45	1603-S-2	–	1	100	YES	NO	CATALOGS AND IN-OUT TRAY
Credenza											
Boxes/Cartons											

CHAIRS / SEATING						MEETINGS / PRIVACY				
Desk Chair						Persons per Meeting				
Steno Chair	1	———			1603-C-2	Mtgs. per Day			Duration	
Side Chair						Confidential Activities				
Drafting Stool						Discussions?			Yes	No
						Papers?			Yes	No
						Private office?			Yes	No
						Semi-private space?			Yes	No

NONE (written diagonally across Meetings/Privacy section)

RICHARD MUTHER & ASSOC. INC. – 166 (Filled-in: RMA-C-2158)

Figure 2.13. Furniture survey. (From R. Muther and L. Hales, *Office Layout Planning*. Kansas City, Missouri: Management and Industrial Research Publications, Inc., 1977.)

DOMINANT CONSIDERATIONS
(Identification & Dominance Rating)

Plant: _Goodall_   Project: _1245_
By: _N. L. Hahn_   With: _L. J. Best et al_
Date: _7 May_   Page: _1_ of ___

EXISTING

Mark "X" if beyond control of company plant · Dominance rating

PHYSICAL — EXTERNAL

#	Consideration	Rating	X
1	Site congested and restricted – already building on leased land	A	
2	Inadequate public roads and railroads	E	X
3	Access roads to and from site are narrow and unsafe	I	X
4	Prevailing wind causes dust problem from local cement mill	I	X
5	Not enough parking space	E	
6	Bridge capacity between plant sites A and B is limited to 5 tons	O	X
7	Boiler house and settling basin not on company-owned land	I	
8	R&D in leased building downtown, very costly to operate	E	

NON-PHYSICAL — EXTERNAL

#	Consideration	Rating	X
1	Local residents upset about parking in streets	I	
2	Limited skilled labor available in the area	E	
3	Government restrictions on pollution and waste	I	
4	City talking about establishing recreation park along NE property line	O	X
5	Percentage of farm-out (outside-contracted) work too high at 60%	I	
6	Pipeline easement runs across NW corner	O	

PHYSICAL — INTERNAL

#	Consideration	Rating	X
1	Not all buildings suitable for current operations	I	
2	Machine shop capacity limited – excessive "farm-outs"	E	
3	Vibration from heavy presses noticeable in adjacent buildings	O	
4	Fire-protection water tanks over buildings restrict upward expansion	I	
5	Blast walls around explosives area require 300 ft. clear space	E	
6	Soil tests "poor" — cause heavy equipment foundation problem	I	X
7	Reservoir – newly constructed – blocks through-road pattern	O	

NON-PHYSICAL — INTERNAL

#	Consideration	Rating	X
1	Production control and scheduling need to be improved	O	
2	Low morale of shop operators	O	
3	Materials handlers' attitude poor – excessive product damage	O	
4	Competitive "pressures" between two divisions	E	
5	Multiple labor union restrictions	O	

FUTURE (through 10 years)

Mark "X" if beyond control of company plant · Dominance rating

PHYSICAL — EXTERNAL

#	Consideration	Rating	X
1	Limited land availability in the area	E	
2	City Plan. Comm. may require company land to widen roads to four lanes	I	X
3	Truck access may be further restricted in the future	E	X
4	Prevailing wind will cause dust problem from local cement mill	I	X
5	Parking space a continuing problem – leased land available down the street	E	
6	Bridge not usable for larger, heavier truck loads	O	X
7	Boiler house and settling basin not on company-owned land	I	
8	Leased R&D facilities or new technical center should be considered	E	

NON-PHYSICAL — EXTERNAL

#	Consideration	Rating	X
1	Local citizens group against major plant expansion in the area	E	
2	Availability of skilled labor will be a continuing problem	E	
3	Gov't regulations of all kinds will increase – pollution, safety, building design	I	X
4	Park may restrict access and land use	I	
5	Should bring work back into the plant – reduce farm-outs to 30%	I	
6	Perhaps move pipeline or restrict NW area to limited use	O	

PHYSICAL — INTERNAL

#	Consideration	Rating	X
1	Some buildings should be demolished (a)	A	
2	(See Non-physical line # 5 above)		
3	Heavy presses should be segregated	O	
4	Fire-protection water tanks over buildings restrict upward expansion	I	
5	Blast walls around explosives area require 300 ft. clear space	E	
6	Soil tests "poor" – heavy equipment foundation problem continuing	I	X
7	Reservoir cannot be moved, practically – will continue to block	O	

NON-PHYSICAL — INTERNAL

#	Consideration	Rating	X
1	Assume these operating problems will be corrected	E	
2			
3			
4	"Pressures" to be reconciled and divisions probably coordinated	O	
5	Labor union restrictions likely to become worse	I	

Reference Notes:
(a) Plant Engineering reports Buildings C-2, C-3, and D-2 to be removed in next ten years

Dominance Rating —
Effect on Facilities Planning

A – Abnormally High
E – Especially Important
I – Important
O – Ordinary
U – Unimportant

Instructions
1. Determine what major considerations affect the existing facilities and future facilities plans.
2. Record these in the appropriate section.
3. Rate each consideration for its relative effect or influence on the project's facilities planning.
4. Mark "X" where consideration is beyond the control of the company plant.

of the project itself. For these reasons it is wise to collect data for each planning phase at the beginning of that phase.

Implications for Computer Aids

Considering all the data that could be collected, and the many survey techniques that could be used, the following questions come to mind:

1. How much data do we routinely need to collect? What should be kept on the computer? Why? Remember, computer memory is cheap, but the time to load it is not.
2. How shall we organize the data? By physical entity--site, tract, building, room? By activity-area for a specific project? If both, is some kind of data base needed, to allow retrieval in many ways?
2. How do we avoid entering the same information several times, if it should be needed for several analyses? Again, is a data base needed?
4. How fast do the data become obsolete? How can they be kept up to date?
5. Are graphic representations really necessary? Of spaces and buildings? Furniture and equipment?
6. Can we standardize our data collection and survey techniques, to make use of standard forms and rating schemes? If not, how will we be able to use the computer?

In Part II we will review a number of computer aids and see how those who use them have answered these questions.

Figure 2.14. List of dominant considerations. Eight categories are covered by considering influences which are external and internal, physical and non-physical, existing and future. Vowel ratings indicate the dominance of each influence on the planning project. An "X" marks those influences beyond the control of the plant or facility. Lists such as this one are useful on all major projects. (From Ref. 8, Vol. 2. Copyright 1980, Richard Muther.)

3

Systematic Planning Techniques

Typical Planning Approaches

Computer programs are nothing more than lengthy sets of detailed instructions. For programs to work properly, there can be no ambiguity in the instructions and no unforeseen steps in the procedure. Planners quickly discover that the successful use of computers rests first on their use of systematic planning techniques.

But the typical planner today is not systematic. In fact, his most commonly used approach might be called "instinct and experience." The planner makes a survey and looks for similarities to past projects. He draws on his experience to formulate solutions. "I think" and "I feel" are common phrases in the development of a plan.

Another approach uses the "cookie cutter." Concepts featured in the latest journals, or presented in recent professional forums, become the proposed solutions for the project at hand. Or, a standard facilities plan is applied, over and over, to each new situation.

The committee or "participative" approach is also popular, especially for planning new offices and labs. The planner leads a large "team" with representatives from each affected area. Solutions evolve from debate and are submitted to a vote to arrive at the best plan.

Communications surveys and flow studies discussed in the previous chapter are an effort by facilities planners to become more sophisticated in their approach. Still, these techniques do not go far enough, limited as they are to a single planning issue.

Systematic Approaches

Leading professionals have long recognized the need for systematic planning. Their concern with techniques and structured approaches predates our current interest in computers by many years (1, 2, 4, 5, 9).

Richard Muther, in his book *Systematic Layout Planning (SLP)*, addresses the heart of facilities planning with the nine-step procedure shown in Figure 3.1 (6). Each box in the pattern covers one or more planning techniques. Taken in sequence, they guide the development of layout plans with clear

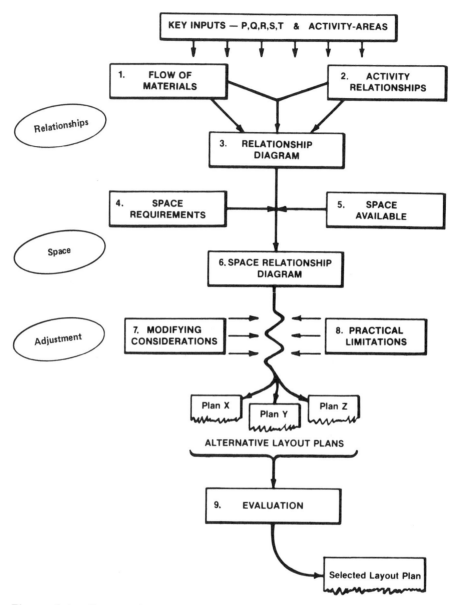

Figure 3.1. Systematic Layout Planning (SLP) Pattern of Procedures. (From Ref. 8, Vol. 1. Copyright 1979, Richard Muther.)

rules and conventions. The pattern of procedures rests on the three "funda-mentals" of relationships, space, and adjustment. These are also the sequen-tial considerations which underlie the balance of this chapter.

Layout, of course, is only one of the five components which make up a fa-cility. Recalling our discussion in Chapter 1, we identified the other four as: handling, communications, utilities, and building. Muther and Hales (your author) have published companion, nine-step planning procedures for each of these components. They appear in the two-volume set *Systematic Planning of Industrial Facilities (SPIF)* (8).

In this chapter we will elaborate on the techniques for layout planning, on the assumption that layout will be the central issue in most planning projects. Even when it is not, layout still serves as a useful starting point for the planner or designer who is interested in computer aids.

Combining Flow and Other-than-Flow Relationships

In Chapter 1 we discussed the recording of material flow, communications, and other relationships between activity-areas. We did not discuss the need to combine material flow and other-than-flow concerns into a single, unified set of relationships. A good, systematic approach to planning must start with the resolution of this question: How important is flow, compared to the other reasons for placing activities close or far apart?

The answer is: it depends upon the type of work performed within the facility. If material flow dominates, the layout can often be made directly from the routings themselves, as shown in Figure 3.2. At the other extreme, consider an executive office area, in which there is virtually no flow. Clearly the layout here must reflect such concerns as appearance, convenience, reception of visitors, personal contact, etc. Between these extremes, the importance of flow is one of degree, as explained in Figure 3.3. Clearly, most planning projects require us to consider (and therefore combine) both flow and other-than-flow relationships.

The mechanics of combining flow (or for that matter, quantitative communications measures) with other types of coded relationships is quite simple, as illustrated in Figure 3.4. In this example, flow and non-flow factors are given equal importance. Material flow data was collected and summarized on a from-to chart. Other-than-flow relationships were summarized on a triangular relationship chart.

Quantitative material flows were converted to vowel-letter ratings by giving an "A" to the highest flow, and then rating the remaining flow in relation to the highest. Next, the vowel letters were converted to points (A = 4, E = 3, I = 1, O = 1, U = 0, X = -1). These were added to the point values of the other-than-flow vowel-letter ratings. The combined totals were then plotted in descending-value sequence, and were calibrated into a new set of vowel-letter ratings. Again, an "A" was given to the highest combined total, with remaining ratings set in relation to the highest. The results are summarized by posting on a "combined" relationship chart (flow and other-than-flow).

Suppose we do not want a 1:1 ratio (equal importance) between flow and the other-than-flow relationships. If so, we add the extra step of multiplying the appropriate vowel-letter point values. If flow is twice as important as other-than-flow, then its vowel-letter point conversions are multiplied by 2 before adding. This makes an "A" rating worth 8 points, "E" worth 6, etc.

Carrying this logic still further, Muther has published the matrices in Figure 3.5 for various ratios of importance between flow and other relationships (6). The appropriate ratio is suggested by the chart shown back in Figure 3.3.

This combining procedure is just the type of rule-based, systematic technique that must be adopted to make meaningful use of computers. It is a numerical technique, and it becomes tedious when done manually on large projects. So the computer is an ideal aid.

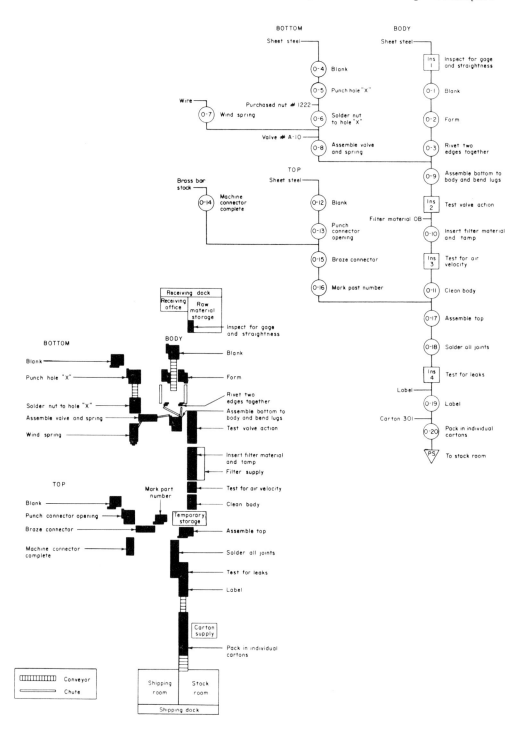

Figure 3.2. The importance of material flow. (From H. B. Maynard and G. J. Stegemerten, *Operation Analysis*. New York: McGraw-Hill.)

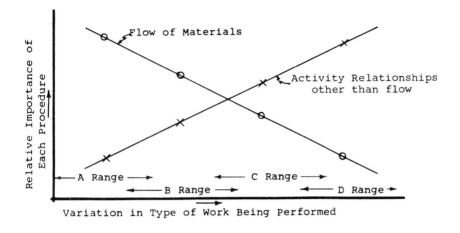

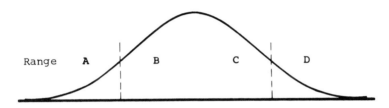

Range A.	Plant layouts involving heavy or awkward products or materials, or relatively large quantities of parts or materials. (Flour Mill; Steel Fabricator; Mass Production Manufacturing)
Range B.	Job shop layouts--having no major clear-cut pattern(s) of material movement. (Tool Making; Custom Plating; General or Specialty Manufacturing)
Range C.	Service Shops with significant amounts of material sequencing or office areas involving heavy flow of papers. (Overhaul and Maintenance Shops; Test Laboratories; High-Volume Office Operations)
Range D.	General Office Areas. (Experimental Labs; Drafting Room; Most Office Areas)

Figure 3.3. The type of work being performed determines the importance of material flow analysis. (From Ref. 6. Copyright 1973, Richard Muther.)

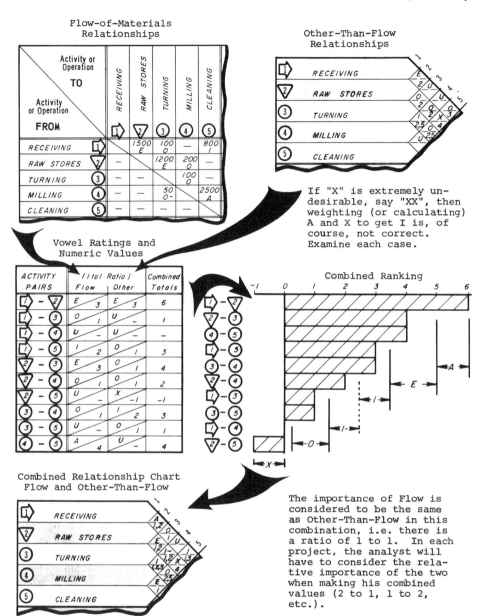

Figure 3.4. Procedure for combining flow and other-than-flow relationships. (From Ref. 6. Copyright 1973, Richard Muther.)

IN VARIOUS SITUATIONS,"FLOW OF MATERIALS" CARRIES DIFFERENT DEGREES OF IMPORTANCE RELATIVE TO FACTORS "OTHER THAN FLOW". ON EACH PROJECT, THE PLANNER OR ANALYST SHOULD DETERMINE THIS DEGREE OF RELATIVE IMPORTANCE, AND THEN ASSIGN A COMBINED CLOSENESS RATING FOR EACH PAIR OF ACTIVITY-AREAS.

THIS CAN BE DONE BY BUILDING A MATRIX OF FLOW AND OTHER-THAN-FLOW; ESTABLISHING A RELATIVE WEIGHT BETWEEN THE TWO; MULTIPLYING THE WEIGHT VALUE BY THE EQUIVALENT CLOSENESS-DESIRED NUMBER VALUE; AND THEN DIVIDING THE TOTALS AT CONSISTENT BREAK-OFF PLACES.

THE DIVISIONS WITHIN ANY MATRIX NEED NOT BE THE SAME FOR DIFFERENT PROJECTS, AS THE PLANNER SHOULD HAVE THE OPPORTUNITY TO END UP WITH A REALISTIC DISTRIBUTION OF NET RATINGS: SAY 1 TO 3% A's; 2 TO 5% E's; 3 TO 8% I's; 5 TO 15% O's; 20 TO 85% U's; 0 TO 10% X's.

FLOW AND OTHER-THAN-FLOW AT RELATIVE IMPORTANCE OF 1 TO 1

OTHER-THAN-FLOW

FLOW		4	3½	3	2½	2	1½	1	½	0	-1	-2
		A	A-	E	E-	I	I-	O	O-	U	X	XX
4	A	8	7½	7	6½	6	5½	5	4½	4	3	2
3½	A-	7½	7	6½	6	5½	5	4½	4	3½	2½	1½
3	E	7	6½	6	5½	5	4½	4	3½	3	2	1
2½	E-	6½	6	5½	5	4½	4	3½	3	2½	1½	½
2	I	6	5½	5	4½	4	3½	3	2½	2	1	0
1½	I-	5½	5	4½	4	3½	3	2½	2	1½	½	-½
1	O	5	4½	4	3½	3	2½	2	1½	1	0	-1
½	O-	4½	4	3½	3	2½	2	1½	1	½	-½	-1½
0	U	4	3½	3	2½	2	1½	1	½	0	-1	-2

FLOW CONSIDERED TWICE AS IMPORTANT AS OTHER-THAN-FLOW.

OTHER-THAN-FLOW

FLOW		4	3½	3	2½	2	1½	1	½	0	-1	-2
		A	A-	E	E-	I	I-	O	O-	U	X	XX
8	A	12	11½	11	10½	10	9½	9	8½	8	7	6
7	A-	11	10½	10	9½	9	8½	8	7½	7	6	5
6	E	10	9½	9	8½	8	7½	7	6½	6	5	4
5	E-	9	8½	8	7½	7	6½	6	5½	5	4	3
4	I	8	7½	7	6½	6	5½	5	4½	4	3	2
3	I-	7	6½	6	5½	5	4½	4	3½	3	2	1
2	O	6	5½	5	4½	4	3½	3	2½	2	1	0
1	O-	5	4½	4	3½	3	2½	2	1½	1	0	-1
0	U	4	3½	3	2½	2	1½	1	½	0	-1	-2

OTHER-THAN-FLOW TWICE AS IMPORTANT AS FLOW

OTHER-THAN-FLOW

FLOW		8	7	6	5	4	3	2	1	0	-2	-4
		A	A-	E	E-	I	I-	O	O-	U	X	XX
4	A	12	11	10	9	8	7	6	5	4	2	0
3½	A-	11½	10½	9½	8½	7½	6½	5½	4½	3½	1½	-½
3	E	11	10	9	8	7	6	5	4	3	1	-1
2½	E-	10½	9½	8½	7½	6½	5½	4½	3½	2½	½	-1½
2	I	10	9	8	7	6	5	4	3	2	0	-2
1½	I-	9½	8½	7½	6½	5½	4½	3½	2½	1½	-½	-2½
1	O	9	8	7	6	5	4	3	2	1	-1	-3
½	O-	8½	7½	6½	5½	4½	3½	2½	1½	½	-1½	-3½
0	U	8	7	6	5	4	3	2	1	0	-2	-4

FLOW CONSIDERED 1½ TIMES AS IMPORTANT AS OTHER-THAN-FLOW. (FLOW 3, OTHER THAN FLOW 2)

OTHER-THAN-FLOW

FLOW		8	7	6	5	4	3	2	1	0	-2	-4
		A	A-	E	E-	I	I-	O	O-	U	X	XX
12	A	20	19	18	17	16	15	14	13	12	10	8
10½	A-	18½	17½	16½	15½	14½	13½	12½	11½	10½	8½	6½
9	E	17	16	15	14	13	12	11	10	9	7	5
7½	E-	15½	14½	13½	12½	11½	10½	9½	8½	7½	5½	3½
6	I	14	13	12	11	10	9	8	7	6	4	2
4½	I-	12½	11½	10½	9½	8½	7½	6½	5½	4½	2½	½
3	O	11	10	9	8	7	6	5	4	3	1	-1
1½	O-	9½	8½	7½	6½	5½	4½	3½	2½	1½	-½	-2½
0	U	8	7	6	5	4	3	2	1	0	-2	-4

OTHER-THAN-FLOW CONSIDERED 1½ TIMES AS IMPORTANT AS FLOW. (FLOW 2, OTHER THAN FLOW 3)

OTHER-THAN-FLOW

FLOW		12	10½	9	7½	6	4½	3	1½	0	-3	-6
		A	A-	E	E-	I	I-	O	O-	U	X	XX
8	A	20	18½	17	15½	14	12½	11	9½	8	5	2
7	A-	19	17½	16	14½	13	11½	10	8½	7	4	1
6	E	18	16½	15	13½	12	10½	9	7½	6	3	0
5	E-	17	15½	14	12½	11	9½	8	6½	5	2	-1
4	I	16	14½	13	11½	10	8½	7	5½	4	1	-2
3	I-	15	13½	12	10½	9	7½	6	4½	3	0	-3
2	O	14	12½	11	9½	8	6½	5	3½	2	-1	-4
1	O-	13	11½	10	8½	7	5½	4	2½	1	-2	-5
0	U	12	10½	9	7½	6	4½	3	1½	0	-3	-6

Figure 3.5. Matrices for combining flow and other-than-flow relationships. (From Ref. 6. Copyright 1973, Richard Muther.)

```
Production Areas:
  Clerk Typists       19 people x  40 sq. ft.    760 sq. ft.
  Key Entry Operators  6    "    x  35 sq. ft.    210 sq. ft.
  Supervisors          1 person x  80 sq. ft.     80 sq. ft.
  Manager              1    "    x 120 sq. ft.    120 sq. ft.
                      Total Production Area      1190 sq. ft.

Service Areas:
  Copier, telex                                    60 sq. ft.
  Main Aisles                                     180 sq. ft.
  Storage, files                                  200 sq. ft.
                      Total Service Area          440 sq. ft.

                      Total Area                 1630 sq. ft.
```

Figure 3.6. The calculation method of determining space requirements. (Courtesy of Richard Muther & Associates, Inc., Kansas City, Missouri.)

Projecting Future Space Needs

Having addressed the issue of relationships, we turn to space, and to systematic techniques for projecting future needs. Muther has identified five techniques which can be used singly or in conjunction, depending upon the project. These are:

1. Rough layout
2. Calculation
3. Space standards
4. Conversion
5. Ratio-trend

When asked how much space is needed, too many planners reply, "I won't know until I lay it out." This is an unfortunate, costly reply. Constructing rough layouts is the slowest and therefore most costly technique of establishing space requirements. There are also many situations in which no layout can be made because furniture or equipment decisions are still pending.

The rough layout technique is legitimate in critical equipment areas with a high cost per square foot and a high degree of fixity. Spaces such as computer rooms, kitchens, and labs, or large, heavy-machining areas are best estimated through rough layout. When using this technique, the planner should avoid premature attention to detail. The objective at first is to determine the space required--not to make a good layout.

When planning for immediate installation, the calculation technique in conjunction with space standards is the most effective way to determine space requirements. A typical example is shown in Figure 3.6. Here, the planner lists the personnel, equipment, and support spaces making up each activity-area. In practice, the list is derived from the data collection and surveys covered in Chapter 2. A space standard is applied to each entry on the list. Extensions are made and totaled to arrive at space required.

The calculation technique takes time, but not as much as rough layout. Also, it requires good, detailed forecasts of manpower and equipment. In most organizations, these are subject to rapid change. So too are the space

WORKPLACE TYPE B

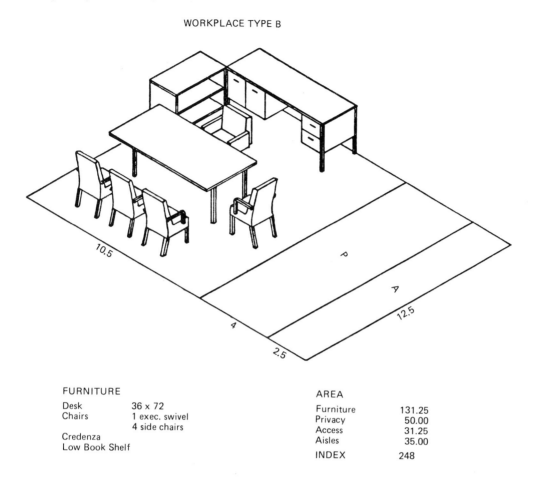

FURNITURE		AREA	
Desk	36 x 72	Furniture	131.25
Chairs	1 exec. swivel	Privacy	50.00
	4 side chairs	Access	31.25
Credenza		Aisles	35.00
Low Book Shelf		INDEX	248

Figure 3.7. Illustrated space standard. Area P is the privacy allowance for an office landscape. Area A is for access. An additional 35 sq. ft. is added for aisles and main circulation not shown.

estimates reached through calculation. Updates add to the already tedious nature of the technique, especially in large facilities. Here again, the computer is an ideal aid.

Space standards are essential in planning administrative and institutional facilities. They are also of some value in planning industrial facilities. When developing and using space standards, the planner must be sure to document the assumptions used. The three most important are: (1) the furniture or equipment used, (2) the aisle width between work places, and (3) the layout or arrangement of work places (e.g., the number of back-to-back stations between aisles). Variations from base assumptions can differ from the standard by 10-25%. Published standards, or those developed by others, should not be used without a clear understanding of their assumptions. The surest understanding is reached with a graphic representation like that shown in Figure 3.7.

```
The accounting area today:   3000 sq. ft.
It should be today:          3500 sq. ft.
  (Because of the production)

Adjustments:
  Increase production          +80%
  Improved methods             -20%
  Night shift                  -10%

New area (space):
3500 x 1.80 x 0.80 x 0.90 = 4550 sq. ft.
```

Figure 3.8. The conversion method of determining space requirements. (Courtesy of Richard Muther & Associates, Inc., Kansas City, Missouri.)

    The rough layout, calculation, and space standard techniques cannot be used without a detailed manpower and equipment forecast. Such forecasts are typically valid for only a few months into the future. Often, the planner must project departmental or activity-area requirements several years into the future. One way to do this is to guess at the future number of clerks, analysts, machines, file cabinets, etc. The results will always be "precisely wrong." In this situation, the conversion technique is best. A brief illustration appears in Figure 3.8.

    The planner surveys each activity-area and makes subjective assessments of current space utilization (tight or loose). The existing square footage is adjusted to what it should be today. Next, the planner interviews the supervisor(s) of the areas in question, and any others who might have knowledge of methods changes, growth or declines in activity, etc. Based on the findings of these interviews, subjective percentage factors are applied to the adjusted current square footage, yielding "ballpark" estimates of future needs.

    Now and then, planners must make long-range, 5-, 10-, and even 20-year projections of space requirements. These are needed for major capacity, location, and site planning decisions. Here, the calculation and conversion techniques are not meaningful. For such space projections, the ratio-trend technique (Figure 3.9) is the most practical. We should note, however, that it only works at the summary levels of space requirements. It does not work at the departmental level and below.

    To use ratio-trends, the planner must first collect historical records of space occupied--in total, and perhaps by 3-5 major classes or types of space. Historical data on business activity must also be collected for the same period-- sales, shipments, employees, etc. The space and business histories are plotted as ratios, with space as the numerator. The ratios will typically exhibit a trend that can be projected into the future. Given forecasts of business activity (the denominators), various estimates of space requirements can be derived from the ratio-trends. This technique lends itself well to statistical regression analysis and curve-fitting--natural uses of computer aids.

Graphical Layout Techniques

With our relationships and space estimates in hand, we are ready for the process of adjustment--converting vowel-letters and numbers into a layout plan or building concept, in the case of architectural design.

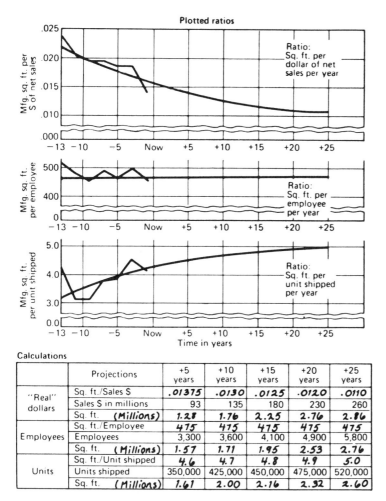

Plotted ratios

Calculations

	Projections	+5 years	+10 years	+15 years	+20 years	+25 years
"Real" dollars	Sq. ft./Sales $	.01375	.0130	.0125	.0120	.0110
	Sales $ in millions	93	135	180	230	260
	Sq. ft. *(Millions)*	1.28	1.76	2.25	2.76	2.86
Employees	Sq. ft./Employee	475	475	475	475	475
	Employees	3,300	3,600	4,100	4,900	5,800
	Sq. ft. *(Millions)*	1.57	1.71	1.95	2.53	2.76
Units	Sq. ft./Unit shipped	4.6	4.7	4.8	4.9	5.0
	Units shipped	350,000	425,000	450,000	475,000	520,000
	Sq. ft. *(Millions)*	1.61	2.00	2.16	2.32	2.60

Figure 3.9. Ratio-trend and projection. In the example above, three ratios are projected based on 13 years of history. Forecasts are obtained for the denominator of each ratio--dollars, employees and units. These, together with the trend-line estimates of future ratio values, lead to the future values for square feet, the numerator in each ratio. Three estimates of square feet result, ranging from 2.6 to 2.86 million in 25 years. Note that the trend lines can also help study changes in the nature of production operations. In the example shown, the ratio trend of sq. ft. per unit shipped would indicate either a growth in product size, increased vertical integration, increased inventories, or combinations of these three. (From Ref. 8, Vol. 2. Copyright 1980, Richard Muther.)

Good planning practice calls for an overall or block layout first--without the details of furniture or equipment. This is the Phase II Overall Plan we discussed in Chapter 1. Even on the smallest projects, a block plan should precede the details. This saves time in the long run, by getting a firm decision early on overall concepts of flow and department or work group position.

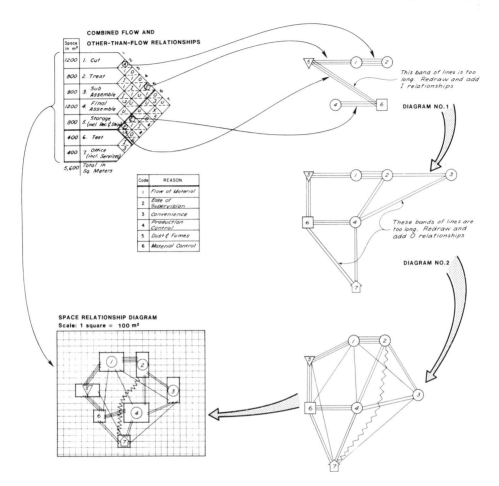

Figure 3.10. Graphical layout technique. (From Ref. 8, Vol. 2. Copyright 1980, Richard Muther.)

There are two systematic ways to develop a block plan. One is the graphical layout or diagramming technique developed by Richard Muther. The other is the use of one or more layout planning algorithms. We will look first at the graphical technique shown in Figure 3.10.

This simple technique converts vowel-letters into a diagram, using a number-of-lines code, and a length-of-line scale. The numbers of lines correspond to the point values used earlier in our combining technique (A = 4, E = 3, I = 2, etc.). The length of the lines is inversely proportional to the strength of the vowel rating. A-ratings are shortest, with others in relation to the length set for an A. The absolute lengths are arbitrary. A common scale in inches is: A = 1/2, E = 3/4, I = 1-1/2, O = 2 and up. U-ratings do not appear, and X-ratings are shown as long, wiggly lines.

The planner scans the triangular relationship chart, building a diagram in the center of a blank piece of paper. "A's" are posted first, then "E's," etc. The diagram is redrawn after each vowel posting, if the length-of-line scale has been exceeded. Symbols are used to represent different types of activity-areas. The resulting diagram represents a theoretically best arrangement, without regard to space.

The finished diagram is "exploded," and symbols redrawn to represent the space of each activity-area. The final step is to fit the activity-areas into the space available, adjusting the shapes as required, while honoring the diagrammed positions of each activity-area--especially those connected by 3 or 4 lines.

This technique, first published in 1960, has been translated into over 10 languages, and is used successfully by several thousand planners around the world. It can be extended to multi-story (and multi-building) plans as shown in Figure 3.11. Clusters are defined by inspecting the finished activity relationship diagram. Space is totaled for each cluster, and the diagram is redrawn to show closeness desired between clusters. The planner makes trial-and-error assignments of clusters to floors, attempting to keep closely related clusters on the same or adjoining floors.

The graphical layout technique gives good results in most single-story situations, but it is very time-consuming in multi-story situations, especially if done manually for very large problems.

Layout Planning Algorithms

In mathematics, an algorithm is a special method of solving a certain kind of problem, generally with repetitive calculations. It is a set of rules expressed in numerical terms and values. By using certain limiting assumptions, we can reduce the development of block layouts to such numerical rules. Since the early 1960s, several dozen algorithms have been published for the purpose of automatically generating block layouts.

In theory, all algorithms are manually executable. In practice, the computer is an essential aid because of the thousands of repetitive calculations involved. Unfortunately, too few planners take the time to explore algorithms manually and thus fully illuminate their logic and limiting assumptions. Because layout planning algorithms are inherently computer aids, we will wait until Part II, Chapter 5, to examine them in full.

Analytical Design Techniques

The block layout gives a new facility its underlying form. It is a basic input to the design of the building enclosure or structure. It also defines the requirements for utilities distribution. At this stage of the facilities planning process, the primary tasks and techniques involve engineering. As such, they must use accepted calculations and adhere to standard procedures. Some of the commonly used analytical techniques grouped by major task are:

1. *Site design*: hydraulic and hydrologic calculations, topographic analysis, earthwork calculations, retaining wall design, etc.
2. *Structural design*: finite element analysis, static and dynamic analysis, seismic and wind force distribution, and beam, column, and slab design.

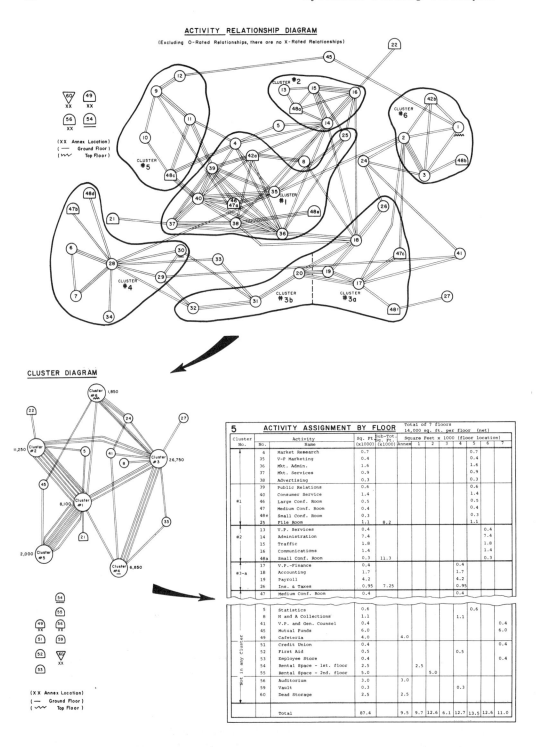

Figure 3.11. Multi-story graphical technique. (From Ref. 6. Copyright 1973, Richard Muther.)

3. *Heating, ventilation and air-conditioning (HVAC) design*: energy load calculation, heat loss, gain, and recovery calculations, duct design
4. *Electrical design*: line sizing and distribution, short circuit analysis, lighting loads, lighting design, etc.
5. *Piping design*: service supply sizing, chilled water sizing, fire sprinkler layout and sizing
6. *Interior design*: acoustical analysis, floor-, wall- and window-covering calculations, ceiling grid calculations

These are but a few of the techniques applied in Phases II and III of the total planning process defined in Chapter 1. The details of these techniques and the use of computers to aid them are part of a field known as computer-aided engineering. This subject is largely beyond the scope of this book, but we should note in passing that the kinds of systematic techniques involved are ideal for computers and in many cases require them for proper application.

Cost Justification

Once plans and designs are complete, the project's cost and economic feasibility can be determined. Assuming an initial feasibility study has already been made, and that funds are available, the planner's task is to choose the most cost-effective plans or designs from among several alternatives. The best are presented to management for approval.

Every organization has its own approach or rules governing cost justification. Most organizations have published these in the form of standard procedures and calculations. All rest on accepted financial and accounting practices and techniques. Payback and discounted cash flow are the most popular techniques. Both can be formulated in a number of ways, but in every case the calculations amount to simple arithmetic. They can be done manually, but for large projects they are best supported with such aids as the "electronic spreadsheet" programs now available on personal computers.

Project Scheduling and Management

Once justified and approved, the facilities plan is implemented using project scheduling and management. These are, by their nature, very systematic techniques. On small projects, installations are scheduled and managed using simple Gantt or bar charts like the one shown in Figure 3.12.

On large projects, PERT (Program Evaluation and Review Technique) or CPM (Critical Path Method) is used to sequence and schedule the many interrelated events. See Figure 3.13. Progress is reported against the schedule. Expense may also be reported and managed if a job or project costing procedure has been established and tied in with the project schedule. Both PERT and CPM are highly systematic, rule-based techniques. They can be applied manually, but as a practical matter they must be computer-aided. We will examine some of the commercially available aids in Part II.

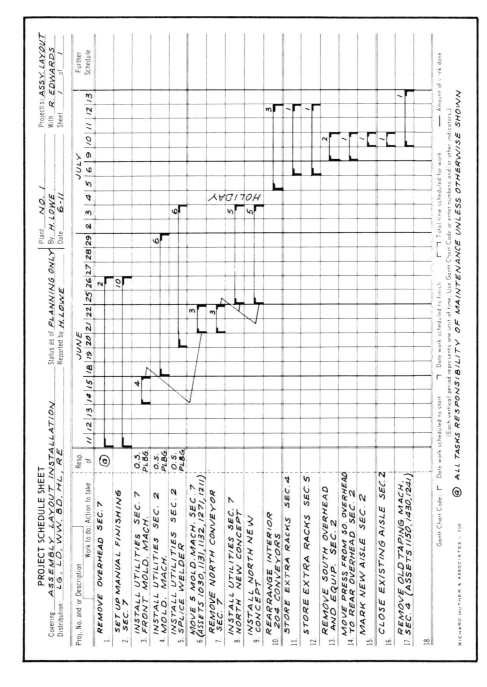

Figure 3.12.   Gantt chart.   (From Ref. 8, Vol. 2.   Copyright 1980, Richard Muther.)

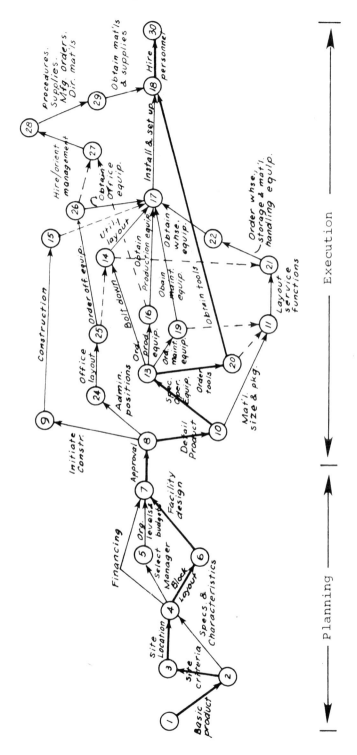

Figure 3.13. Critical path network diagram. (From Ref. 8, Vol. 2. Copyright 1980, Richard Muther.)

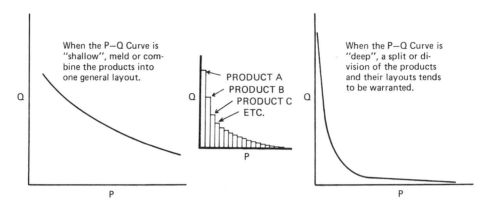

Figure 3.14. Product-Quantity (P-Q) Analysis. (From Ref. 6. Copyright 1973, Richard Muther.)

Analytical Techniques for Industrial Facilities

The techniques covered so far are applicable to all types of facilities. Industrial planners often need additional, special techniques to address the complexities of manufacturing and production control. We will consider four particularly systematic techniques which are useful in a variety of industrial facilities. The four are:

1. Product-Quantity (P-Q) Analysis
2. Group Technology
3. Distance-Intensity (D-I) Plot
4. Quantified Flow Diagram

Product-Quantity or P-Q Analysis is a phrase coined by Richard Muther to cover facilities applications of Pareto's curve, often called the 80/20 rule (6). In every industrial plant, activity is devoted to many products or materials. When the quantities of each product or material are plotted in rank order, the result is a P-Q curve like the ones shown in Figure 3.14. A shallow curve (no dominant products or materials) suggests that the facility should be planned as a general purpose or job-shop operation, with the internal layout organized around the general process sequence, e.g., forming, treating, fabrication, assembly, test, etc. Most or all products are expected to pass through each process area.

A deep curve (one or a few dominant products or materials) suggests that the facility be split into a dedicated, special area for high volume, dominant products, and a general purpose or job-shop area for the rest of the low-volume products or materials. The special, high-volume area often employs production line techniques.

The P-Q curve also suggests which technique of flow analysis should be used in planning the layout and material handling methods. The use of the curve appears in Figure 3.15. The "A" range includes one or a few high-

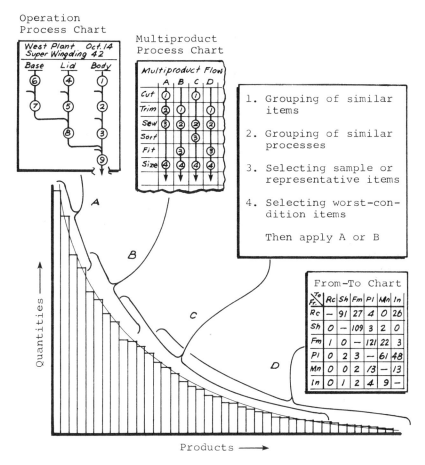

Figure 3.15. Method of analysis varies with product-mix. (From Ref. 8, Vol. 2. Copyright 1980, Richard Muther.)

volume products. These are analyzed with operation process charts. The "B" range includes a few more products or materials and requires the use of multiproduct process charts. The "C" range has too many entries to analyze each explicitly. So, simplifying steps are taken and the process chart methods are applied. In the "D" range the best approach is a statistical summary using the from-to chart.

Group technology is a collection of techniques that identify and bring together related or similar components in a batch manufacturing process. Its use improves the production economies for medium- and low-volume items (3). These are the items in the "C" and "D" range of the typical P-Q curve, or possibly the entire curve if it is shallow. Group technology can also reduce work in process inventory, and aid in standardizing and reducing the number of parts in machined products.

The results of group technology are often called cells or flexible manufacturing systems, although the latter can also connote the application of robotics for material handling and machine operation. Instead of grouping all like equipment by individual process step (e.g., all lathes, all grinders, all broaches),

Diameter \ Length	0 0-20	1 20-50	2 50-100	3 100-150	4 150-240	5 240-400	6 400-600	7 600-1000	8 1000-2000	9 2000
0 · 0-20	Capstan work 1. Round bar 2. Hex, bar		Not supported, m/c'd from one side			Shafts 1. Round bar 2. Hexagonal bar Machine between centers Probably on center-lathe				
1 · 20-50										
2 · 50-100			Cylinders (Medium-size lathes)							
3 · 100-150	Disks Suitable for short-bed lathes									
4 · 150-240										
5 · 240-400										
6 · 400-600	Vertical lathes and borers Workpieces of very large diameters not possible to machine on standard lathes					Very large rotational parts Capacity failure				
7 · 600-1000										
8 · 1000-2000										
9 · >2000										

Figure 3.16. Length/diameter grouping of rotational parts used in machine tool production. (From Ref. 3.)

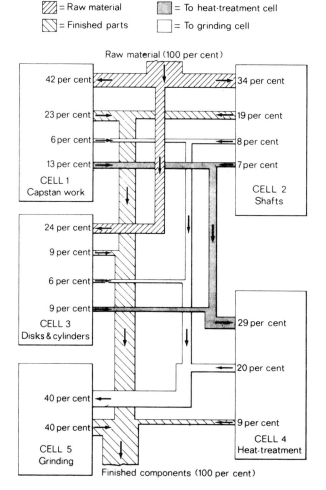

Figure 3.17. Work flow analysis for the rotational parts grouped in Figure 3.16. (From Ref. 3.)

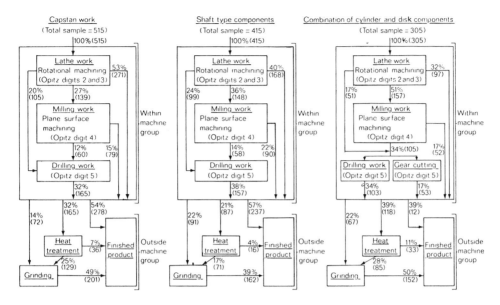

Figure 3.18. Detailed work flows for three parts families. (From Ref. 3.)

each cell contains the equipment necessary for a series of process steps on a family of parts or products.

A number of parts coding and classification schemes have been published to aid in the analysis and identification of cells. Traditional techniques of flow analysis are also used. The application of group technology to a machine shop layout is shown in Figures 3.16 through 3.19.

The Distance-Intensity Plot, published by Muther and Haganaes, is a simple yet powerful technique for evaluating layout plans and designing material handling systems (7). The intensity (rate) of material flow for each product over each route is plotted against the length or distance of the route. The resulting scatter plot can be interpreted as shown in Figure 3.20.

The arithmetic product of intensity times distance is defined as transport work and is generally proportional to the ultimate cost of moving the materials in question. Notice that the area "behind" each point on the plot represents transport work. The sum of all the areas is the total transport work associated with the underlying facility plan. This is an excellent scoring device for comparing alternative layouts. It forms the basic criteria used to measure results in many layout planning algorithms. Obviously, the objective is to keep the distance short on high-intensity routes.

The quantified flow diagram is another graphic technique for visualizing transport work. It is done directly on the layout plan using the width of a flow line to represent intensity of material flow. The area enclosed by the flow line (intensity or width times distance or length) represents transport work. An example appears in Figure 3.21.

In this chapter, we have seen a representative sample of systematic planning techniques--the kinds of step-by-step procedures which are a necessary foundation for computer-aided facilities planning. In Part II, we will see how these and other techniques have been converted into a wide range of commercially available computer aids.

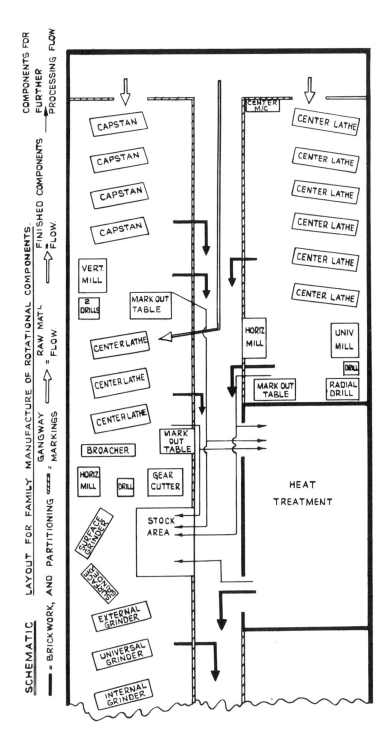

Figure 3.19. Machine shop layout, including the three cells defined in Figure 3.18. (From Ref. 3.)

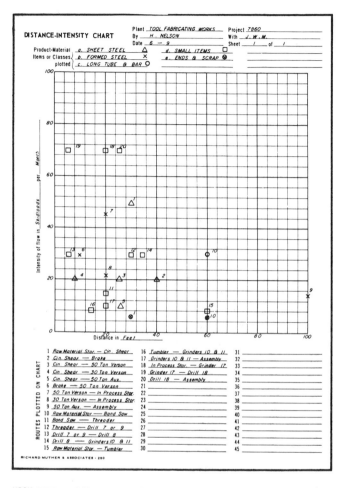

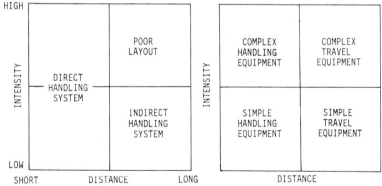

Figure 3.20. Distance-Intensity (D-I) Plot showing the flow of each material class on each route. Symbols indicate material class. The number beside each symbol refers to the route. The position of the various points suggests the types of handling systems and equipment to be used. Points in the upper right quadrant suggest layout improvements. (From Ref. 7. Copyright 1969, Muther International.)

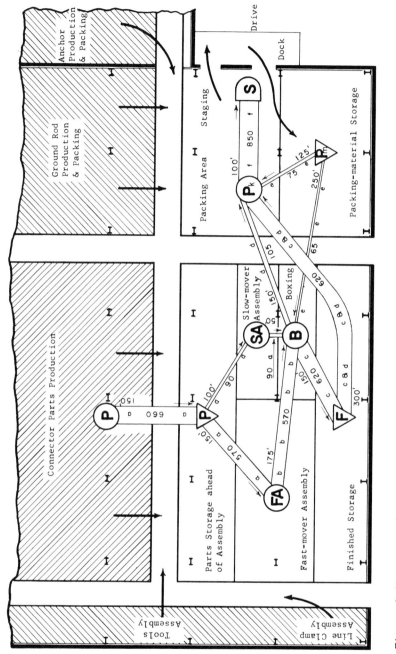

Figure 3.21. Quantified flow diagram of assemble, store, pack, and ship for electrical connectors plant. The width of flow lines corresponds to intensity. Direction and actual route distances are shown alongside each flow line. Lower-case letters indicate material class. (From Ref. 7. Copyright 1969, Muther International.)

Part II

HOW COMPUTERS CAN HELP

4

Uses of Computer Technology

Available Computer Technologies

The terms "computer, computer system, and computer aid" are almost meaningless today without some further definition. As facilities planners, we need to define and understand these five computer technologies:

1. Decision Support Systems (DSS)
2. Computer-Aided Engineering (CAE)
3. Computer-Aided Design (CAD)
4. Word Processing (WP)
5. Management Information Systems (MIS)

Each of these technologies can benefit facilities planners. To apply them successfully we must first understand their general purpose and their relationships to different types of hardware, software, and configurations.

Decision Support Systems (DSS)

This use of computers applies statistics, mathematical algorithms, and various calculations to address frequent planning issues and decisions. The term "DSS" is most frequently used to describe interactive financial planning models. In the context of facilities planning, decision support systems can be applied to such issues as: Whether to modernize or build new? How much space to provide? Own or lease? Also included would be the issue of block layout or best arrangement of departments and spaces. Each of these decisions can be reached with structured or analytical approaches. While there may be no single "right" or "best" answer, rules and calculations can be coded into the computer to arrive at "good" solutions or estimates. Or, the computer can be used to calculate the implications of different alternatives presented by the planner--a "what if?" approach to facilities planning decisions.

DSS is an applications software concept. Any computer, from personal to mainframe, can be used depending upon the scope of the specific application involved (19, 46).

A wide variety of decision support models and calculation routines are commercially available to the facilities planner. We will see several examples in the next two chapters. We should note, however, that decision support systems may need custom-tailoring to fit specific situations. And in some cases--simulation models, for example--a DSS must be built and not bought, as one might purchase a word processing or even a CAD system.

Computer-Aided Engineering (CAE)

This computer technology uses extensive calculation routines to model intended structures and designs. The purpose of the models is twofold: first, to insure compliance with professional standards and codes, and second, to compare the costs and performance of different design concepts. In the latter regard, CAE may also use simulation models to show the probable response of a design to differing physical forces. CAE routines also calculate the proper sizes and numbers of structural members, components, and materials--including such items as ductwork, piping, insulation, lighting, electrical service, etc.

Sophisticated CAE uses interactive graphics as an input medium. Intended designs are drawn at a terminal using a CAD system. The relevant characteristics of the finished design are fed into the CAE routines. The results may even be graphically displayed. In structural analysis, for example, the deflection of a structure under various stresses can be displayed on a graphics terminal or plotted on a drawing.

Like DSS, computer-aided engineering is largely a software concept that can be realized on a variety of devices. However, the complexity of some engineering calculations may require the use of large minicomputers or mainframe installations. CAE is a well-developed field. Hundreds of programs are commercially available for a wide range of specific tasks. These programs will run on a wide variety of computers and are also available through a number of time-sharing services.

Computer-Aided Design (CAD)

This technology employs interactive graphics to visually develop plans and designs at a computer terminal. The finished work can then be converted automatically to drawings and bills of materials. With facilities-oriented CAD, special calculation routines are used to compute areas, lengths, and cubes of elements in a plan. As noted above, there may be links to CAE routines for structural analysis or other engineering-related calculations.

CAD is a well-developed technology and has been successfully employed by many facilities planners, architects, and engineers. Over 40 companies sell complete CAD systems (hardware and software included). We will review these systems completely in Chapter 7. It is also possible to purchase CAD software only, for installation on existing equipment.

Word Processing (WP)

This readily available computer technology aids the planner in documentation--bills of material, bid lists, specifications, and reports. Word processing software can be purchased for use on almost any existing hardware. And, of course, complete word processing systems can also be acquired. Word processing is included here for the sake of completeness. Because its use is already widespread, and it is not central to facilities planning, we will not discuss it further.

Management Information Systems (MIS)

After computation itself, information storage and retrieval was the next historical use of computers. Over the years, the construction of files, and the input, update, security, and reporting of data from these files, has come to be called MIS. This technology usually incorporates a number of simple calculation routines to aid in the summarization and statistical analysis of stored information.

Most facilities planners employ some form of MIS technology. Lists of corporate assets--buildings, furniture, machinery, and equipment--are the most common form of MIS. Typically, an MIS is used to produce regular reports on asset or facilities utilization. These might be monthly, quarterly, or annual. An example of a standard quarterly report would be a listing of:

> all assets disposed of in the prior quarter, including book
> value, replacement cost, age, and condition

A good MIS also permits the planner to ask for unanticipated information such as:

> the replacement costs of all vertical boring mills in the aero-
> space division which are over 3 years old

> locations of buildings with 5000 sq. ft. of unused space which
> is carpeted, with good views, off-street parking, and access
> for the handicapped

In later chapters we will see examples of commercially available information systems for facilities planners. In practice, these products will need some tailoring to meet individual planners' needs. All types of hardware can be used--from personal computers to mainframes--the key determinants being the amount of good data already stored in an existing system, and the ultimate storage capacity required to accommodate the data.

Hardware Issues

Because hardware is the most visible aspect of computer technology, it often dominates the planner's decisions with respect to computer aids. This is unfortunate, since hardware today is often a matter of convenience. Far more emphasis needs to be placed on software and its impact on system performance. Still, the planner needs to be aware of these four devices:

1. Personal computers
2. Microcomputers
3. Minicomputers
4. Mainframe computers

Today's personal computers (PCs) can perform many tasks required in facilities planning. They are ideal for small DSS applications. They can do much of what is routinely required for CAE--all but the strenuous structural analysis computations and large simulations. There are even limited CAD systems available that employ personal computers. PCs are ideal for word processing and for MIS applications where the need for data storage is limited. A growing number of PCs can act as terminals to larger computers--micros, minis, or mainframes. They can also be linked in networks with other PCs to share files and programs, as well as printers. In the absence of a network, data and programs can be easily transferred by copying onto disks or tape cassettes and loading them into another computer.

Personal computers are not designed to give simultaneous data access to multiple users. Each must have his own copy. In some applications this may be a limitation. Memory is another limitation when compared to larger microcomputers. Memory indirectly determines speed and capacity for certain types of applications. There is a risk that initial uses of a personal computer may grow beyond the power of the machine. If so, the user may have to start over with new programs on a larger device--micro, mini, or mainframe. Personal computer hardware will be discussed further in Chapter 6.

Microcomputers range from what might be called powerful personal computers on the low end to small minicomputers on the high end. In the middle range, microcomputers offer more computing power and internal memory than the typical PC. They can also provide several users with simultaneous access to the same data. Most can act as terminals to larger minicomputers and mainframes. Networks of microcomputers can also share data, programs, and peripheral equipment, such as printers and graphics plotters.

Like PCs, microcomputers can do almost all the tasks of facilities planning--all but the structural analyses of CAE. Micros can be linked to disk and tape drives for mass storage, so there is essentially no limitation imposed on their use by large amounts of data. The limitation of CAE can be overcome by using the micro as a terminal to a minicomputer or mainframe. Microcomputers are also very popular in CAD and word processing systems.

Minicomputers (and mainframes) differ in a fundamental way from personal and microcomputers. Minis are a powerful, shared resource for dozens of simultaneous users. PCs and most micros, however, are dedicated resources for one user at a time. A minicomputer, coupled with mass storage devices (tapes or disk), can provide easy access to large central files and common analytical routines. Such systems are simply scaled-down versions of the typical mainframe computer which supports hundreds of terminals in a time-shared or remote-job environment.

Minicomputers are powerful enough to perform all facilities-related tasks, including the structural analyses of CAE. They are also popular as the computing resource in large multi-station CAD systems, although this use is being challenged by the growing popularity of intelligent CAD workstations using microcomputers.

Mainframe installations, of course, can handle all facilities requirements with ease. They can also be used to run CAD software, provided the proper terminals and input and output devices are also present.

The problems in practice with minicomputers and mainframes are access, response, downtime, and, indirectly, cost. Because of their powerful but expensive nature, these machines must be shared by multiple users. Even minicomputer-based CAD systems are designed to be used in a multiple-station arrangement. As more users seek to perform more tasks, access must be managed and often restricted. Response times lengthen, reducing perform-ance. And, if the machine fails, all users are deprived of their resource.

As a practical matter, most corporate facilities planners cannot get enough unrestricted access to load all their desired applications on existing mainframes or minis. Fortunately, the personal computer and microcomputer provide a cost-effective way to meet the planner's needs. Today, there is little reason (other than available capacity) to use a mini or mainframe for facilities applica-tions, provided the chosen PC or micro can "grow" through easy links to stor-age devices, other micros, or even minis and mainframes. This will reduce the chance of outgrowing the device and any device-specific programs that run on it.

In your author's opinion, each type of hardware has its place as follows:

Personal computers: Best for limited calculation routines and small models needing no link to other devices, and no simultaneous data access. Use for applications that will not outgrow the device.

Microcomputers: Good for most facilities applications including CAD. Use as a means to integrate a variety of technologies--DSS, CAE, CAD, WP, MIS--into a single facilities planning system. A micro is the ideal "de-partmental" computer.

Minicomputers and mainframes: Use when already available, provided: (1) the new applications will not overload the system, and (2) the facilities planner can have ready, relatively unrestricted access to the machine.

The relationship between our five computer technologies and the hardware just discussed is summarized in Figure 4.1.

Software Issues

Hardware is useless without the software to make it run. Three types of soft-ware are required for effective facilities planning applications. These are:

1. *The operating system (OS)*: "central nervous system" of a computer which coordinates the actions of keyboards, displays, central process-ing unit (CPU), main memory, disk drives, printers, etc. Also, the interface between applications programs and the CPU.
2. *The data base management system (DBMS)*: an optional collection of programs which control the creation, modification, and use of a cen-tral or common data base. The DBMS is an interface between the OS and the applications programs.
3. *Applications programs*: the coded routines employed by the user to enter, retrieve, and perform operations on his data.

All computers come with an operating system. Often the buyer has a choice between a proprietary operating system of the manufacturer's own design, or a "standard" system written by a separate software firm. The initial choice is im-portant, since all applications programs must be written to run under a specific OS. If there is ever a need to change from one machine to another, using a

COMPUTER TECHNOLOGY	HARDWARE			
	PERSONAL COMPUTER & FLOPPY DISK	MICROCOMPUTER & HARD DISK	MINICOMPUTER DISK & TAPE	MAINFRAME SYSTEMS
DSS DECISION SUPPORT SYSTEMS	Adequate for many routine calcula-tions and some algorithms. May have trouble access-ing existing data files on minis and mainframes	Adequate for all tasks. May need extra memory for large algorithms.	Adequate for all tasks. Not needed as a dedicated device.	Adequate for all tasks. Not needed as a dedicated device.
CAE COMPUTER AIDED ENGINEERING	Adequate for most sizing calcula-tions. Not adeq-uate for heavy structural analy-sis or large energy simula-tions. May have trouble communic-ating with larger devices for this purpose.	Adequate for most sizing calcula-tions. Can comm-unicate with minis and main-frames for heavy structural analy-sis and large simulations.	Adequate for all tasks. (Large minis required for heavy struc-tural analysis)	Adequate for all tasks. Not needed as a dedicated device. May en-counter conten-tion for lengthy analysis and simulation.
CAD COMPUTER AIDED DESIGN	Adequate for small installa-tion drawings and layouts. Limited commands. Slow, restricted capac-ity.	Adequate for all tasks when sup-ported by extra disk storage. Sophisticated commands. Fast response. No de-gradation with multiple users.	Adequate for all tasks. Supports multiple users but degrades response. Must be dedicated to CAD.	Adequate for all tasks. Not needed as a dedicated device, but may encounter con-tention with non-CAD users.
WP WORD PROCESSING	Adequate. Some inconvenience due to limited storage.	Adequate for all tasks. Unecessary as a dedicated device.	Adequate for all tasks. Not needed as a dedicated device.	Adequate for all tasks. Not needed as a dedicated device.
MIS MANAGEMENT INFORMATION SYSTEMS	Limited due to storage. Little capacity for large-scale computations.	Adequate with access to extra storage.	Adequate for all tasks. Not needed as a dedicated device.	Adequate for all tasks. Not needed as a dedicated device.
PRICE CPU & STORAGE	$1 - $5K	$10 - $50K	$25 - $200K	$200K and over

Figure 4.1. The relationship between technology and hardware.

different operating system, the applications programs may have to be rewritten. And any data base may have to be reconfigured and reloaded.

It would be nice if all computers used the same operating system. Unfortunately this is not the case. The situation today is somewhat similar to the early spread of electricity with its great variety of voltages, phases, and cycles. As a result, electrical machinery and appliances were not nearly as portable then as they are today.

While a common operating system has not appeared, a de facto standard has emerged at least for smaller personal computers. Those built around the 8-bit, Z-80 microprocessor from Zilog either come with or can be made to use the CP/M operating system. CP/M stands for Control Program for Microcomputers. It is a product of Digital Research. Apple and Tandy--two leading makers of 8-bit PCs--use proprietary operating systems. A card can be purchased, however, to convert them to CP/M.

A variety of operating systems are used on larger, 16-bit PCs. The IBM-PC and others come equipped to use PD/DOS or MS/DOS from MicroSoft. The UCSD p-System from SoftTech Microsystems and the Pick Operating System from Pick Systems are also used, as is CP/M-86, an extension of CP/M for 16-bit microprocessors. Apple and Tandy have remained with proprietary operating systems on their 16-bit machines. UNIX, a product of Bell Labs, is the emerging standard for microcomputers using the Motorola family of 16- and 32-bit microprocessors. All of these operating systems are designed to deliver multiple-task, multiple-user capabilities, taking advantage of the larger memories and greater speed of 16- and 32-bit devices.

Several products allow one operating system to "talk" to others. For example, UNIX can operate as a "host" for CP/M, MS/DOS, and the p-System. Others allow one computer to operate either of two systems such as CP/M and MS/DOS if both are loaded onto the disk. Some 16-bit machines come with an 8-bit Z-80 too, allowing the user to run software for each.

The word "standard" is used with caution in any discussion of operating systems. Planners can expect to find different versions and releases of every system just named. This confusion is compounded with different disk formats, and even different editions of the same model computer. The result is extreme difficulty moving any program or data from one computer to another. Where personal computers are concerned, hardware is portable; software generally is not.

Minicomputers and mainframes are sufficiently complex to require proprietary operating systems. Most manufacturers insure portability of programs across different models in their own product lines. However, there is little direct portability among different makes.

A data base management system (DBMS) represents an optional but useful software investment. In facilities planning work, we find ourselves using the same basic data over and over in different ways--square footage of buildings, occupancies, lists of furniture, machinery and equipment, head counts by department--all have multiple uses. On a large facilities planning staff, several people may need recourse to the same basic data at the same time. And, more importantly, this data may be updated or modified by more than one person.

As we develop computer aids, we might choose to create separate data files for each applications program. However, we will find ourselves entering and storing the same basic data in several places. This is tedious and wasteful and leads inevitably to updating problems.

To improve on this situation, we might make up a list of basic data files, name them, and "call" them as needed from our applications programs. This eliminates the redundancy of entry and storage and solves much of the updating problem. However, the inevitable results are contention by simultaneous users and also maintenance problems with the applications programs. Any change in file structure requires us to modify all the applications programs using the file.

If we are going to have a common data base, the only practical way to do so is with a data base management system. A DBMS is really a tool of the applications programmer, not the facilities planner (unless they happen to be the same person). Still, the presence of a DBMS offers some valuable features to the planner, as outlined in Figure 4.2.

```
1. Permits changes in the organization and structure
   of the data-base without the need to re-write the
   applications programs which use it.

2. Edits all inputs to the data-base and rejects likely
   errors for review.

3. Provides for the re-establishment of the current
   data-base if storage disks are destroyed or damaged.

4. Allows multiple users to have simultaneous access
   to the data-base without creating errors, conflicts,
   or confusion.

5. Lets the user or programmer write applications in
   any of several languages -- whichever is most
   convenient.

6. A standard (non-proprietary) system provides some
   portability of the data-base and applications
   software; i.e. it may let the user move from one
   manufacturer's device to another.
```

Figure 4.2. What a DBMS does for the end-user.

A variety of data base management systems are commercially available for use under several operating systems and a wide variety of processing units. Several computer manufacturers offer their own proprietary DBMS as well. The use of these, however, limits portability to the manufacturer's product line. This may or may not be a problem for the facilities planner.

With an operating system and DBMS in place, we can install or write our applications software. To do so, we must use a programming language. At this point, we can use only those languages recognized by the CPU, operating system, and DBMS. Most systems today will support any of the following: BASIC, FORTRAN, Pascal, C, PL/1, COBOL, RPG, and Assembler. However, the support of a programming language requires "overhead" resources--primarily memory. For this reason the number of languages supported simultaneously is limited on personal computers. The planner needs to also be aware that each language has different versions. Some manufacturers provide what is essentially a proprietary version of BASIC, for example. These variations are a further impediment to portability, requiring typically minor modifications to make programs work on different brands of hardware.

Figure 4.3 summarizes the relationships we have been discussion between hardware and software. The issues involved suggest that considerable planning and foresight be devoted to the investment in computer aids. The facilities planner or manager should have a clear idea of where he wants to go with data bases and applications, and some idea of how to get there.

Configurations

Up to now, we have assumed that the facilities planner either has access to an existing computer or plans to obtain one for dedicated use. This is in fact the most common situation. However, the development of sound computer aids should not be defined solely on the basis of what is already available. The planner who wants the greatest benefits from computer usage should have a long-term (3-5 year) configuration in mind. There are five basic choices:

CPU	PERSONAL COMPUTER	MICROCOMPUTER	MINICOMPUTER	MAINFRAME
	Typically 8-bit microprocessor	Typically 16-bit microprocessor	16- or 32-bit processors	32- or 64-bit processors
OPERATING SYSTEMS	Proprietary or standard. CP/M is most popular standard.	Proprietary or standard. UNIX and MS/DOS are most popular standards.	Typically proprietary but will communicate with other systems such as CP/M, MS/DOS and UNIX.	
DATA BASE MANAGEMENT SYSTEMS	Proprietary or standard. Most standard DBMS will work in several CPU/OS environments.			
PROGRAMMING LANGUAGE	Typically BASIC. Often in proprietary version.	BASIC, PASCAL, C, FORTRAN, PL/1		
APPLICATIONS PROGRAMS	Proprietary, third-party, or user-written. Must be written in a language recognized by the DBMS and operating system.			

Figure 4.3. The relationship between hardware and software.

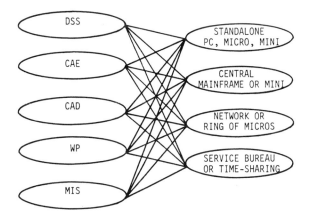

Figure 4.4. Configuring today's technology.

1.   Standalone PCs, micros, or mini
2.   Central mainframe or mini
3.   Network or ring of intelligent (micro) terminals
4.   Service bureau or time-sharing service
5.   Combination of the above

As shown in Figure 4.4, any one of these configurations can be used with any of our five technologies. In short, there is no "one best way." Choice of configuration is generally a function of:

1.   Current availability
2.   Budget

STAGES	PHASES	PLANNER'S TASKS AND DECISIONS	COMPUTER AIDS	TECHNOLOGY
PLAN	0	• Forecasting requirements	• Calculations; curve-fitting and extrapolations	DSS
		• Product-mix analysis	• P-Q plot	
	I	• Preliminary site planning		CAE
		- Drainage analysis	• Hydrologic and hydraulic calcuations	
		- Cut and fill analysis	• Topographic and earthwork calculations	
		- Site layout	• Coordinate geometry calculations	
			• Graphic layout procedure	
	II	• Block layout planning		DSS
		- Communications analysis	• Scoring calculations; cluster analysis	
		- Material flow analysis	• Flow diagrams; process charting	
		- Block layout	• Graphic procedures; layout algorithms	
		• Material handling analysis	• Process simulation; calculations	
		• Preliminary building design	• Structural analysis; beam, column, and slab calculations; energy calculations	CAE
			• 2-D; 3-D drafting	CAD
	II-III	• Preliminary utilities design	• Load and sizing calculations; circuit and routing analysis	CAE
			• 2-D drafting	CAD
	I-III	• Cost justification	• Life-cycle costing; interactive cash flow models	DSS
	III	• Detailed layout planning	• Graphic procedures; 2-D drafting	CAD
		• Detailed building design	• 2-D; 3-D drafting; solids modeling	
		• Detailed utilities design	• 2-D; 3-D drafting; solids modeling	
		• Interior design	• Area and surface calculations; acoustical analysis	CAE
		• Cost estimating	• Bill of materials take-offs; pricing; report generator	CAD
		• Specifications	• Text editing	WP
		• Project scheduling	• PERT/CPM	MIS
PROVIDE	IV-V	• Bidding	• On-line data entry and edit; data base management system; report generator	MIS
		• Puchase/work order generation		
		• Project management		
REVIEW	ONGOING	• Asset management	• Data base management system; report generator	MIS
		- Inventory control		
		- Financial reports		
		- Utilization studies		
		• Lease management	• On-line data entry and edit; data base management system; report generator	MIS
		- Expiration review		
		- Special studies		

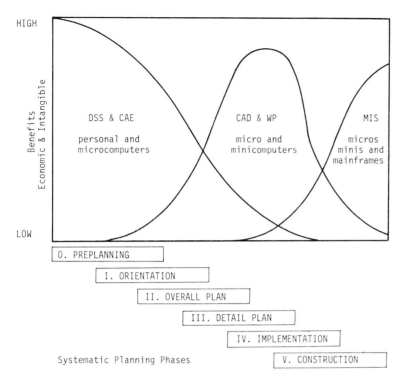

HIGH

Benefits
Economic & Intangible

DSS & CAE

personal and
microcomputers

CAD & WP

micro and
minicomputers

MIS

micros
minis and
mainframes

LOW

0. PREPLANNING

I. ORIENTATION

II. OVERALL PLAN

III. DETAIL PLAN

IV. IMPLEMENTATION

Systematic Planning Phases

V. CONSTRUCTION

Figure 4.6. Project benefits of computer technology. Different technologies benefit different phases of a facilities planning project. The data bases of MIS can also be the data sources for DSS. (Adapted from Ref. 19.)

   3.   Frequency, urgency, and extent of use
   4.   Long-term plan

   In the face of this complexity, the advice of an outside consultant or an objective DP professional may be of value. So too is the information that can be gained from newsletters, journals, and conferences and other technical gatherings. (See Appendix A.)

Aids to Facilities Planning

Until now, we have kept our discussion of computer aids at a very general level. Our purpose has been to first define five computer technologies, each

Figure 4.5. A list of computer aids, organized by stage of the facilities management cycle, and by the systematic phases of planning. Note the progressive application of five computer technologies. This list is somewhat abbreviated. The tasks required for process plant design are covered only indirectly. Many of the documentation tasks where word processing is used have been omitted. (Adapted from Ref. 19.)

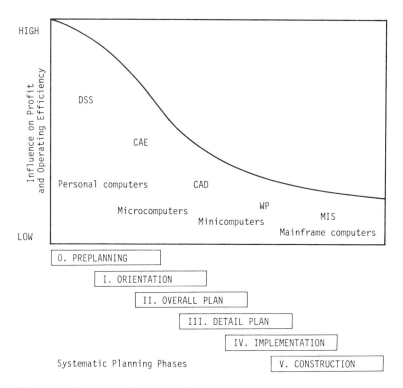

Figure 4.7. Project impact of computer technology, derived from the curves in Figure 1.6 and Figure 4.6. The later phases of a project have less impact on a facility's profit and operating efficiency. The same is true of those technologies which benefit the later phases. We should note however, that the data bases used for MIS can have great benefit in the early phases when linked to decision support systems. (Adapted from Ref. 19.)

having a role in facilities planning. Let us now relate each technology to the specific needs of facilities planners. In Figure 4.5, we have listed the tasks of facilities planning in rough order by the phases defined in Chapter 1 (Pre-planning—Phase O through Construction—Phase V). Specific computer aids are proposed for each task. The "technology" of each aid is also defined. Figure 4.5 may be used as a checklist or guide to the full range of computer aids which could be acquired or developed.

   In the early phases (O, I, and II) the computer aids consist primarily of calculations and schematic graphics. The technologies of interest are DSS and CAE. In the middle phase (III), most aids are applications of CAD. In the later phases of implementation, construction, and beyond, the computer aids are primarily reporting or MIS applications.

The Impact of Computer Technology

Close examination of Figure 4.5 reveals that different technologies benefit different phases of planning. This can be seen more clearly in Figure 4.6. The relationship between technologies and planning phases is of great importance since it suggests to us the emphasis we should place on each technology and on specific computer aids.

Back in Chapter 1 (Figure 1.6) we discussed the economic consequences of facilities planning--suggesting that our influence on profit and operating efficiency is greatest in preplanning and declines thereafter. Reintroducing this concept here, we can immediately see in Figure 4.7 that the impact of computer technology is also greatest in preplanning and declines thereafter. The conclusion is clear. Our first priority should be the development of decision support and computer-aided engineering. These technologies will have the greatest impact on our results. Coincidentally, they are generally the least expensive to acquire or develop, running as they generally do on personal or microcomputers.

Figure 4.7 does not suggest that CAD and MIS are insignificant. In fact, they are of great benefit to the planner. But if we look at the big picture and our fundamental objective of more profitable, efficient facilities, we can see that DSS and CAE should be fully exploited first, followed by CAD, word processing, and MIS.

We cannot address all five of these computer technologies in a single volume. For this reason we will limit our discussions to DSS, CAD, and MIS. These three technologies are of greatest interest to facilities planners. Chapters 5 and 6 will cover a variety of decision support applications. Chapter 7 will cover CAD and Chapter 8, MIS.

5

Computer-Aided Layout Planning

Approaches to Layout Planning

In Chapter 1 we established the layout as one of a facility's five components, the others being: handling, communications, utilities, and buildings. In general, layout is the "lead component" in the planning process, giving the facility its overall form and functional features. (The design of speculative offices and warehouses is a notable exception. Here, the building itself is the lead component, with layout planning done later by each tenant.)

Also, in Chapter 1, we introduced the concept of planning phases, noting the difference between an overall plan or layout and a detailed plan. The overall or block layout typically shows little or no equipment. It is constructed first to get approvals and agreements on the position of major activities.

In Chapter 2, we reviewed the collection of data on material flow, communications, and other relationships between activity-areas or departments in the facility. The planner uses this information to develop or validate layouts--most commonly at the block planning level.

In Chapter 3, we outlined typical approaches to facilities and layout planning and presented the graphical technique developed by Richard Muther. We mentioned the use of mathematical layout algorithms as an alternative to the Muther technique. We noted that such algorithms are necessarily computer-based because of the computation they require. In this chapter, we will elaborate on algorithms and related computer aids. Our discussion will be limited to block layout, since this is the focus of the aids available to us. We will assume that the planner has collected data on relationships and space, as prescribed in Chapter 2.

Before we assess the available computer aids, it may be useful to review what it is that the layout planner is trying to accomplish. In its most general terms, layout planning can be described as the attainment or satisfaction of multiple objectives, subject to a variety of constraints. The objectives (which may be conflicting) typically include:

1. Effective movement of materials and personnel
2. Effective utilization of space
3. Adaptability to unforeseen changes

4. Easy expansion
5. Safety
6. Control of noise
7. Easy supervision and control
8. Good appearance
9. Security
10. Low cost

The most common planning constraints include:

1. One or more fixed activities
2. Activities which must be separated
3. Architectural limitations
4. Material handling limitations
5. Utility limitations
6. Organizational restrictions
7. Budget
8. Code restrictions
9. Time

Attainment of layout objectives while satisfying the relevant constraints is a challenging task, especially for major rearrangements and new facilities. Typically the planner begins by applying classical concepts and principles to the following issues:

1. *Definition of activities and departments*: layout by product, by process, by stage of manufacture, or some combination thereof (such as the group technology approach outlined in Chapter 3)
2. *Flow patterns*: layout for flow--straight-through, U-shaped, L-shaped, star-shaped, or some variation thereof
3. *Expansion plans*: layout for future growth in zones, extensions, or duplications
4. *Design concepts*: truck docks at the rear, executives on top / cafeteria on bottom, offices along the windows, etc.

The interplay of these issues, adapted to the relationships and spaces at hand, will quickly yield a half dozen or more acceptable concepts. These are then firmed up into workable block layouts and compared in some way to see which is best.

Over the years, some planners and students of planning have found this approach too subjective and lacking in comprehensive technique. Several more rigorous approaches have been applied. These are:

1. *Bubble diagramming*: This is not really a technique but rather a means of illustrating an intended arrangement of activities. It is popular among architects and interior designers.

2. *Graphic layout technique*: As defined by Richard Muther and introduced in Chapter 3. This is a graphical means of resolving multiple objectives and constraints. It is most popular among industrial engineers.
3. *Scoring techniques*: Which calculate the relative worth of a completed
   block layout. These are popular among industrial engineers and office
   space planners.
4. *Clustering techniques*: A numerical approach that focuses on grouping
   of activities, often without reference to size, place, or layout.
5. *Layout algorithms*: Mathematical routines that automatically generate
   block plans from data on relationships and space.

In the sections that follow, we will use the graphic technique as a point of
reference. Scoring and clustering will be discussed, but the main emphasis
will be on layout algorithms.

Graphical Layout Revisited

We described this technique and presented two illustrations in Chapter 3. A
simplified application is included here for subsequent comparison with algorithm
results. The example is the Tool Fabrication Works first published by Richard
Muther and John Wheeler in 1962. The inputs of relationships and space re-
requirements are shown in Figure 5.1. The space available appears in Figure
5.2. The relationship diagram and one of three resulting layouts are shown in
Figure 5.3. Finally, in Figure 5.4, the layouts are compared using Richard
Muther's evaluation technique. The chosen plan is then detailed as a final step.
   The reader should not be misled by the small scale (2585 sq. ft.) of the ex-
ample. It could just as easily cover 25,000 or even 250,000 sq. ft. A few extra steps would be necessary, but the underlying procedure would be the same.
The basic characteristics of the example are:

1. twelve activity-areas with three of them fixed
2. five reasons for closeness between activities
3. one reason for separation
4. sixty-three relevant relationships of which 27 are unimportant, and only
   2 are absolutely necessary
5. six planning factors or objectives, each with its own weight (three factors are decidedly intangible)

To use the graphical technique successfully on problems of this and greater
complexity, the planner must fully understand the reasons for closeness and
separation. This in turn requires an understanding of the activities that occur
within each activity area. The planner should also know the evaluation factors
or objectives in advance, along with their relative importance. But all this
data, understanding, and knowledge still do not produce a layout. The missing ingredients are the planner's imagination, experience, and drafting skill.
   Your author has used the Muther graphical technique with good results on
more than 50 occasions. Based on this experience, the chief limitation usually

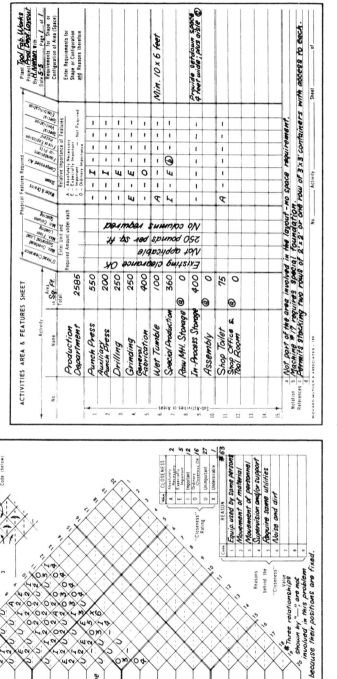

Figure 5.1. Basic input on relationships and space. (From R. Muther and J. Wheeler, *Simplified Schematic Layout Planning*. Kansas City, Missouri: Management and Industrial Research Publications, Inc., 1962.)

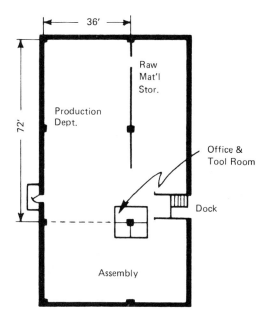

Figure 5.2. Space available for the Tool Fabrication layout problem. The Production Dept. is the area to be laid out. Raw Mat'l. Stor., Assembly, and the Office & Tool Room correspond to activities 8, 10, and 12 in the preceding figure. (Courtesy of Richard Muther & Associates, Kansas City, Missouri.)

proved to be time. The manual process of posting from relationship chart to diagram, interpreting the diagram, adjusting it, and manually preparing the block plans restricts the planner to a few alternatives--typically five to ten. There is a natural tendency to stop when a "satisfactory" alternative is achieved, possibly overlooking improvements. Critics of this technique also point to the subjective nature of the evaluation, preferring a more quantitative and cost-based approach (43). Computer-based scoring, clustering, and algorithms are all responses to these perceived limitations and the subjectivity of the graphical technique.

Scoring Techniques

Closeness scoring techniques are a simple first step to computer-aided layout planning. Depending on how far the planner wishes to go in automating his tasks, scoring may be the only step taken. Closeness scoring techniques do not generate layouts; they evaluate the "goodness" of those put to them. All these techniques assume that the planner seeks to maximize the closeness of highly interrelated activities. Well-designed scoring techniques will highlight desirable adjustments to a layout and will thus guide the planner toward better results. Scores are typically based on a weighted measure of distance between activities in a plan. Such scores do not explicitly recognize or value the application of classical principles and concepts. In other words, there is no value

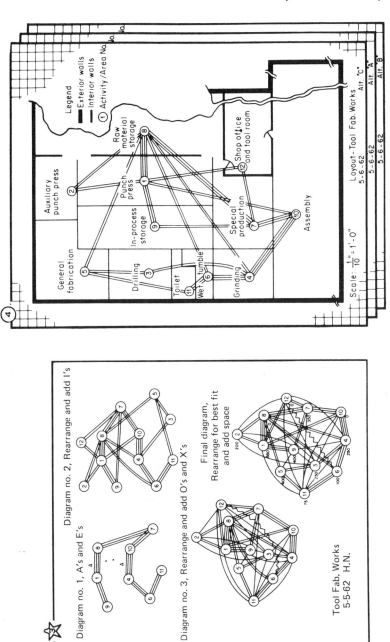

Figure 5.3. Relationship diagram developed from the relationship chart shown in Figure 5.1. At the right, the diagram is superimposed upon one of three block layouts. Single-line, "O" ratings are omitted for clarity. (From R. Muther and J. Wheeler, *Simplified Systematic Layout Planning*. Kansas City, Missouri: Management and Industrial Research Publications, Inc., 1962.)

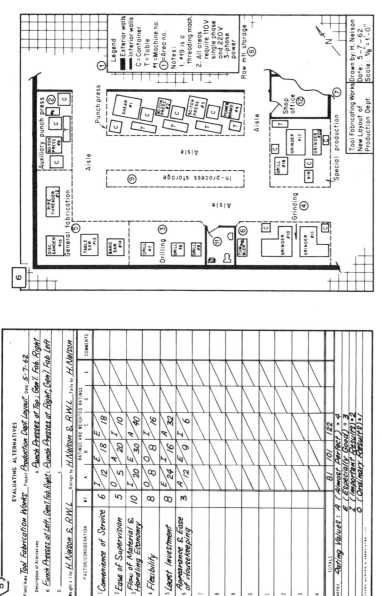

Figure 5.4. Final steps in the layout process. On the right, three alternatives are evaluated using a weighted-factor approach. Ratings are made using vowel-letters. These are converted later to points and multiplied times a factor weight, to arrive at a total score. On the right, the best block plan (Alt. C.) is detailed. (From R. Muther and J. Wheeler, *Simplified Systematic Layout Planning*. Kansas City, Missouri: Management and Industrial Research Publications, Inc., 1962.)

placed on U-shaped flow, or keeping all storage together, or putting offices along windows. Depending on the planner's philosophy and the situation at hand, this can be a benefit or a limitation of scoring techniques.

Distance-based scores are inherently cost-minimizing approaches to layout planning. It is assumed that costs are incurred in proportion to the distance traveled by people or materials. Therefore, it is assumed that distances between activities should be minimized, especially on heavily traveled routes.

The Distance-Intensity (D-I) Plot mentioned in Chapter 3 can be used as a very effective closeness scoring technique. It uses transport work (distance times intensity of flow) as a surrogate for cost. By calculating and summing the transport work on all routes, a total "score" is obtained for each layout. By interpreting the D-I Plot, the planner can identify which routes need to be shortened and adjust his layout accordingly. Each successful adjustment will lower the score, driving down the transport costs associated with the plan.

This technique lends itself well to the personal computer. Figure 5.5 was produced on a small Hewlett-Packard 7470 plotter driven by a Hewlett-Packard 85 computer. Distances are scaled from a site plan and entered via the keyboard on the HP-85. Intensities are entered from a movement summary like the one illustrated earlier in Chapter 2. The scoring routine automates the tedious multiplications and plotting that would be required manually.

Some planners prefer a more explicit treatment of costs than that provided by transport work. This is achieved with cost tables covering the specific handling methods or personnel traffic involved with each route. Costs are expressed per unit of distance. They are then multiplied by measured distances and totaled to give a "score," as shown in Figure 5.6. This table is taken from a larger program called PAINT which also runs on a personal computer--the Exidy Sorcerer.

To load a scoring program, the distances must be entered for each route. Generally these are obtained in one of three ways:

1. *Center-to-center*: Distances are measured in straight lines between the center of each activity area.
2. *Rectilinear*: Center points are connected with right-angled lines.
3. *Specific route*: Distances are measured over a specific travel path from pick-up to set-down.

Specific routes are more trouble but give a closer estimate of "real" distance-related costs. If the layout is developed using interactive computer graphics, a program can be written to automatically load distances into the scoring routine. The PAINT program employs this approach, using a 12-inch Video 100 monitor. Pick-up and set-down points for each potential route are noted with a symbol as shown in Figure 5.7. Distances are rectilinear between pick-up and set-down. The graphic capabilities of the computer and monitor can also be used to illustrate material flow as shown in Figure 5.8. Output in each case is achieved with a low-cost, dot-matrix printer.

The D-I Plot and PAINT programs are not real-time computer aids. A batch run is made when all distances and intensities have been loaded. A score is then produced. More powerful, graphic systems are capable of real-time, interactive scoring. As changes are made to the layout through the graphics terminal, distances and flow are recomputed, and the total score is displayed on the screen.

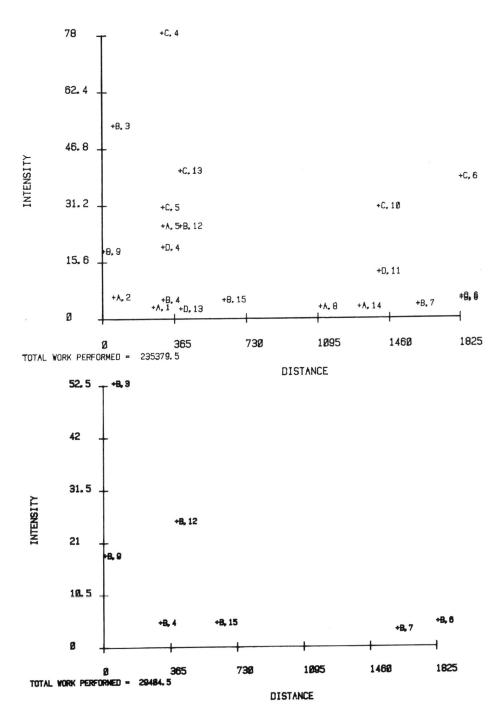

Figure 5.5. Distance-Intensity Plot. Above, a plot of all material classes. Below, class B only. (Courtesy of Micro-Vector, Armonk, New York, and Richard Muther & Associates, Inc., Kansas City, Missouri.)

```
TITANTIC TANK COMPANY

Evaluation Of Layout Number  3
Move Cost Detail

DEPT                    DEPT                 DAILY   EQUIPMENT        MOVE
FROM                    TO                   TRIPS     TYPE          COST $

-SHIP. & RECEIV.  -FLAT STORAGE        .3    Fork Lift          35
-SHIP. & RECEIV.  -COIL STORAGE        .3    Fork Lift          36
-SHIP. & RECEIV.  -SIDES-R F & P       5     Fork Lift          903
-FLAT STORAGE     -SQUARE SHEAR        70    Slide              3684
-SQUARE SHEAR     -OBLONG FLANGE       80    Slide              14589
-SQUARE SHEAR     -2 PB ROLL & FORM    4     Tow Truck          322
-SQUARE SHEAR     -CRC SHEAR & FLG     10    Tow Truck          1495
-OBLONG FLANGE    -FOUNTAIN FAB.       10    Slide              1342
-OBLONG FLANGE    -BOTTOM STORE        70    Slide              3753
-FOUNTAIN FAB.    -ASS FOUN TO TANK    10    Tow Truck          679
-2 PB ROLL & FORM -2 PIECE SWAGE       4     Tow Truck          252
-2 PIECE SWAGE    -CRC SHEAR & FLG     4     Tow Truck          259
-CRC SHEAR & FLG  -BOTTOM STORE        14    Tow Truck          827
-BOTTOM STORE     -ASS. & SPOT WELD    100   Slide              12831
-COIL STORAGE     -DECOIL & SHEAR      .4    Fork Lift          42
-DECOIL & SHEAR   -SIDES-R F & P       100   Slide              3003
-SIDES-R F & P    -ASS. & SPOT WELD    5     Tow Truck          329
-ASS. & SPOT WELD -IN PROCESS TANKS    100   Slide              2807
-IN PROCESS TANKS -INSIDE CREASE       100   Slide              2610
-INSIDE CREASE    -WELD SIDE BRACE     10    Slide              447
-INSIDE CREASE    -STAGE FOR SOLDER    90    Slide              4472
-WELD SIDE BRACE  -STAGE FOR SOLDER    10    Slide              388
-STAGE FOR SOLDER -SOLDER SEAM         100   Slide              4674
-SOLDER SEAM      -PAINT & SEAL        100   Slide              3396
-PAINT & SEAL     -SEALANT CURE        100   Slide              3495
-SEALANT CURE     -ASS FOUN TO TANK    10    Slide              801
-SEALANT CURE     -F G STORAGE         90    Slide              5268
-ASS FOUN TO TANK -F G STORAGE         10    Slide              811
-F G STORAGE      -SHIP. & RECEIV.     100   Slide              10276
                                                                -------

Annual Comparative Move Cost Layout    3  =                     83843.6
```

Figure 5.6. Scoring a layout based on comparative material handling costs.
(Courtesy of Corliss V. Little & Co., Raytown, Missouri.)

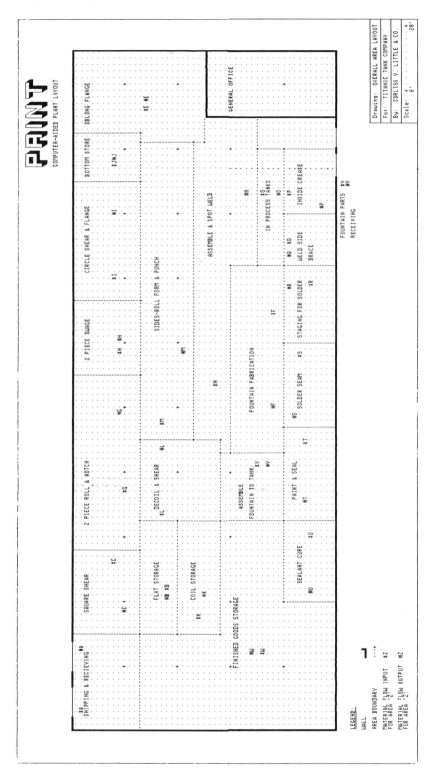

Figure 5.7. Block layout with pick up "*" and set down "#" points. Developed using the cursor control on an Exidy Sorcerer computer. Distances are computed automatically and provided to the scoring routine shown in the preceding figure. (Courtesy of Corliss V. Little & Co., Raytown, Missouri.)

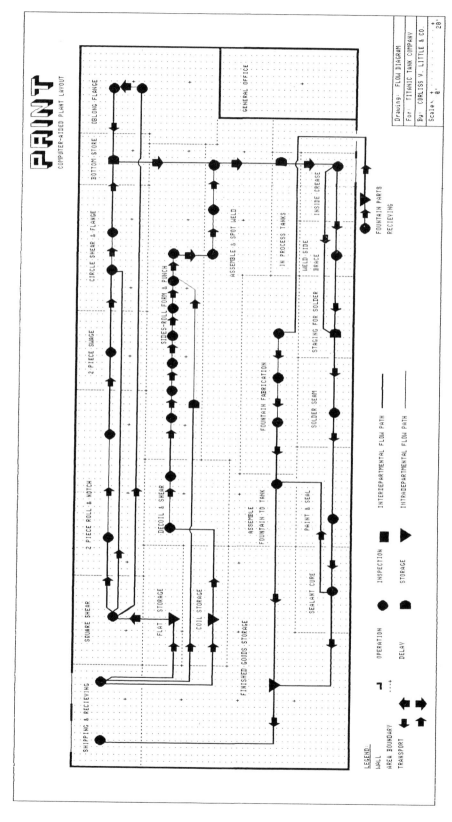

Figure 5.8. Material flow diagram. (Courtesy of Corliss V. Little & Co., Raytown, Missouri.)

SCORE/CORELAP is a scoring program developed by Moore Productivity Software, Blacksburg, Virginia. It uses the scoring logic of the CORELAP layout algorithm discussed later in this chapter. SCORE/CORELAP is a weighted distance approach that uses the numerical equivalents of vowel-letter ratings (A = 4, E = 3, I = 1, etc.) instead of quantitative flow data as required by the D-I Plot. This makes it useful as a scoring mechanism where non-flow factors are significant (37). SCORE/CORELAP is available on a time-shared basis from the General Electric Information Services Company. The memory and computational requirements of the program limit it to mainframe installations.

Clustering Techniques

Clustering is used to generate groupings of people or activities. It is an analytical tool which can be used in three ways:

1. To define activity-areas prior to layout planning. Here, it is an alternative to judgment or reliance on existing organizational units.
2. To study relationships between activity-areas, as an alternative to the Muther relationship and cluster diagrams.
3. To help reveal the overall sizes or blocks of space required for closely related activities. Here, clustering is a prelude to assignment or placement of activities.

In Chapter 2, we presented the Herman Miller questionnaire for studying communication and interaction. The results of this questionnaire are shown in Figure 5.9. The output is achieved with a mathematical technique known as hierarchical decomposition. In laymen's terms this is clustering. The repetitive mathematics involved requires a computer. The Herman Miller programs were originally developed at the Illinois Institute of Technology by Dr. Charles Owen (41). His programs VTCON and RELATN generate similar results, as shown in Figures 5.10 through 5.12. The GENERATE and INACT programs developed by Dr. Larry Ritzman at Ohio State University also fall into this class (23, 39).

These programs were written in the 1970s for use on mainframe computers. Because of their approach and the computation involved, several minutes of CPU time are required, even on the larger and faster mainframes. Minicomputers such as the VAX 11/780 or the Prime 400 can be used but run time is significantly lengthened. (In the next chapter we will review a similar clustering program developed specifically for micro- and personal computers.)

Clustering programs do not generate a definitive arrangement of activities, as do the Muther diagramming techniques. However, they do consider a wide range of flow and non-flow factors, as well as constraints in arriving at clusters of closely related departments. These programs are a way of gaining a better understanding of a particular layout problem. The planner can make successive runs, varying the importance of flow and non-flow factors. The program output will reveal the sensitivity of the clusters to these changes. Attempting this manually in large design or planning problems would consume too much time. Clustering techniques appear best suited to large office and institutional layout plans, where material flow patterns and other classical issues are less of an organizing influence on the block plan.

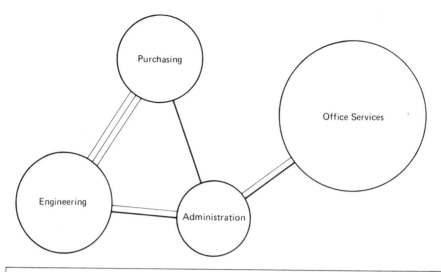

```
DEPARTMENTAL INTERACTION REPORT                                    Page    1

ADMINISTRATION

Links within Administration          GROUP                              STRG

    1)Adams, Jeffrey                   5
       2)Anderson, Melissa             5                                 81

    2)Anderson, Melissa                5
       1)Adams, Jeffrey                5                                 81

Links to Engineering

    1)Adams, Jeffrey                   5
       7)Browning, David               1                                 53

    2)Anderson, Melissa                5
       8)Bryan, Michael                1                                 42

Links to Office Services

    1)Adams, Jeffrey                   5
       3)Andrews, Marian               3                                 54
      10)Collins, Everett              3                                 37

Links to Purchasing

    1)Adams, Jeffrey                   5
       5)Black, Thomas                 2                                 58
```

Figure 5.9. Results of the Herman Miller communications survey. The diagram is a hand-drawn interpretation of the departmental interaction report. Other reports are produced to show interaction at the individual and company level. (Courtesy of CORE, Herman Miller, Inc., Grandville, Michigan.)

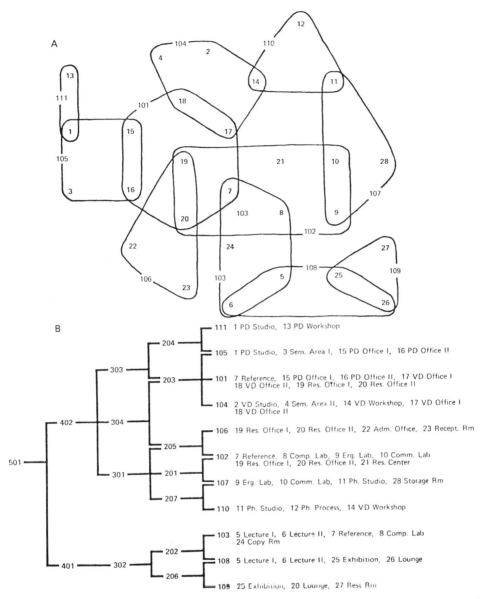

Figure 5.10. Hand-drawn, first level clusters (A) and complete hierarchy (B), based on printouts of the RELATN and VTCON programs. These examples are based on the desired proximity between activities. Companion clusters based on other factors are shown in Figures 5.11 and 5.12. Note that one activity, such as "19 Res. Office I," may appear in several clusters. (From Ref. 41. Copyright 1980, Association for Computing Machinery. Reprinted by permission.)

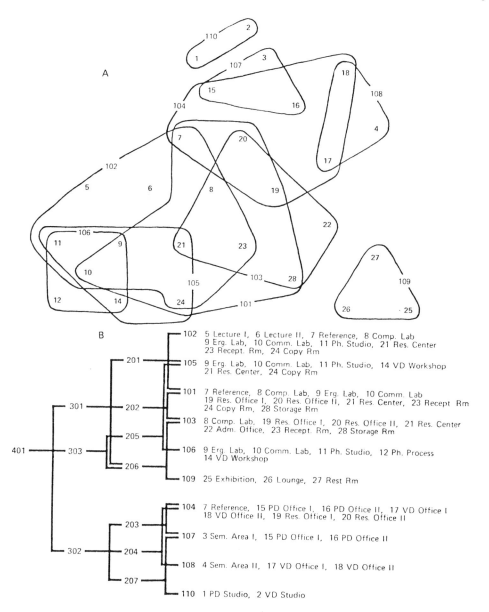

Figure 5.11. Clusters (A) and hierarchy (B) based on the similarity of activities. (From Ref. 41. Copyright 1980, Association for Computing Machinery.)

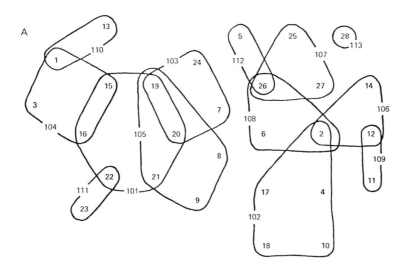

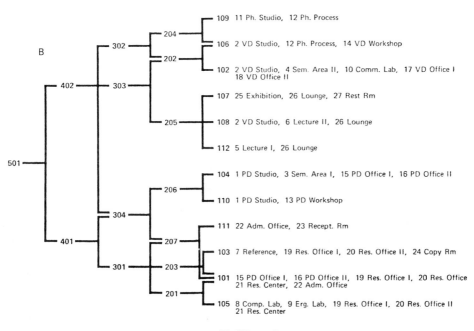

Figure 5.12. Clusters (A) and hierarchy (B) based on the flow or movement of people and materials. (From Ref. 41. Copyright 1980, Association for Computing Machinery.)

Layout Planning Algorithms

The field of computer-aided layout planning began 20 years ago, with publication of the CRAFT layout algorithm by Elwood Buffa, Gordon Armour, and Thomas Vollman (10). Since then, the merits of algorithms over traditional approaches have been hotly debated, especially among industrial engineers. Today, there are perhaps 50 published layout algorithms available for use on a variety of computer systems. Each of these generates a "finished" layout in the sense that departments or activity-areas are represented to scale and graphically adjoined. Algorithms fall into two categories:

1. *Improvement routines*: take an existing or proposed layout and make changes that reduce movement costs
2. *Construction routines*: take relationships between activity-areas and generate a block layout.

Multistory and multibuilding planning requires a special type of algorithm often called vertical stacking. This contains routines similar to clustering. Some produce a block layout for each floor; others merely assign activities to each floor and leave the layout to manual methods. Some industrially-oriented algorithms include explicit treatment of material handling methods. Those oriented toward offices do not. Several facilities-oriented CAD systems include some form of layout algorithms as part of their standard software offering.

Improvement Routines

Improvement routines are extensions of the scoring routines already discussed. Their basic approach is to minimize transportation or movement cost by reducing the distances on heavily traveled routes.

CRAFT is the oldest and most well-known improvement routine. It has been copied and extended by many researchers over the past 20 years. However, its underlying algorithm remains representative of all improvement routines. For this reason, we will use it as a convenient example and as a basis for some reasonable generalizations.

CRAFT (and other improvement routines) begin with the following inputs: (1) from-to chart, (2) move costs, and (3) initial layout. The from-to chart or volume array is the same device presented in Chapter 2. Volumes may be expressed in absolute units (pieces, tons), equivalent units (loads, skids) or in trips. However, use of trips in an industrial application may beg the issue of volume if the loads moved are not uniform. The alert reader will recognize volume in this context as synonymous with intensity of flow. CRAFT accepts flow between as many as 40 activity-areas. Some improvement routines can accept up to 100.

Move costs are expressed as a cost per unit of volume per unit of distance, for example: $0.05 per ton-foot, or $0.10 per trip-foot. In CRAFT, costs may vary by direction of flow and by route. However, no cost differences may be given to different methods on the same route and direction. As a practical matter, many users load their cost array with 1.00 as a uniform cost in both directions on all routes. This begs the issue of costs and defaults the logic to distance times intensity or transport-work.

An initial layout provides the dimensions of the overall space, the number of activity-areas, their size, and, indirectly, their center points. The overall shape must be a rectangle, but dummy activities can be declared to create different effective shapes within the rectangular envelope. Activity-areas may be fixed in their initial positions by declaring them unavailable for subsequent manipulation.

In addition to these three inputs, the planner must also set various control parameters and give instructions governing output. CRAFT inputs are visualized in Figure 5.13. The routine works in the following manner:

1. The center of each activity-area is found. Rectilinear distances between centers are computed.
2. Distances are multiplied by flow volume and move cost to arrive at a total movement cost for the layout.
3. For each successive activity-area, the program identifies all possible exchanges with other activity-areas. To be considered, the activities must adjoin or border on a common activity. They must also be the same size. A parameter setting allows the size restriction to be relaxed to varying degrees. Both two-way and three-way exchanges may be considered. This is also set by control parameter. The move cost reductions of all possible two- and/or three-way exchanges are compared. The greatest cost reduction is identified.
4. If the greatest cost reduction represents a significant reduction in total move cost, the appropriate activity-areas are exchanged. If not, the routine ends. The significance level is a control parameter.
5. The process is repeated with the exchanged activities constituting a new initial layout. Output controls will print the best layout found at each iteration or simply the final best layout.

The logic of CRAFT has been incorporated by James Tompkins into COFAD, another well-known improvement routine (43, 44). COFAD is an extension of CRAFT that addresses industrial materials handling costs in a more explicit fashion. For each route, COFAD evaluates the cost of alternative handling methods. A cost-minimizing set of handling methods is selected and the required pieces of handling equipment are calculated. The result is a handling system for the layout, and a potentially new set of move costs, which triggers another iteration of the CRAFT-like layout routine.

The logic of CRAFT has also been employed by Larry Ritzman for office layout planning (23). In the LAYOUT routine, from-to flow data is replaced by an interaction matrix (other-than-flow relationship chart). Relationships are weighted by distance and treated in the same manner as CRAFT treats movement costs. (The LAYOUT routine is a companion to the GENERATE and INTACT routines used for clustering activities.)

Improvement routines have a straightforward, appealing logic. In practice, however, their results may leave something to be desired. The CRAFT

CRAFT VOLUME ARRAY												
Activity or Department Number												
01	02	03	04	05	06	07	08	09	10	11	12	13
0000	3000	2000	0000	0000	0000	4000	0000	0000	2000	0000	0000	0000
0000	0000	0000	3000	0000	0000	0000	3000	0000	2000	0000	0000	0000
0000	0000	0000	3000	0000	0000	2000	0000	0000	1000	1000	1000	1000
0000	0000	0000	0000	3000	0000	1000	0000	0000	1000	1000	1000	1000
0000	0000	0000	0000	0000	2000	2000	0000	0000	0000	0000	1000	1000
0000	0000	0000	0000	0000	0000	4000	0000	0000	2000	0000	1000	1000
0000	0000	0000	0000	0000	0000	0000	0000	0000	0000	0000	0000	0000
0000	0000	0000	0000	0000	0000	0000	0000	4000	1000	0000	0000	0000
0000	0000	0000	0000	0000	0000	0000	0000	0000	4000	0000	0000	0000
0000	0000	0000	0000	0000	0000	0000	0000	0000	0000	0000	1000	1000
0000	0000	0000	0000	0000	0000	0000	0000	0000	0000	0000	0000	0000
0000	0000	0000	0000	0000	0000	0000	0000	0000	0000	0000	0000	0000
0000	0000	0000	0000	0000	0000	0000	0000	0000	0000	0000	0000	0000

(Rows labeled 01–13 for Activity or Department Number)

CRAFT INITIAL LAYOUT

```
10 10 10 10 10 10 10 10 10 10
10 10 10 10 10 10 10 10 10 10
08 08 12 13 13 13 13 13 06 06
09 08 12 12 13 13 13 13 06 06
09 02 01 01 01 07 07 07 06 06
02 02 01 01 01 07 07 07 06 06
02 02 01 01 01 07 07 07 06 06
02 02 03 03 03 07 07 07 06 06
02 02 04 04 04 04 04 05 06 06
02 02 04 04 04 04 04 05 05 11
04 04 04 04 04 04 04 05 05 11
04 04 04 04 04 04 04 05 05 11
```

CRAFT COST ARRAY												
Activity or Department Number												
01	02	03	04	05	06	07	08	09	10	11	12	13
1000	1000	1000	1000	1000	1000	1000	1000	1000	1000	1000	1000	1000
1000	1000	1000	1000	1000	1000	1000	1000	1000	1000	1000	1000	1000
1000	1000	1000	1000	1000	1000	1000	1000	1000	1000	1000	1000	1000
1000	1000	1000	1000	1000	1000	1000	1000	1000	1000	1000	1000	1000
1000	1000	1000	1000	1000	1000	1000	1000	1000	1000	1000	1000	1000
1000	1000	1000	1000	1000	1000	1000	1000	1000	1000	1000	1000	1000
1000	1000	1000	1000	1000	1000	1000	1000	1000	1000	1000	1000	1000
1000	1000	1000	1000	1000	1000	1000	1000	1000	1000	1000	1000	1000
1000	1000	1000	1000	1000	1000	1000	1000	1000	1000	1000	1000	1000
1000	1000	1000	1000	1000	1000	1000	1000	1000	1000	1000	1000	1000
1000	1000	1000	1000	1000	1000	1000	1000	1000	1000	1000	1000	1000
1000	1000	1000	1000	1000	1000	1000	1000	1000	1000	1000	1000	1000
1000	1000	1000	1000	1000	1000	1000	1000	1000	1000	1000	1000	1000

(Rows labeled 01–13 for Activity or Department Number)

Figure 5.13. Sample CRAFT inputs. (Courtesy of Richard Muther & Associates, Inc., Kansas City, Missouri.)

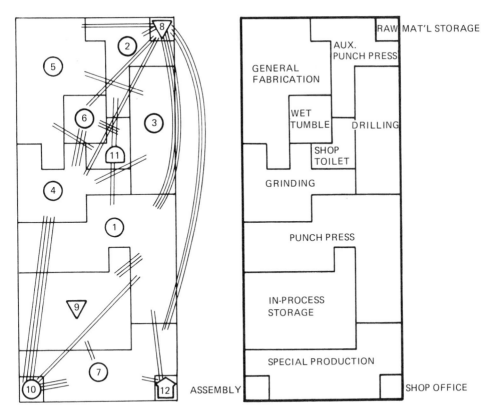

Figure 5.14. CRAFT output (right) for the Tool Fabrication problem presented earlier. For purposes of evaluation, a relationship diagram is presented on the left. Note the poor placement of activities 4 and 10. Also the irregular activity shapes. Activity 8 appears to be poorly placed; however, this is not the case. The reader will recall that activity 8 actually spans the right side of the area to be planned. Remember, this is one alternative. Others, perhaps better, can be produced by varying the initial layout. (Copyright 1980, Richard Muther & Associates, Inc.)

output for our sample problem is shown in Figure 5.14. Close inspection reveals that one of only two "absolutely necessary" relationships has not been honored. Such oversights are not uncommon and stem from inherent limitations in the algorithm approach. The most troublesome limitations are:

1. Different initial layouts give different "final" solutions. The number of possible exchanges of activity-areas depends upon the starting layout. To have any confidence in the results, the planner must use several different starting layouts. Even so, there is an element of randomness involved in arriving at the final result.
2. Outputs contain unrealistic locations, shapes, and alignments. Manual adjustment is always required. However, in some cases it is extensive enough to raise doubts about the value of the approach.

3. The improvement algorithms cannot generally consider a negative "X" relationship. In some cases this may not matter, but in others the manual adjustment required to separate two activities may seriously amend the rest of the layout.

4. Improvement algorithms do not deal easily with other-than-flow relationships. To be sure, there are some, such as the LAYOUT program just mentioned, that substitute "interaction" for flow. But in the CRAFT program, closeness data must be expressed as flow of materials or movement of people.

5. Architectural influences are very difficult to consider. Uses of windows and views, the issues of floor loadings or bay heights, or types of space, or utility service, can only be addressed indirectly by fixing the location of some activities, or by adding dummy flows to drive activities together.

Improvement programs require a building shape as input. This is not really a limitation. If the planner desires a program that will generate a building shape along with the internal layout, then a construction routine should be used.

Construction Routines

Construction routines are automated graphical techniques. Their basic approach is to find a starting point or initial activity placement and then add remaining activity-areas in accordance with logical rules. In some routines the rules are similar to Muther's vowel-letter sequencing; in others the rules are a blend of this sequence and improvement-like, distance minimization.

CORELAP is the oldest and perhaps best-known construction routine. It was developed by James Moore and published in 1967 (26). Like CRAFT, it has been copied and adapted over the years. As a class, construction routines show more variety than improvement routines. Still, their results show many of the same characteristics and limitations. CORELAP is reasonably representative of these, so we shall use it as our example. Our discussion of its features is taken largely from the writings of James Tompkins and James Moore (43, 44).

CORELAP and other construction routines begin with two major inputs: (1) a relationship chart and (2) space requirements. The relationship chart (interaction or preference matrix) is typically the same device presented in Chapter 2. In industrial engineering circles, it has become customary to use the Muther vowel-letter ratings as an expression of closeness desired or interaction. In the field of office space planning and architectural design, a 1-9 numerical scale is more common.

CORELAP accepts relationships between as many as 40 departments or activity-areas. It also contains a "driver" routine which helps to fix certain activity-areas, such as shipping, to an outside wall or corner. CORELAP works in the following manner:

1. Vowel-letter ratings are converted to their numerical equivalents (A = 4, E = 3, etc.). The ratings for each activity-area are summed

and compared to identify the activity with the highest "total-closeness-rating" (TCR). This activity is then placed in the center of the layout.

2. Remaining activity-areas are examined. The one having the greatest desired closeness to the initially placed activity is placed next to it in the developing block plan.

3. Each remaining activity-area is examined for its relationship to those already placed. Placement is made in descending order of closeness desired to the activities already placed. If there is a tie, (equally desired closeness) between two or more activity-areas, the TCR calculated in step 1 is used to break it.

4. The routine ends when all activities have been placed. The resulting layout is scored using the shortest distance between each pair of activities, weighted by the desired closeness between them.

The CORELAP output for our sample problem appears in Figure 5.15. Several limitations are apparent. The most obvious is the irregular shape of the layout and most of the departments. While CORELAP allows the planner to specify a length-to-width ratio for the final layout, the results may still leave something to be desired. Significant manual adjustment may be necessary to correct for irregular activity shapes. The "driver" discussed earlier is not an ability to truly fix activity locations. This may also lead to major adjustments. In our sample problem, several important relationships are not well satisfied. In truth, this problem contains constraints in shape and activity location which work against the success of CORELAP and similar construction routines.

The ALDEP routine, developed by Jerrold Seehof and Wayne Evans, takes a different approach to the construction task (42). ALDEP reduces closeness desired to "important" and "unimportant." It then proceeds to pick an initial activity at random and places it in the "northwest corner" of the layout. The next and successive activity-areas are placed in order of their relationship to those already placed. Ties are broken by random selection. Placement uses a technique called sweep, illustrated in Figure 5.16. Once all departments or activities have been placed, another layout is generated beginning again with a randomly selected activity-area. The routine continues until a predetermined number of layouts have been generated. Each layout is scored based on the number of related activities that are adjacent, weighted by the relative closeness initially desired between them.

ALDEP works within a fixed building or boundary. It allows the location of activities to be precisely fixed. It can also address basic architectural features such as halls, stairs, and core areas by blocking them out of the activity placement routine. Up to three floors may be considered. The ALDEP output for our sample problem appears in Figure 5.17. The shapes of the activity-areas are much more regular than those of CRAFT and CORELAP. This is a by-product of the sweep technique. This particular run failed to satisfy many important relationships. This is to be expected, however, with a randomly seeded, brute-force approach. In theory, a satisfactory layout should eventually appear. As a practical matter, the planner may run out of time before it does.

In your author's judgment, the algorithms used in construction routines contain an obscure but fundamental limitation in their selection and placement of activity-areas. By terminating when all activities have entered the layout, it is possible and in some cases probable, that significant closeness relationships will not be considered. This may explain the frequent inability of these routines to

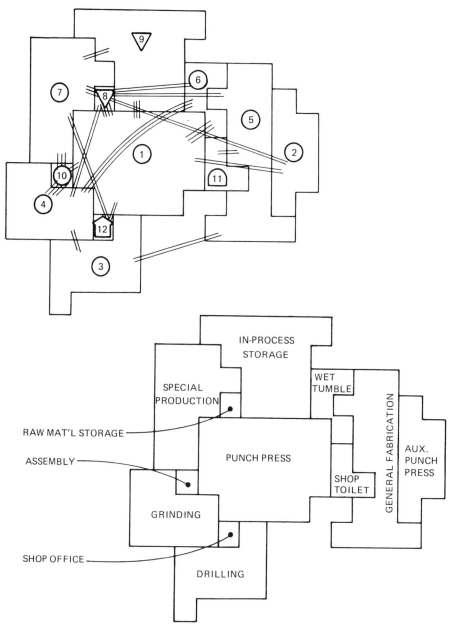

Figure 5.15. CORELAP output (bottom) for the Tool Fabrication problem. Note the inability to deal with fixed activity locations and the dimensions of an existing space. As a construction routine, CORELAP is not well suited to constrained problems such as Tool Fabrication. (Copyright 1980, Richard Muther & Associates, Inc.)

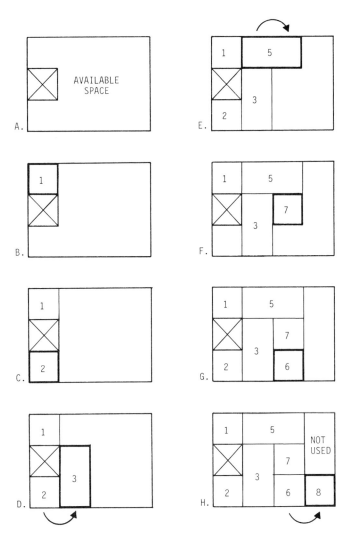

Figure 5.16. The ALDEP sweep technique. All but the blocked-out space is available. Activity 1 is placed first in the "northwest corner." Activity 2 is most closely related to Activity 1. It is placed next, as close as possible to Activity 1, in the first sweep "column." Activity 3 is placed next, sweeping "up." Activity 5 must extend into the third sweep column. Placement continues in this fashion until all activities have entered the layout. The sweep width is controllable.

satisfy even a small number of important relationships. To see why this could be so, we need to compare placement logic with the graphical approach already presented. When manually constructing a relationship diagram, the Muther procedure calls for posting all the "A" relationships first, proceeding from the top of the chart to the bottom. All "E" ratings are posted next, followed by "I's," "O's," and "X's." In this way, every relationship is considered in order of its priority or importance. Algorithms on the other hand, post *activities* instead of relationships. Proceeding in top-down fashion, on the relationship chart, all activities may be placed before all relationships have

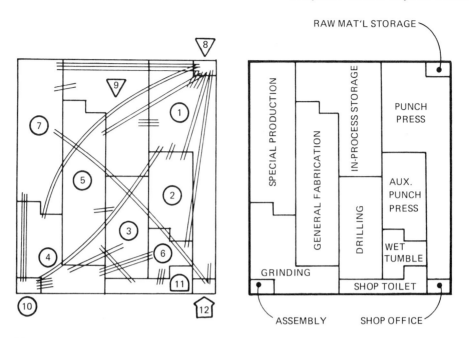

Figure 5.17. ALDEP output (right) for the Tool Fabrication problem. (Copyright 1980, Richard Muther & Associates, Inc.)

been examined. In other words, a portion of all relationships will be sufficient to place all departments. If this portion does not contain all critical relationships, then the resulting layout may include significant flaws. The presence of numerical scores and the prescription of making multiple program runs may not overcome this limitation.

Multistory Routines

The assignment of activities to floors in a multistory building is a formidable planning problem. If approached in a systematic way, the amount of calculation is overwhelming on projects of significant size. In recent years, a number of algorithms have been developed for planning large multistory facilities. A number of these computer aids stop with activity assignment, leaving the final block plan to manual methods. This is sufficient in many multistory office settings for which these routines have generally been developed. Building floors are often relatively small, with relatively few activities assigned to them. Given the inevitable mandatory placements for one or two activities and the obvious choices for others, the planner can manually achieve an effective floor layout in a short time.

Multistory routines employ a variety of algorithms and approaches. Some resemble the clustering techniques and outputs already discussed. Others, like like SPACECRAFT, are based on improvement algorithms. This routine, offered by Decision Products, Inc., of Santa Clara, California, is an extension of CRAFT that incorporates vertical travel costs. The planner can specify separate travel times for upward and downward movement, as well as an expected waiting time for each interfloor route. In addition to fixing the precise location of a given activity, the planner can specify a range of acceptable floors, any one of which will do.

Planning ADES, developed by Allan Cytryn in the mid-1970s, employs a construction routine approach to multistory planning (15). Closeness desired is expressed numerically as a degree of desired face-to-face contact. It is assumed that workers will travel to achieve their desired degree of contact. Their travel time is estimated and multiplied by appropriate wage data to give a travel cost for each alternative set of floor assignments. The program seeks the minimum travel cost. Once floor assignments have been made, the program generates a bubble or space relationship diagram that can be adjusted manually to arrive at a block layout for each floor. Some diagrams need more adjustment than others as seen in Figure 5.18.

Planning ADES was a set of card input programs for use on IBM 360 and 370 mainframe computers. A successor to these programs has been developed for the Apple IIe using interactive graphics and more effective design algorithms. We will examine this newer personal computer program in the next chapter.

The assumption that people will walk as far as necessary to achieve their desired level of contact is central to vertical stacking algorithims. It also applies to the single-level algorithms already discussed. Thomas J. Allen of the Sloan School, Massachusetts Institute of Technology, has compiled a considerable body of research suggesting that this assumption is false. He has shown that in numerous settings people stop seeking face-to-face contact when the distance between them is more than 80 feet in single-level facilities and less in multistory facilities.

This behavior can be modeled in an algorithm by adjusting the cost functions and closeness desired ratings to give a severe cost penalty on placements that exceed the "80-foot rule." The need for face-to-face contact is being lessened by new telecommunications technology--store and forward voice mail, for example. It is up to the planner to consider such issues in his use of computer aids. As delivered, computer aids are neutral and may be subject to misapplication. As a general rule, they are only as perceptive as the people who developed and coded them.

The SABA routine (also published as the Space Planning System) is a unique combination of improvement and construction algorithms (27). It has been developed by William Mitchell and the Computer-Aided Design Group of Santa Monica, California. Mitchell and his associates, Charles Reeder and Jeff Hamer, have incorporated a number of features into their routine, which addresses the historical limitations of algorithms.

In the SABA routine, a construction algorithm assigns activities to floors and, if desired, to zones within floors. The improvement algorithm then takes over and attempts to improve the arrangement of each floor and zone. In addition to travel costs, the algorithm also considers move or rearrangement costs. This allows the algorithm to pass up exchanges of activities where the savings in travel would be less than the cost of rearranging. This is a considerable improvement over simply fixing, a priori, the positions of certain departments. Other useful features of the SABA routine include:

1. Fixing activities to floors and zones as well as specific locations.

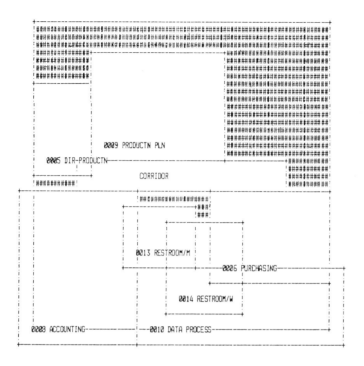

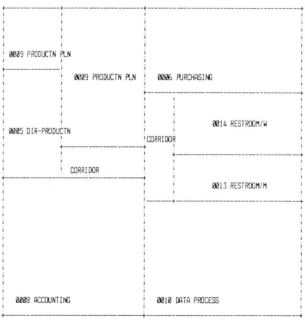

Figure 5.18. Bubble or space relationship diagram (above) and adjusted block layout (below). The diagram is produced automatically in accordance with relationship inputs; however, the results may include overlapped departments and "unused" space. The diagram is redrawn on the screen using a joystick. (Courtesy of Allan Cytryn, Tenafly, New Jersey, and Richard Muther & Associates, Inc., Kansas City, Missouri.)

2. Restricting activities from floors or zones with the use of penalty re-arrangement costs. This and the previous feature allow the routine to consider architectural and interior design issues such as use of views, floor loadings, types of space, existing features, bay sizes, and utility service.

3. The ability to split activities between zones and floors with different penalties for the type of split involved.

4. A placement strategy in the construction routine that places largest activities first, thus reducing the number of splits.

5. A coherence ratio for each activity, in addition to the specification of length and width. This helps to reduce irregular activity shapes in the block planning phase. See Figure 5.19.

6. Consideration of straight-line or rectilinear distances in the travel cost computation. In some settings, the straight-line distance is the best representation.

SABA and other multistory routines require a great deal of input data, usually obtained by survey. There is a temptation, when routines can accept up to 100 activities, to make full use of this capacity. The planner should resist this temptation and continue to define problems with the minimum number of departments for a realistic solution. True, the additional run time is insignificant but the data collection and initialization are not. Remember, a 45-activity problem contains 990 paired relationships. A 100-activity problem contains 4950.

CAD-Based Block Layout Planning

The output of most layout algorithms is a static plot or printout. Adjustments must be made manually by the planner who typically redraws the output. The adjusted, redrawn layout must then be re-keyed for the next program run, should one be desired. This is a tedious approach left over from the 1960s and early 1970s when most algorithms were developed. At that time, the cost of interactive computer graphics was beyond the reach of algorithim developers.

Today, facilities planners' growing use of computer-aided design is prompting some CAD vendors to introduce layout algorithms as part of their software product. With these systems, the algorithm's results are displayed on a graphics terminal. Using a light pen or cursor control, the planner can modify or adjust the displayed result without manual drafting. Then, if another run is desired, the input is automatically loaded from the adjusted plan. There is no need to re-key input data.

The PEAC system by Decision Graphics, Southborough, Massachusetts, was the first CAD system to offer an interactive block layout routine (over 10 years ago). In its current form, the routine displays a score associated with the closeness of activity symbols in a bubble diagram. The symbols are placed automatically by the algorithm. As the planner converts symbols and adjusts the spaces into a block plan, the center points of the activity spaces are tracked. The score is recomputed dynamically as activities are moved about the display. This feedback gives the planner useful guidance throughout the adjustment process, so that the final plan may approach or even improve the bubble diagram score. (See Figure 5.20.)

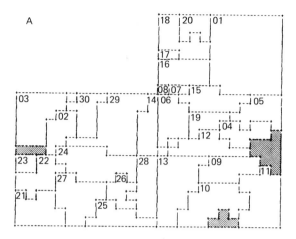

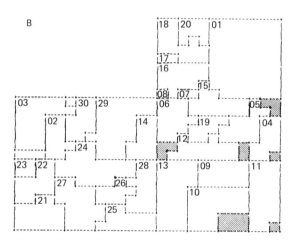

Figure 5.19. Controlling irregular shapes. Final block plan before (A) and after (B) shape adjustment. A coherence ratio measures the degree to which a rectangular shape has been achieved. The program modifies the shapes of adjacent activities until improvements in one diminish the coherence of the other. In this example, significant shape improvements are made in activities 9, 10, 11, and 13. On the other hand, the coherence of activities 18 and 20 could not be improved. (From Ref. 27.)

The routine by Cytryn discussed earlier offers interactive graphic adjustment on the Apple IIe, using a joystick. Scoring is based on the adjusted layout rather than on the intermediate space relationship diagram generated by the algorithm.

At General Motors, in the late 1970s, special routines were written for an Applicon CAD system. These allowed the planner to display a layout and then trace a forklift route upon it with a light pen. Using a keyboard, data about

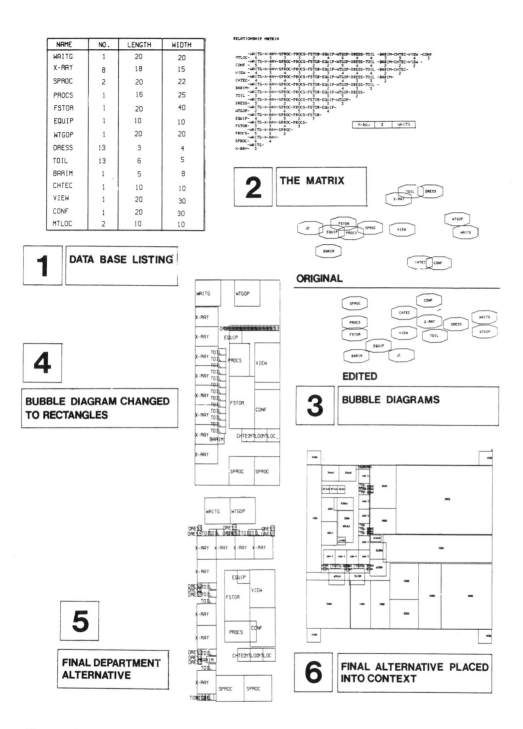

Figure 5.20. CAD-based block layout on the PEAC system. In Step 1, the
shape and size of each activity is pre-defined. Desired closeness is entered
in Step 2, using a 1-9 scale. A bubble diagram is generated (Step 3) automatically, representing a graphic resolution of matrix values. The diagram is
edited using the cursor control. Bubbles are converted to rectangles (Step 4)
and manipulated into a desired arrangement (Step 5). The entire arrangement
is then moved as an entity into its larger context (Step 6). (Courtesy of Decision Graphics, Inc., Southborough, Massachusetts.)

the route could be entered and used for subsequent flow diagrams, calculations, and simulation of materials handling requirements.

In the next few years, these and other progressive uses of interactive graphics will be common offerings by those CAD vendors who are focused on facilities planning. The underlying algorithms will be similar if not the same as those discussed in this chapter.

Practical Considerations

Surveys in the 1970s by James Moore found that few industrial planners were using layout algorithms (12, 18, 32, 33, 35, 36). Further, most of those who did found the results to be of limited value. The reasons that this is so may be apparent from our preceding discussions of the various available routines.

Contrary to popular belief, algorithms cannot produce a demonstrably best or optimum layout. At best they can provide a good solution. While they do consider a great many alternatives, it does not follow that their results will exceed those of an astute planner working unaided.

Most of the algorithms described require a large mainframe computer. Virtually every facilities planner has access to such a machine and enough funds to experiment with the better known programs. The oldest algorithms--CRAFT, COFAD, CORELAP and ALDEP--are available at a nominal charge from the SHARE Program Library, Triangle University Computation Center, Research Triangle Park, North Carolina. CRAFT, COFAD, and CORELAP are also available from Moore Productivity Software as part of the PLANPAK Library. This collection includes P-Q plotting and scoring. It is offered on a time-sharing basis through the General Electric Information Services Company. The other programs discussed in this chapter are available from the developers. In the case of universities, the fee is often nominal. The programs by Larry Ritzman, for example, are available for $300 from the Faculty of Management Sciences, Ohio State University.

As this is written (September 1983), the Institute of Industrial Engineers has announced CRAFT and CORELAP programs for personal computers. These and two other programs will be offered as a set for under $200. Scoring programs are ideal for personal computers. But it remains to be seen whether the run time will be acceptable for the full-scale CRAFT and CORELAP algorithms (21, 37). Experiments have shown that several hours to a full day or more may be required, depending on the programmer's approach, the speed of the disk drive, and the internal capacity of the computer. During this time, the computer is vulnerable to disk errors, power surges, and any other accidental disruptions.

Rather than accept this chapter's discussion as definitive, the interested planner should set a goal of experimenting with at least two or three algorithms. Then, compare results with a structured manual approach such as the Muther graphical technique. In this way, the planner will gain familiarity with a well-developed class of computer aids and be able to form his own conclusions.

6

Personal Computer Applications

Personal and Microcomputers

This chapter will survey the uses of small computers. In Chapter 4, we distinguished between micro- and personal computers (PC's) by defining the micro to be "larger." Later in this chapter we will attempt a more definitive distinction between these two classes of "small" computers. In the meantime, we will emphasize personal computer applications--those that generally run on single-task, single-user, floppy disk systems. For continuity with our discussion in Chapter 4, we will limit our survey of applications to the following computer technologies:

> Decision Support Systems (DSS)
> Computer-Aided Design (CAD)
> Management Information Systems (MIS)

A variety of programs are commercially available within each of these technologies. Our emphasis will be on DSS applications, as these are particularly well suited to personal computers--more so perhaps than CAD and MIS. The examples presented in this chapter are meant to be representative rather than inclusive. For a more complete picture of available software, the reader should use the directories listed in Appendix A.

DSS Applications

Decision support systems involve a mix of calculation, modeling, algorithms, and data handling--often with graphical output. Personal computers perform these functions effectively and economically. For many facilities decisions, the complexity of calculations, modeling, and simple algorithms is within the power of 8- and 16-bit machines.

DATE 06/30/83   CARPET ESTIMATION

ROOMS:

ROOM	LENGTH	WIDTH	AREA
1	15'0	10'0	150.00
2	15'0	10'0	150.00
3	25'0	15'0	375.00
4	15'0	15'0	225.00
5	20'0	15'0	300.00
6	20'0	20'0	400.00
7	35'0	15'0	525.00
8	45'0	35'0	1575.00
9	210'0	150'0	31500.00
10	95'0	65'0	6175.00
11	35'0	30'0	1050.00
12	135'0	125'0	16875.00
13	75'0	45'0	3375.00
14	35'0	30'0	1050.00
15	25'0	15'0	375.00
16	15'0	10'0	150.00
17	25'0	20'0	500.00
18	35'0	20'0	700.00
19	20'0	20'0	400.00
20	15'0	15'0	225.00
21	20'0	15'0	300.00
22	35'0	20'0	700.00
23	40'0	20'0	800.00

```
            TOTAL AREA REQUIRED          67875.00
            SQUARE YARDS REQUIRED        `7542.00
```

COST:

GRADE	COST PER SQ YD	TOTAL COST
A	8.37	63126.54
B	11.39	85903.38
C	16.77	126479.34
D	21.46	161851.32

Figure 6.1. Printout from a carpet estimating program. The planner supplies lengths, widths, and costs. (Courtesy of BASICOMP, Inc., Mesa, Arizona.)

The simplest application of a PC is in calculations of the type shown in Figures 6.1 and 6.2. Here, the computer is serving as a high-priced, programmable calculator. It is favored over a calculator because it is faster, can drive a printer, and has a better display. Programs such as those shown for carpet and wallcovering estimates can be purchased for under $200. They can also be written with ease by the facilities planner himself.

The next level of complexity in DSS is the spreadsheet. In reality, spreadsheet programs such as VisiCalc and others perform series of chained calculations. An excellent application is shown in Figure 6.3. Here the arithmetic

DATE 06/30/83   WALLCOVERING ESTIMATION

WALLS:

WALL	WIDTH	HEIGHT	AREA
1	12'0	8'6	102.00
2	14'0	8'6	119.00
3	14'0	8'6	119.00
4	12'0	8'6	102.00
5	20'0	8'6	170.00
6	20'0	8'6	170.00
7	25'0	8'6	212.50
8	25'0	8'6	212.50

                       TOTAL   1207.00

WINDOWS:

WINDOW	WIDTH	HEIGHT	AREA
1	7'0	5'2	36.16
2	14'0	5'2	72.33

                       TOTAL   −108.50

DOORS:

DOOR	WIDTH	HEIGHT	AREA
1	3'0	6'8	20.00
2	5'0	6'8	33.33

                       TOTAL   −53.34

               TOTAL AREA REQUIRED   1045.16
                       TOTAL ROLLS   35

COST:

GRADE	COST PER ROLL	TOTAL COST
A	12.37	432.94
B	14.12	494.19
C	15.86	555.10
D	18.29	640.14

Figure 6.2. Printout from a wall covering estimating program. The planner supplies widths, heights, and costs. No consideration is made for patterns and repeats. (Courtesy of BASICOMP, Inc., Mesa, Arizona.)

extension of a from-to chart has been accomplished using the VisiCalc spreadsheet from VisiCorp. The formatting of the from-to chart is accomplished with a "template" from Moore Productivity Software, Blacksburg, Virginia. The complete set of templates for the Muther approach to layout planning is called

```
          F R O M - T O          C H A R T

Plant...ABC Mechanical Works        Project: New Layout
By......RHH            With.....DL        Date.....Jan. 9

Items(s) Charted.All Items      Basis of Values...Unit Loads/day
```

From	To	Press	Weld	Machine	Assembly	Paint	SteStor	Parts	Fin Store	Driveway	Totals
Press	....		15	12							27
Weld	....					11			2		13
Machine	....		3		15			2	12		32
Assembly	....					3			18		21
Paint	....			7	13						20
Steel Storage		12	5	1							18
Parts & Supply			1	4	3	1					9
Finished Storage				2						20	22
Driveway	....						8	11			19
Totals		12	24	26	31	15	8	13	32	20	181

Figure 6.3. From-to chart using VisiCalc, and a pre-formatted from-to "template" contained in the SLPCALC package. The example is the same one shown earlier in Figure 2.2. A change in any entry prompts a recalculation of its row, column, and the grand total. (Courtesy of Moore Productivity Software, Blacksburg, Virginia.)

SLPCALC. The cost is $250. To this must be added the cost of the VisiCalc software and a personal computer, if these are not already available. SLPCALC will run on the Apple II, Apple III, Radio Shack TRS-80 Model III, and the IBM-PC.

The departmental space report shown in Figure 6.4 is typical of a variety of software packages now available for space projections. The report multiplies the headcount projections by a space standard, down-totals for each period, and factors-up the sum for two types of circulation. These tasks could be accomplished with a spreadsheet program, but the output would not be as well formatted. And, it would not be as easy to transmit the resulting data to other files and programs where they might also be used. This program was developed by Micro-Vector, Inc., of Armonk, New York. Several versions are available. The smallest runs on the Hewlett-Packard 85 and 87. Other versions run on the IBM-PC and a variety of microcomputers supporting the UNIX operating system. Specialized programs of this type are priced between $2000 and $5000.

A higher-order DSS application is shown in Figure 6.5. This report is produced by a trend analysis program. It is more complicated than the previous examples. The program employs several curve-fitting routines, applies them to historical data sets, interprets the results, and make a projection using the best curve-fit (highest correlation coefficient). To create such a program on one's own requires the appropriate statistical functions to be present. The planner must also know how to call them from his program or reproduce their code within his own. This is considerably more difficult than

```
NOV 22 1982        A B C   I N V E S T M E N T   B A N K I N G      Page 6

        DEPARTMENT              CORPORATE                    2000
        SECTION                 CEO                          1170
        JOB STATUS FACTOR       MOVED IN                     22
        STATUS DATE             2/15/82                      12


TITLE                    CODE  OFF.  SQ.FT.  1982   1983   1984   1985   1987   1989
                               TYPE  REQ'D.

CHAIRPERSON              CHRP  G      180      1      1      1      1      1      1
PRESIDENT                PRES  A      200      1      1      1      1      1      1
VICE PRESIDENT           VPRE  A      200      2      2      2      2      2      2
SECRETARY                SECY  F       64      4      4      4      4      4      4
COFFEE TABLE             *CFT          64      1      1      1      1      1      1
CONFERENCE TABLE         *CTB  D       80      1      1      1      1      1      1
BOARD ROOM               -BRM          80      1      1      1      1      1      1
CONFERENCE ROOM          -CRM  H      180      1      1      1      1      1      1
RECEPTION ROOM           -RCP         180      1      1      1      1      1      1

 * * * *   SECTION                  * SUMMARY TOTALS FOR PROJECTION YEARS *
 * * * *   1170                       1982   1983   1984   1985   1987   1989

      TOTAL PERSONNEL REQUIRED         8      8      8      8      8      8
      TOTAL SQ.FT.REQ'D. WITHOUT CIRC. 1620  1620   1620   1620   1620   1620
        13% INTER-DEPT CIRC.           211    211    211    211    211    211
        18% INTRA-DEPT CIRC.           292    292    292    292    292    292
      TOTAL SQ.FT.REQ'D. WITH CIRC.    2123  2123   2123   2123   2123   2123
      SQUARE FEET/EMPLOYEE             265    265    265    265    265    265
 * * * * * * * * * * * * * *   G R O W T H   * * * * * * * * * * * * * * * * *
      PERSONNEL GROWTH FOR 1982 - 1989:           0%
      AVERAGE PERSONNEL GROWTH PER ANNUM:         0%
      AVERAGE SQUARE FEET GROWTH PER ANNUM:       0%
      PROJECTED SQUARE FT. REQ'D. FOR 1989:    2,123
```

Figure 6.4. Departmental space projections. Job codes, office types, and space standards are entered once into master files and used as needed for each departmental projection. Results can be fed to related space planning programs. (Courtesy of Micro-Vector, Inc., Armonk, New York.)

working up a spreadsheet application. The example shown is another Micro-Vector program. It too runs on the HP-85 and 87, and IBM-PC, and certain microcomputers supporting UNIX.

The addition of graphic, plotted output appears in Figure 6.6. The application is the P-Q curve discussed in Chapter 3. The example shown is the plotted output of a simple sorting routine. It could be produced with modest effort using good business graphics software. ("Good" software will plot correctly from randomly entered P-Q data.) The output shown here was produced on a special P-Q program from Micro-Vector. The computer used was an HP-85 connected to an HP-7470 plotter. The value of a special program like this is found in such features as automatic scaling of the Y-axis, and pre-formatting of the title blocks and labels.

The Class of Space Projection shown in Figure 6.7 combines statistical trend analysis with plotted output, also on the HP-85 and HP-7470. This program was written several years ago, before the widespread availability of low-cost business graphics software. The same results could be achieved to-day with standard statistical packages that incorporate business graphics. Some re-keying of data may be required. However, if the statistics and graphics are fully integrated this should not be necessary. Such software products are called "integrated" because they allow the user to take statistical results

PAGE   1

ABC INVESTMENT BANKING CORPORATION

STATISTICAL COMPARISON REPORT OF
PROJECTION DATA VS. TREND ANALYSIS

DEPARTMENT   1000:   ADMINISTRATION
SECTION   1010:   SECRETARIAL

PERSONNEL   (TREND ANALYSIS ON HISTORICAL DATA SHOWS A CORRELATION COEFFICIENT = .7823)

*********** HISTORY DATA FOR YEARS: ****************

1977	1978	1979	1980	1981
20	23	23	35	36

******************** PROJECTION DATA FOR YEARS: ********************

	1982	1983	1984	1985	1987	1989
TREND	42	50	59	69	95	131
(-)INTERVIEW	45	50	53	56	65	74
(=)DIFFERENCE	-3	0	6	13	30	57
% DIFFERENCE	-7	0	11	23	46	77

SQUARE FOOTAGE   (TREND ANALYSIS ON HISTORICAL DATA SHOWS A CORRELATION COEFFICIENT = .9307)

*********** HISTORY DATA FOR YEARS: ****************

1977	1978	1979	1980	1981
4,000	4,212	4,719	5,000	5,399

******************** PROJECTION DATA FOR YEARS: ********************

	1982	1983	1984	1985	1987	1989
TREND	5,742	6,100	6,459	6,818	7,535	8,252
(-)INTERVIEW	5,612	6,022	6,256	6,586	7,384	8,166
(=)DIFFERENCE	130	78	203	232	151	86
% DIFFERENCE	2	1	3	4	2	1

Figure 6.5. Statistical comparison report, used to compare historical trends of manpower and space against the estimates obtained in interviews. In the example shown, the personnel projections obtained through interviews are not consistent with past trends. The planner should clarify this issue before going further. In practice, this analysis should be applied only at the company, division, or large department level, where reasonably accurate historical records are available. (Courtesy of Micro-Vector, Inc., Armonk, New York.)

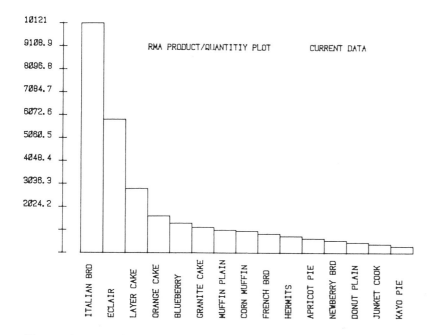

Figure 6.6. P-Q curve. The planner supplies product name and production quantities. The Y-axis is automatically scaled. A deep curve like this suggests that high volume products be split off from others and produced with specialized, dedicated equipment. (From an example by James Moore. Courtesy of Richard Muther & Associates, Inc., Kansas City, Missouri.)

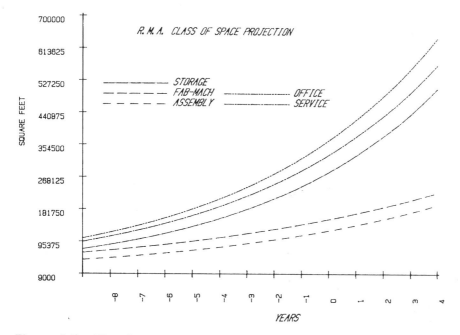

Figure 6.7. The planner supplies historical data. The rest is automatic. (Courtesy of Micro-Vector, Inc., Armonk, New York.)

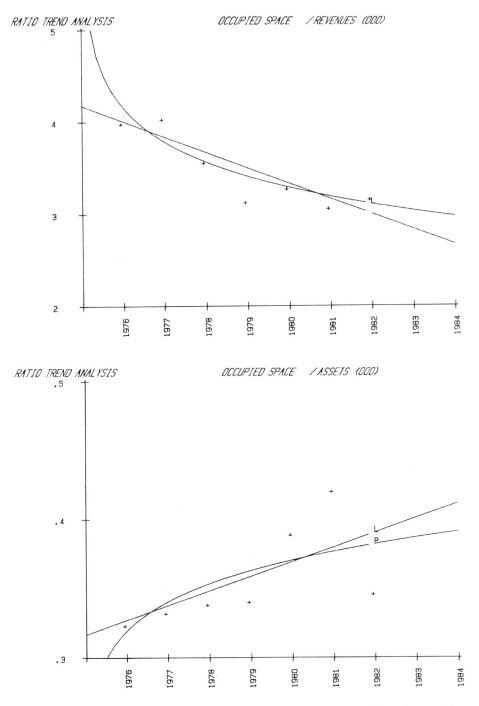

Figure 6.8. Ratio-trend plots. The planner supplies historical data. The program finds the best curve fit (linear, hyperbolic, logarithmic, etc.), plots the ratio, and extends it to a specified future period. By examining a series of such trends, a predictive model can be constructed as shown in Figure 6.9. (Courtesy of Micro-Vector, Inc., Armonk, New York.)

```
MICRO-VECTOR, INC.   MULTIPLE LINEAR REGRESSION

      Y = C + A1X1 + A2X2 + ... AnXn

WHERE  Y = DEPENDENT VARIABLE
       C = CONSTANT
       A1...An ARE COEFFICIENTS OF INDEPENDENT VARIABLES X1...Xn
             FOUND BY THE METHOD OF LEAST SQUARES

DEPENDENT VARIABLE = OCC SPACE (000)
         CONSTANT = -218609.406081
               A1 = +1.101025E+02   YEAR
               A2 = -1.529201E-03   ASSETS (000)
               A3 = +2.824548E-02   STOCK EQUITY (000)
               A4 = -8.287342E-04   LOANS (000)
               A5 = +4.226443E-03   DEPOSITS (000)
               A6 = +3.635791E-02   GROSS REVENUE(000)
               A7 = -3.268480E-02   OP+NET IN REV(000)
               A8 = -2.954432E-02   EXPENSES (000)
               A9 = -9.956942E-03   NON-IN EXP (000)
               A$ = -9.880156E-02   NET OP INCOME(000)
               A$ = -1.959203E-01   PERSONNEL

COEFFICIENT OF DETERMINATION         = .999999942522
COEFFICIENT OF MULTIPLE CORRELATION  = .99999997126
STANDARD ERROR OF ESTIMATE           = .055308227236


ORIGINAL DATA:

SAMPLE # 1
     YEAR                =     1972.000
     ASSETS (000)        =   662390.000
     STOCK EQUITY (000)  =    80410.000
     LOANS (000)         =   310058.000
     DEPOSITS (000)      =   532810.000
     GROSS REVENUE(000)  =    41323.000
     OP+NET IN REV(000)  =    28553.000
     EXPENSES (000)      =    31183.000
     NON-IN EXP (000)    =    18413.000
     NET OP INCOME(000)  =     8426.000
     PERSONNEL           =      909.000
     OCC SPACE (000)     =      220.000

SAMPLE # 2
     YEAR                =     1973.000
     ASSETS (000)        =   744037.000
     STOCK EQUITY (000)  =    80828.000
     LOANS (000)         =   371405.000
     DEPOSITS (000)      =   560350.000
     GROSS REVENUE(000)  =    54886.000
     OP+NET IN REV(000)  =    31820.000
     EXPENSES (000)      =    44470.000
     NON-IN EXP (000)    =    21404.000
     NET OP INCOME(000)  =     8327.000
     PERSONNEL           =     1008.000
     OCC SPACE (000)     =      237.000
```

Figure 6.9. Multiple linear regression, used to constuct a model in which space is prescribed by a set of financial variables. (Courtesy of Micro-Vector, Inc., Armonk, New York.)

directly into business graphics without key-entry. The screen is divided into
windows, with the statistics visible in one while the projections graph is constructed in another. In practice, certain limitations may apply both to statistical capabilities and to business graphics. For this reason a specialized, facilities-oriented program may be desirable and worth the extra expenditure.
Again, the chief benefits are automatic scaling and formatting, including standardized line types for each class of space. An example of modeling appears in
Figures 6.8 and 6.9. Here, a statistical model has been constructed to predict
long-range space requirements from a series of ratio trends. The programs
(from Micro-Vector) incorporate the simple trend analyses and plotting seen
earlier with the use of a multiple linear regression model. The same effort
could be accomplished (without plotted graphics) using standard statistical
packages.

Up to now, our examples have focused on space projections and preliminary analysis. These support Phase I--Orientation tasks in our phased approach to planning. The personal computer can also be used to support a
variety of Phase II--Overall Plan decisions, especially those involving layout.
In Figure 6.10, we see a different type of modeling program for scoring block
layouts. In the preceding chapter, we discussed scoring routines at some
length and presented an example of the PAINT program running on the Exidy
Sorcerer personal computer. The program described here in Figure 6.10 is
called quite simply Plant Layout Program. It was developed at Purdue University by Professors Suresh Khator and Colin Moodie. Written in BASIC, it contains roughly 300 lines of code and runs on the Radio Shack TRS-80 and the
IBM-PC.

The Plant Layout Program uses much of the CORELAP logic (total-closeness-rating) described in the preceding chapter. The required inputs are a
relationship chart (A, E, I, O, U, X format) and an existing or proposed layout plan. Two results are provided. The first is an activity selection order to
be used when constructing a layout. If the departments or activities are placed
in accordance with the suggested order, the important adjacencies should be
honored. The second result is a score that rates the overall worth of a completed layout based on the number of important adjacencies achieved. For a
complete discussion of the logic and the code itself, the reader should refer
to *Industrial Engineering*, Volume 15, No. 3, March 1983 (24).

In Figures 6.11 through 6.15 we move up the scale of complexity from
scoring to clustering. The examples shown are demonstrations of a program
called CAM, Communications Analysis Matrix. This is a trademarked product of
BASICOMP, Inc., of Mesa, Arizona. Beginning with the questionnaire shown
in Figure 6.11, the program produces sorted reference lists of those responding
and a series of adjacency resports at the individual, department, and company
level. First developed on the Commodore personal computer, the CAM program will run on any computer supporting the CP/M-86 or MS/DOS operating system. With 96K of memory, up to 2600 survey respondents can be
clustered. This is a feat possible only on the largest mainframe computers
as recently as 5 years ago. (The INACT program mentioned in the preceding
chapter uses 450K of memory on an Amdahl or IBM mainframe and is limited
to 100 survey inputs per run.) CAM can be used as an aid to multistory
planning, block layout, and even detailed layout of individual workstations,
subject of course to the reservations we expressed in Chapter 2 concerning
the survey approach.

The CAM program represents several man-years of effort by experienced
programmers. Yet like all the personal computer aids reviewed thus far, it

sells for less than $3000. With this kind of computer power available at such
low prices, there is no point in writing one's own programs.

After clustering, the next level of complexity is the generation of block
layouts. As noted in the previous chapter, the Institute of Industrial Engineers has announted both CRAFT and CORELAP for several makes of personal computers. However, for our example here, we will return to the program by Allan Cytryn introduced in the preceding chapter. This program
uses algorithms with joystick cursor control to produce a "finished" block
layout. The process is illustreated in Figures 6.16 through 6.21, using the
Tool Fabrication example from our earlier discussion of layout planning.

Inputs consist of a department list and an affinity map. The department
list includes a growth factor and usage code for each department. Affinities
are entered on a -9 to 99 scale. A rough block layout is generated automatically and adjusted with a joystick. An Iso-Cost Profile is produced
for each layout. This is actually a pseudo D-I Plot in which the numerical affinity expressions (-9-99) are used as a surrogate for the intensity
of flow. (This is the same approach used in the LAYOUT program by
Larry Ritzman.)

A total cost is produced for each layout. This is the product of the distances between departments times the numerical affinity expressed as dollars.
Three scores are given. The first is the simple average of the distances between all related departments. The second is weighted by the affinities. Finally, a ratio between the two averages is given.

The Cytryn program runs on the Apple IIe computer with 64K of RAM.
Results can be plotted on a small 80/132-column printer. The program is
capable of multistory applications of up to 44 floors. Like the CAM program just reviewed, several man-years of effort lie behind the example
shown here. The result is a very powerful planning aid on a very small
computer. It does not follow that such programs are simple to duplicate.

Our final example of Phase II--Overall Plan applications is a vertical stacking (multistory) planning program by Micro-Vector. Instead of block floor
plans, this program plots a scaled, vertical profile showing the assignment
of activities to floors. Program inputs include the gross and net available
space on each floor, the relationships between activities, and the space required by each. Two algorithms are invoked. One minimize the distances
between activity cents; the other minimizes the number of split activities.
Distance minimization takes precedence unless the user specifies the "no-split
rule" to be of greater importance. Activity locations can be fixed or pre-
assigned as needed. Results are plotted as shown in Figure 6.22. With a
program like this, a task that might take several days is reduced to a few
minutes. As with any algorithm, there is no guarantee of an "optimum" or best
layout, yet the results are graphic enough to reveal any impractical or unacceptable assignments at a glance.

In the past few pages we have reviewed a dozen DSS applications ranging
from carpet calculations to multistory layout and stacking. All run on equipment costing less than $5000. The specific devices employed include the most
popular makes of personal computers: Apple, Radio Shack, IBM, Hewlett-
Packard, and Commodore. Prices for individual applications range from $200
to $5000. These are modest outlays by any standard. Yet they afford big pay-
offs through better, faster analysis and the consideration of more alternatives
that might otherwise be possible. The examples we have just seen reinforce
the prescription made in Chapter 4 to place first priority on decision support
applications in Phase O, I, and II of the facilities planning process.

```
RUN

                    PLANT LAYOUT PROGRAM
              SCHOOL OF INDUSTRIAL ENGINEERING
                    PURDUE UNIVERSITY
              **********************************


ENTER:
        1   FOR FINDING SELECTION ORDER OF THE DEPARTMENTS
        2   FOR EVALUATING A GIVEN LAYOUT CONFIGURATION
? 1

NO. OF DEPARTMENTS? 7


GET READY TO ENTER RELATIONSHIP MATRIX IN UPPER TRIANGULAR FORM
PERMISSIBLE RELATIONSHIPS ARE: A,E,I,O,U(OR SPACE) AND X

ENTER RELATIONSHIP BETWEEN

              DEPARTMENT 1 AND DEPARTMENT 2 ? E
              DEPARTMENT 1 AND DEPARTMENT 3 ? O
              DEPARTMENT 1 AND DEPARTMENT 4 ? I
              DEPARTMENT 1 AND DEPARTMENT 5 ? O
              DEPARTMENT 1 AND DEPARTMENT 6 ? U
              DEPARTMENT 1 AND DEPARTMENT 7 ?

ENTER RELATIONSHIP BETWEEN

              DEPARTMENT 2 AND DEPARTMENT 3 ?
              DEPARTMENT 2 AND DEPARTMENT 4 ? E
              DEPARTMENT 2 AND DEPARTMENT 5 ? I
              DEPARTMENT 2 AND DEPARTMENT 6 ? I
              DEPARTMENT 2 AND DEPARTMENT 7 ?

              DEPARTMENT 4 AND DEPARTMENT 7 ?

ENTER RELATIONSHIP BETWEEN

              DEPARTMENT 5 AND DEPARTMENT 6 ? A
              DEPARTMENT 5 AND DEPARTMENT 7 ? I

ENTER RELATIONSHIP BETWEEN

              DEPARTMENT 6 AND DEPARTMENT 7 ? E

RELATIONSHIP MATRIX
DEPT NO.         1   2   3   4   5   6   7

   1             S   E   O   I   O   U   U
   2             E   S   U   E   I   I   U
   3             O   U   S   U   U   O   U
   4             I   E   U   S   I   U   U
   5             O   I   U   I   S   A   I
   6             U   I   O   U   A   S   E
   7             U   U   U   U   I   E   S


WANT TO CHANGE ANY RELATIONSHIP (Y/N)? N

CLOSENESS RANK OF THE DEPARTMENTS:

   6   5   2   4   1   7   3

SELECTION ORDER OF THE DEPARTMENTS:

   6   5   7   2   4   1   3

WANT TO EVALUATE A LAYOUT CONFIGURATION (Y/N)? Y
```

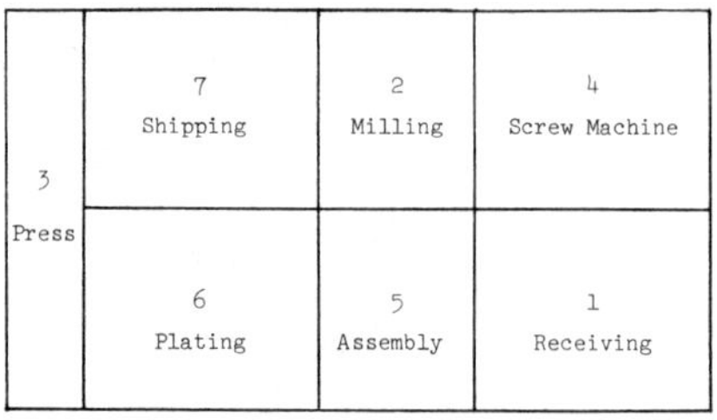

	7 \ Shipping	2 \ Milling	4 \ Screw Machine
3 \ \ Press	6 \ Plating	5 \ Assembly	1 \ Receiving

Figure 6.10. Plant Layout Program. For evaluating single-floor, block layouts.
Also for sequencing the placement of departments in a manual layout process.

```
WANT TO CHANGE CLOSENESS RELATIONSHIP VALUES (Y/N)
(DEFAULT VALUES ARE: A=8, E=4, I=2, O=1, U=0, X= -8)? N

ENTER DEPARTMENT NUMBER (O TO STOP)? 3

          ENTER NEIGHBORING DEPARTMENT(O TO STOP)? 7
          ENTER NEIGHBORING DEPARTMENT(O TO STOP)? 6
          ENTER NEIGHBORING DEPARTMENT(O TO STOP)? O

ENTER DEPARTMENT NUMBER (O TO STOP)? 7

          ENTER NEIGHBORING DEPARTMENT(O TO STOP)? 2
          ENTER NEIGHBORING DEPARTMENT(O TO STOP)? 6
          ENTER NEIGHBORING DEPARTMENT(O TO STOP)? O




ENTER DEPARTMENT NUMBER (O TO STOP)? O


CLOSENESS MATRIX
DEPT NO.       1  2  3  4  5  6  7

1              0  0  0  1  1  0  0
2              0  0  0  1  1  0  1
3              0  0  0  0  0  1  1
4              1  1  0  0  0  0  0
5              1  1  0  0  0  1  0
6              0  0  1  0  1  0  1
7              0  1  1  0  0  1  0

WANT TO CHANGE ANY CLOSENESS VALUE(Y/N)? N


TOTAL SCORE FOR THE GIVEN LAYOUT CONFIGURATION= 22

WANT TO TRY ANY VARIATIONS OF THIS PROBLEM (Y/N)? Y



ENTER:
     1   FOR FINDING SELECTION ORDER OF THE DEPARTMENTS
     2   FOR EVALUATING A GIVEN LAYOUT CONFIGURATION
? 2

WANT TO CHANGE ANY RELATIONSHIP (Y/N)? N

WANT TO CHANGE CLOSENESS RELATIONSHIP VALUES (Y/N)
(DEFAULT VALUES ARE: A=8, E=4, I=2, O=1, U=0, X= -8)? N

WANT TO EVALUATE A NEW LAYOUT (Y/N)?   Y
```

3 Press	1 Receiving	2 Milling	4 Screw Machine
7 Shipping	6 Plating	5 Assembly	

```
CLOSENESS MATRIX
DEPT NO.       1  2  3  4  5  6  7

1              0  1  1  0  0  1  1
2              1  0  0  1  1  1  0
3              1  0  0  0  0  0  1
4              0  1  0  0  1  0  0
5              0  1  0  1  0  1  0
6              1  1  0  0  1  0  1
7              1  0  1  0  0  1  0

WANT TO CHANGE ANY CLOSENESS VALUE(Y/N)? N


TOTAL SCORE FOR THE GIVEN LAYOUT CONFIGURATION= 27

WANT TO TRY ANY VARIATIONS OF THIS PROBLEM (Y/N)? N
Ok
```

The logic used is that of the CORELAP algorithm. Graphics are for illustration only. They are not produced by the program.

```
PERSON-TO-PERSON RESPONSES                              PERSON-TO-PERSON RESPONSES
INFORMATION
PREPARED BY:
COMMUNICATIONS (ADJACENCY) ANALYSIS QUESTIONNAIRE            QUESTION SHEET
PERSON-TO-PERSON RESPONSES                                   QUESTION SHEET
**NOTE**..........Be sure the correct answer sheet is being used............**NOTE**
Choose the most appropriate answer for each question and place your response on the
answer sheet.
A. Does this person give you information which you use in your daily responsibilities?
   5-Very Often    4-Often    3-Sometimes    2-Not Very Often    1-Almost Never

B. Do you receive direction from this person for your daily activities?
   5-Very Often    4-Often    3-Sometimes    2-Not Very Often    1-Almost Never

C. Does this person's information give you alternatives to your daily activities?
   5-Very Often    4-Often    3-Sometimes    2-Not Very Often    1-Almost Never

D. Do you keep the information given to you by this person to refer to in the future?
   5-Very Often    4-Often    3-Sometimes    2-Not Very Often    1-Almost Never

E. Does this person's information outline your daily activities procedures?
   5-Very Often    4-Often    3-Sometimes    2-Not Very Often    1-Almost Never

F. Do you ask this person's opinion even if it is different than your opinion?
   5-Very Often    4-Often    3-Sometimes    2-Not Very Often    1-Almost Never
```

```
M. Rate the importance of your face-to-face communication with this person as it re-
   lates to your activities and responsibilities.
   5-Very Important   4-Important   3-Some Importance   2-Little Importance
   1-Not Important

N. How often do you communicate with this person as it relates to your daily activi-
   ties and job responsibilities?
   5-Constantly   4-Many Times A Day   3-Many Times A Week   2-Many Times A Month
   1-Many Times A Year

PERSON-TO-PERSON RESPONSES                              PERSON-TO-PERSON RESPONSES
                                                              capa/cam-104
                                                              Updated 2/28/83
```

Figure 6.11. Communications questionnaire and handwritten response. In the current version of the program, handwritten responses must be keyed into the computer. Use of optical character recognition (OCR) and a scanner would eliminate this step. (Courtesy of BASICOMP, Inc., Mesa, Arizona.)

```
PERSON-TO-PERSON RESPONSES                          PERSON-TO-PERSON RESPONSES
CLIENT NAME:
PROJECT NUMBER:
```

```
COMMUNICATIONS (ADJACENCY) ANALYSIS QUESTIONNAIRE            ANSWER SHEET
PERSON-TO-PERSON RESPONSES                                  ANSWER SHEET
**NOTE**..........Be sure the correct answer sheet is being used.............**NOTE**
```

Employee
Code#/.1.2./.0.3./.0.2./.0.8/   Employee Name: ROSE M. ERICKSON

Department MARKETING   Job Title VP- BIS. DEVELOPMENT

```
**************************************************************************
Co-Worker's
Code(8-digits)/.1.2./.0.3./.0.2./.0.3./    A. 1   B. 1   C. 1
Co-Worker's
Name  JANET BARNES                          D. 2   E. 1   F. 1
Co-Worker's
Job-Title  ADMIN. ASST.                     G. 2   H. 1   I. 2        M. 4
Co-Worker's
Department  MARKETING                       J. 5   K. 5   L. 5        N. 5
**************************************************************************
Co-Worker's
Code(8-digits)/.1.2./.0.3./.0.2./.1.2./    A. 4   B. 1   C. 3
Co-Worker's
Name  RANDY SMITH                           D. 3   E. 1   F. 3
Co-Worker's
Job-Title  MARKETING MANAGER                G. 3   H. 1   I. 2        M. 5
Co-Worker's
Department  MARKETING                       J. 5   K. 5   L. 4        N. 3
**************************************************************************
Co-Worker's
Code(8-digits)/.1.0./.0.0./.0.0./.0.1./    A. 5   B. 3   C. 4
Co-Worker's
Name  N. J. HUDSON                          D. 4   E. 3   F. 3
Co-Worker's
       PRESIDENT                            G. 4   H. 4   I. 3        M. 5
```

```
DATE: 06/30/83        ABC POWER SYSTEMS EMPLOYEE RESPONSES LISTING

EMPLOYEE    CO-WORKER    A B C D E F G H I J K L M N

01000001    01000002     2 1 1 2 1 1 2 1 2 5 5 5 4 5
01000001    01000004     4 2 3 4 2 3 1 1 2 4 5 4 5 3
01000001    01000006     4 2 3 4 2 3 1 1 2 4 5 4 5 3
01000001    01000010     4 2 3 4 2 3 1 1 2 4 5 4 5 3
01000001    03000001     4 2 3 4 2 3 1 1 2 4 5 4 5 3
01000001    05020101     2 3 4 3 3 3 2 3 4 2 2 2 1 2
01000002    01000001     5 4 4 4 4 3 3 4 4 2 1 2 5 5
01000002    01000003     3 1 3 2 1 2 2 1 1 5 5 5 5 4
01000002    01000004     3 3 3 3 3 3 3 3 3 1 1 2 4 3
01000002    01000005     3 1 3 2 1 2 2 1 1 5 5 5 5 4
01000002    01000006     3 3 3 3 3 3 3 3 1 1 2 4 3
01000002    01000007     3 3 2 2 2 3 3 3 3 1 1 2 4 3
01000002    01000008     3 1 3 2 1 2 2 1 1 5 5 5 5 4
01000002    01000009     3 1 3 2 1 2 2 1 1 5 5 5 5 4
01000002    01000010     3 3 3 3 3 3 3 3 1 1 2 4 3
01000003    01000002     4 5 4 4 5 4 4 5 5 2 1 1 4 4
01000004    01000001     4 5 5 5 4 4 4 4 5 3 1 3 5 3
01000004    01000002     2 1 1 1 1 1 1 1 1 3 3 4 4 4
01000004    01000006     0 0 0 0 0 0 0 0 0 0 0 0 0 0
01000004    01000007     3 1 3 3 2 2 1 1 2 4 5 4 4 3
01000004    01000011     3 1 3 2 2 1 1 1 2 4 5 4 4 3
01000004    02020001     3 1 2 1 2 1 1 1 2 4 5 4 4 3
```

```
DATE: 06/30/83              ABC POWER SYSTEMS PERSONNEL CODES

   CODE                    NAME                          TITLE

01000001  JACKSON B.                        PRESIDENT
01000002  THOMAS S.                         SECRETARY TO PRESIDENT
01000003  ANDERSEN T.                       EXECUTIVE RECEPTIONIST
01000004  HENDERSON B. R.                   V.P. - CORPORATE PLANNING
01000005  ZILOSKY M. T.                     EXECUTIVE SECRETARY
01000006  ANDERSON T.                       V.P. CORPORATE FACILITIES
01000007  ERICKSON D. R.                    CONTROLLER
01000008  HEMMINGWAY M.                     EXECUTIVE SECRETARY
01000009  COLLINGSWORTH R.T.                CORPORATE SECRETARY
01000010  MABRY R. M.                       V.P.-DIV. CUSTOMER SVC.
01000011  MARTIN S.                         DIRECTOR-SPEC.-COPPS.
02010001  ALEXANDER THOMAS                  MANAGER
02010002  AUSTIN T.                         ASSISTANT MANAGER
02010003  STAPLETON B.                      SENIOR ENGINEER
02010004  ROBINSON F.                       ENGINEER
02010005  STARKWEATHER D.                   INTERNSHIP
02020001  BECK                              DIRECTOR
03000001  ELDREDGE                          MANAGER
03000002  SARDINICK U. P.                   SENIOR EDP OPERATOR
```

```
DATE: 06/30/83            ABC POWER SYSTEMS SORTED PERSONNEL

   CODE                    NAME                          TITLE

02010001  ALEXANDER THOMAS                  MANAGER
01000003  ANDERSEN T.                       EXECUTIVE RECEPTIONIST
04030001  ANDERSON                          DIRECTOR
01000006  ANDERSON T.                       V.P. CORPORATE FACILITIES
02010002  AUSTIN T.                         ASSISTANT MANAGER
02020001  BECK                              DIRECTOR
04010001  BECKERTON P. L.                   PROJECT MANAGER
06000005  BENSTER D. W.                     RESEARCH ANALYST I
04050001  BIGLOW T.E.                       DIRECTOR
04010002  BILINGSLEY T. T.                  COST SCHED. SPECIALIST
03000004  BURKHARDT A. N.                   STAFF FINANCE AUDITOR
06000006  BUS J.                            RESEARCH ANALYST II
06000007  CHINN A. N.                       RESEARCH CLERK
04050003  COBB O.                           MANAGER
01000009  COLLINGSWORTH R.T.                CORPORATE SECRETARY
04040004  CRAIN W.                          INTERNSHIP
04050002  CRAVENS R.E.                      SECRETARY
04040005  CRAWFISH V.                       EIT
04040003  CRAWFORD C.                       SECRETARY
```

Figure 6.12. Sorted listings of questionnaire respondents. (Courtesy of
BASICOMP, Inc., Mesa. Arizona.)

```
DATE: 06/30/83              INDIVIDUAL ADJACENCY REPORT                PAGE   1

DEPARTMENT      010000    EXECUTIVE

01 JACKSON B.
   PRESIDENT
                                  RECEIVES           GIVES
                            INFO INST INFL    INFO INST INFL FREQ  IMP POWER
      02 THOMAS S.           40   20   33      90   90   87   90   100   95
         SECRETARY TO PRESIDENT
         EXECUTIVE

      06 ANDERSON T.         60   27   57      83   93   87  100    60   80
         V.P. CORPORATE FACILITIES
         EXECUTIVE

      04 HENDERSON B. R.     60   27   57      83   93   87  100    60   80
         V.P. - CORPORATE PLANNING
         EXECUTIVE

      10 MABRY R. M.         30   17   27      40   50   40   50    30   40
         V.P.-DIV. CUSTOMER SVC.
         EXECUTIVE
```

Figure 6.13. Individual adjacency report, showing the nature of the interaction between B. Jackson, President, and others in the executive department. The various scores are interpreted by the planner in a detailed layout plan. (Courtesy of BASICOMP, Inc., Mesa, Arizona.)

```
DATE: 06/30/83              DEPARTMENT ADJACENCY REPORT

DEPT. 010000 EXECUTIVE

LINKS TO

      DEPT. 030000 INTERNAL AUDIT
      REPORTED LINKS                         AVERAGE
           3                                  73.3

      DEPT. 040500 ENVIRONMENTAL
      REPORTED LINKS                         AVERAGE
           1                                  70.0

      DEPT. 040400 QUALITY ASSURANCE
      REPORTED LINKS                         AVERAGE
           1                                  70.0

      DEPT. 050102 STATE
      REPORTED LINKS                         AVERAGE
           1                                  70.0
```

Figure 6.14. Department adjacency report, showing the number and relative strength of the links between the executive department and others in the organization. (Courtesy of BASICOMP, Inc., Mesa, Arizona.)

```
DATE: 06/30/83              COMPANY DEPARTMENT ADJACENCY REPORT


    DEPT. 010000 EXECUTIVE
    DEPT. 020200 COMMUNICATIONS SYSTEM
      REPORTED LINKS                              AVERAGE
           2                                       75.0

    DEPT. 010000 EXECUTIVE
    DEPT. 030000 INTERNAL AUDIT
      REPORTED LINKS                              AVERAGE
           6                                       71.7

    DEPT. 010000 EXECUTIVE
    DEPT. 060000 PLANNING AND ADMINISTRATION
      REPORTED LINKS                              AVERAGE
           1                                       70.0

    DEPT. 010000 EXECUTIVE
    DEPT. 050102 STATE
      REPORTED LINKS                              AVERAGE
           1                                       70.0

    DEPT. 020100 POWER PLANNING
    DEPT. 050101 FEDERAL
      REPORTED LINKS                              AVERAGE
           2                                       70.0

    DEPT. 020200 COMMUNICATIONS SYSTEM
    DEPT. 050101 FEDERAL
      REPORTED LINKS                              AVERAGE
           1                                       70.0
```

Figure 6.15. A complete listing of all departmental relationships, in order of descending strength. (Courtesy of BASICOMP, Inc., Mesa, Arizona.)

```
VERSION 1.0, JAN-1982 ***'TOOL-WORKS DEPARTMENT LIST'*** PAGE 1
```

TOOL-WORKS DEPARTMENT LIST

```
    DEPT ....NAME.... FL .AREA.. GRO .1982.. .USE.

    0001 PUNCH PRESS   1     550   0%     550   1000
    0002 AUX PUNCH PR  1     200   0%     200   1000
    0003 DRILLING      1     250   0%     250   1000
    0004 GRINDING      1     250   0%     250   1000
    0005 GEN FABRICAT  1     400   0%     400   1000
    0006 WET TUMBLE    1     100   0%     100   1000
    0007 SPEC. PROD    1     360   0%     360   1000
    0008 RAW MTL STRG  1       1   0%       1   2000
    0009 IN-PROC STRG  1     400   0%     400   1000
    0010 ASSEMBLY      1       1   0%       1   2000
    0011 SHOP TOILET   1      75   0%      75   3000
    0012 SHOP OFFICE   1       1   0%       1   2000
    0012 *** TOTAL *** 2588         2588
```

Figure 6.16. Department or activity list, showing the activity number and space for each activity. No growth factor is used in this example. "Use" is a type-of-space code. Here "2000" indicates the three fixed activities. Data corresponds to that given in Figure 5.1. (Courtesy of Allan Cytryn, Tenafly, New Jersey.)

TOOL—WORKS AFFINITY MAP

```
                          0001
0001 PUNCH PRESS         -0002          KEY TO SYMBOLS
0002 AUX PUNCH PR        +-0003         ─────────────
0003 DRILLING           -!-0004         #  61-99  ESSENTIAL
0004 GRINDING           --+-0005        *  41-60  VERY IMPPRTANT
0005 GEN FABRICAT       -+-0006         +  21-40  IMPORTANT
0006 WET TUMBLE         ---*-0007       !  01-20  DESIRABLE
0007 SPEC. PROD         ------0008      -   0     NOT MEANINGFUL
0008 RAW MTL STRG       #+!+++*-0009    X  (0     UNDESIRABLE
0009 IN-PROC STRG       *!!!!-+-0010
0010 ASSEMBLY           +-!#-*--
                                  0011
0011 SHOP TOILET        !-!!+*!-! -0012
0012 SHOP OFFICE        +!!!-X+-- !-
```

TOOL—WORKS AFFINITY LIST

DEPT ....NAME....	DEPT/AF	DEPT/AF	DEPT/AF	DEPT/AF	DEPT/AF	DEPT/AF	DEPT/AF
0001 PUNCH PRESS	0001/99	0002/25	0008/95	0009/55	0010/25	0011/10	0012/25
0002 AUX PUNCH PR	0001/25	0002/99	0003/10	0008/25	0009/10	0012/10	
0003 DRILLING	0002/10	0003/99	0004/25	0005/25	0008/10	0009/10	0010/10
	0011/10	0012/10					
0004 GRINDING	0003/25	0004/99	0006/55	0008/25	0009/10	0010/95	0011/10
	0012/10						
0005 GEN FABRICAT	0003/25	0005/99	0008/25	0009/10	0011/25		
0006 WET TUMBLE	0004/55	0006/99	0008/25	0011/55	0012/-9		
0007 SPEC. PROD	0007/99	0008/55	0009/25	0010/55	0011/10	0012/25	
0008 RAW MTL STRG	0001/95	0002/25	0003/10	0004/25	0005/25	0006/25	0007/55
	0008/99						
0009 IN-PROC STRG	0001/55	0002/10	0003/10	0004/10	0005/10	0007/25	0009/99
0010 ASSEMBLY	0001/25	0003/10	0004/95	0007/55	0010/99	0011/10	
0011 SHOP TOILET	0001/10	0003/10	0004/10	0005/25	0006/55	0007/10	0010/10
	0011/99	0012/10					
0012 SHOP OFFICE	0001/25	0002/10	0003/10	0004/10	0006/-9	0007/25	0011/10
	0012/99						

Figure 6.17. Affinity map and list uses a -9-99 closeness scale, calibrated as shown in the key. In this example, the vowel-letter ratings in the original problem (Figure 5.1) were converted to A = 95, E = 55, I = 25, O = 10, and X = -9. (Courtesy of Allan Cytryn, Tenafly, New Jersey.)

TOOL—BLDG—1/B01V0

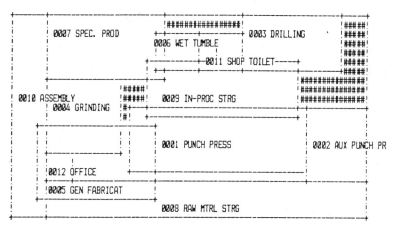

Figure 6.18. Space relationship diagram, generated automatically from the department and affinity lists previously entered. The algorithm results in overlaps and gaps which are resolved by redrawing with a joystick. (Courtesy of Allan Cytryn, Tenafly, New Jersey.)

*** VERSION 1.0, JAN-1982 ***' TOOL-BLDG-1/B01V1' AT SCALE 1"=16F ***

TOOL-BLDG-1/B01V1

USAGE SUMMARY (SQ-F)

GROSS-AREA	3520	100%
PERIMETER WALL	0	0%
INTERIOR WALL	0	0%
UTILITY/SERVICE	0	0%
NET AVAILABLE	3520	100%
ALLOCATED	3520	100%
AVAILABLE	0	0%

ALLOCATION SUMMARY (SQ-F)

(1)	0010 ASSEMBLY	352	(1)	0008 RAW MTRL STR	576	(1)	0012 OFFICE	36
(1)	0001 PUNCH PRESS	550	(1)	0002 AUX PUNCH PR	200	(2)	0004 GRINDING	252
(1)	0005 GEN FABRICAT	399	(1)	0006 WET TUMBLE	182	(1)	0007 SPEC. PROD	360
(1)	0009 IN-PROC STRG	374	(1)	0011 SHOP TOILET	72	(1)	0003 DRILLING	248

Figure 6.19. Final block layout and related space summaries. With the exception of activity 0005, very little adjustment was required from the previous step. (The reader is encouraged to compare this result with those of Chapter 5.) (Courtesy of Allan Cytryn, Tenafly, New Jersey.)

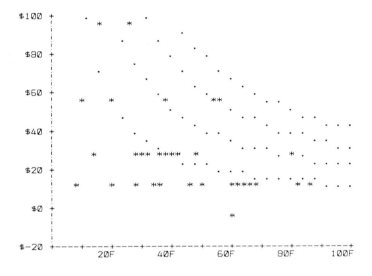

Figure 6.20. Iso-Cost Profile (psuedo D-I plot) in which the numerical affinities (95, 55, 25, 10, -9) are plotted as costs against the distances derived from the final block layout. (Courtesy of Allan Cytryn, Tenafly, New Jersey.)

***

VERSION 1.0, JAN 1982 *** ISO-COST PROFILE FOR PROJECT=TOOL-WORKS   AND PLAN=TOOL-BLDG-1/B01V1   *** PAGE 2 ***

PROJECT: TOOL-WORKS
PLAN...: TOOL-BLDG-1/B01V1

*** 0001 PUNCH PRESS ***

```
0002 AUX PUNCH PR   30F   X 25$/F =  $750      0008 RAW MTL STRG   12.3F X 95$/F = $1163
0010 ASSEMBLY       45.8F X 25$/F = $1143      0011 SHOP TOILET    31.3F X 10$/F =  $312
                                               0009 IN-PROC STRG   17.8F X 55$/F =  $976
                                               0012 SHOP OFFICE    34.3F X 25$/F =  $856
```

*** 0002 AUX PUNCH PR ***

```
0001 PUNCH PRESS    30F   X 25$/F =  $750      0003 DRILLING       31F   X 10$/F =  $310
0009 IN-PROC STRG   47.8F X 10$/F =  $477      0012 SHOP OFFICE    64.3F X 10$/F =  $642
                                               0008 RAW MTL STRG   38.3F X 25$/F =  $956
```

*** 0003 DRILLING ***

```
0002 AUX PUNCH PR   31F   X 10$/F =  $310      0004 GRINDING       77.8F X 25$/F = $1943
0008 RAW MTL STRG   57.8F X 10$/F =  $577      0009 IN-PROC STRG   43.3F X 10$/F =  $432
0011 SHOP TOILET    18.3F X 10$/F =  $182      0012 SHOP OFFICE    83.8F X 10$/F =  $837
                                               0005 GEN FABRICAT   11.8F X 25$/F =  $293
                                               0010 ASSEMBLY       79.8F X 10$/F =  $797
```

*** 0004 GRINDING ***

```
0003 DRILLING       77.8F X 25$/F = $1943      0006 WET TUMBLE     52.3F X 55$/F = $2873
0009 IN-PROC STRG   34.5F X 10$/F =  $345      0010 ASSEMBLY       24F   X 95$/F = $2280
0012 SHOP OFFICE    6F    X 10$/F =   $60      0008 RAW MTL STRG   34F   X 25$/F =  $850
                                               0011 SHOP TOILET    59.5F X 10$/F =  $595
```

*** 0005 GEN FABRICAT ***

```
0003 DRILLING       11.8F X 25$/F =  $293      0008 RAW MTL STRG   46F   X 25$/F = $1150
0011 SHOP TOILET    27F   X 25$/F =  $675      0009 IN-PROC STRG   31.5F X 10$/F =  $315
```

*** 0006 WET TUMBLE ***

```
0004 GRINDING       52.3F X 55$/F = $2873      0008 RAW MTL STRG   35.8F X 25$/F =  $893
0012 SHOP OFFICE    58.3F X -9$/F = $-525      0011 SHOP TOILET    7.3F  X 55$/F =  $398
```

Figure 6.21. Detailed listing of iso-cost data. (Courtesy of Allan Cytryn, Tenafly, New Jersey.)

```
ADJACENCY  EVALUATIONS  FOR  VERTICAL  STACK
```

```
PRIMARY  DEPT:   AMS-B              AREA:        3315

SECONDARY DEPT.              AREA            ADJ.

    AMS-CS                   8897             9
    AMS-IA                   14261            9
    TS-IS                    11668            9



PRIMARY  DEPT:   AMS-CS             AREA:        8897

SECONDARY DEPT.              AREA            ADJ.

    AMS-IA                   14261            9
    TS-IS                    11668            9



PRIMARY  DEPT:   AMS-IA             AREA:        14261

SECONDARY DEPT.              AREA            ADJ.



PRIMARY  DEPT:   CIS-SVP            AREA:        2800

SECONDARY DEPT.              AREA            ADJ.

    CIS-NCF                  8232             5
    CIS-IM                   2387             5
```

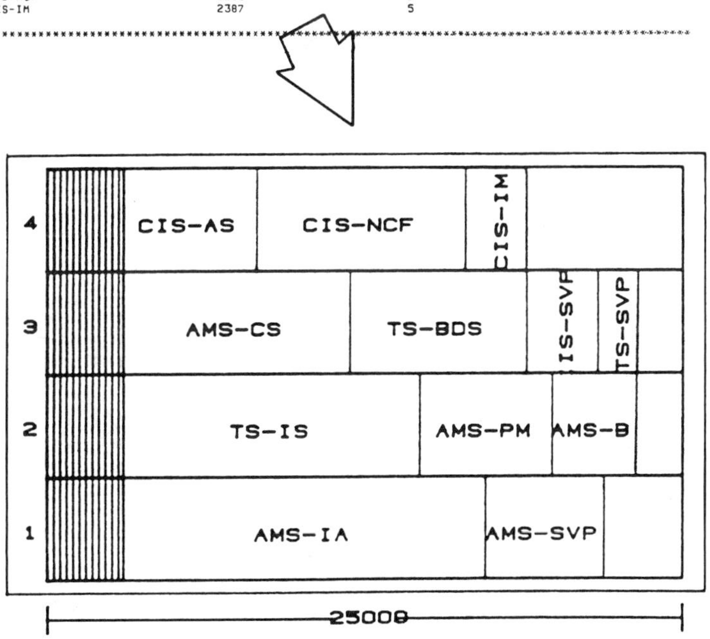

Figure 6.22. Vertical stack input and output. Shaded areas represent the building core. Blank areas are unassigned. (Courtesy of Micro-Vector, Inc., Armonk, New York.)

CAD Applications

The usefulness of personal computers has recently been extended to Phase III--Detail Plans, with the advent of low-cost software for computer-aided drafting. The scope and power of such software is limited but can be considered for certain small applications. Since the next chapter is devoted entirely to CAD, our discussion here will be somewhat abbreviated. A typical personal computer drafting system is illustrated in Figure 6.23. Such a system can be obtained today for under $10,000, often using an existing Apple or IBM machine.

While they may outwardly look alike, there are in fact two classes of personal CAD systems. One uses off-the-shelf computers; the other uses enhancements. Off-the-shelf systems have the following general characteristics:

1. Standard 8- or 16-bit CPU, running at a speed of 4-8 MHz
2. 64K to 256K of memory
3. Dual floppy disk drives
4. Small (9-13 in.) monochromatic display with low to medium resolution
5. Small (11 x 11 in.) digitizing tablet for cursor control
6. Small plotter
7. Single-task operating system (no plotting while drafting or vice versa)
8. Software limited to general purpose 2-D drafting

Software for these systems ranges from $250 to $2000. Typical software offerings include: Vector Sketch from GTCO Corporation of Rockville, Maryland, CADAPPLE from T & W Systems, Inc., of Fountain Valley, California, and PC-Draw from MICROGRAFX of Dallas, Texas.

The interested planner must typically acquire his software from one source, his computer and peripherals from another. It is then his responsibility to connect the equipment and install the software to achieve a working system.

Enhanced systems add memory, hard disk drives, and usually a graphics processor to the standard-issue personal computer. This processor is in fact a microcomputer on its own card or circuit board. If not added to the computer, it may be incorporated instead into the display device. These enhancements increase speed and the quality of the graphic display. Drafting is an intensive task by comparison with other uses of computers. Consequently, standard-issue PC's are slow on many drafting applications and are not effective drivers of higher resolution and larger color displays. The general characteristics of enhanced systems include:

1. 16-bit CPU and graphics processor (often 16/32 bit) running at a speed of 8-10 MHz
2. 256K to 512K of memory
3. Hard disk drive
4. Small, possibly color display with medium to high resolution (sometimes dual displays)
5. Small digitizing tablet
6. Medium-speed, C- or D-sized plotter
7. Single-task operating system

Figure 6.23. Personal computer CAD system includes small plotter, digitizing tablet, disk drive, CPU, monitor, and keyboard. (Courtesy of GTCO Corporation, Rockville, Maryland.)

8.   General purpose, 2-D drafting software

These enhancements bring total costs to $25,000 or more. Typical of the software for these systems are: AutoCAD from AutoDESK of Mill Valley, California, Versa CAD from T & W Systems, Inc., and Draft-Aide from United Networking Systems, Inc., Houston, Texas.

In some cases, software vendors or their local dealers will act as turnkey systems houses--buying the computer, graphics processor, and peripherals, making the necessary enhancements, and installing the software. The planner receives a fully-operational system with one source assuming responsibility for malfunctions. Otherwise the planner may have to buy the computer from one source, graphics processor from another, peripherals from a third and fourth, and software from perhaps a fifth. While this saves several thousand dollars, it is time-consuming and can lead to annoyances and delays in the event of malfunctions. It also takes the self-confidence to put it all together.

Typical accomplishments with personal-computer-based CAD systems are shown in Figures 6.24 through 6.26. These plots are in many respects indistinguishable from those of the large CAD systems reviewed in the next chapter. But it does not follow that personal computers will satisfy all our drafting needs. A man can move a pallet of bricks if he makes enough trips. So too can a personal computer produce a set of working drawings. It does not follow that this approach is cost-effective. To use an analogy, the large CAD systems covered in the next chapter represent the power of word processing systems, when compared to the electric typewriters we are discussing here. Remember, a finished document may look the same no matter which device was used.

PC-based systems, both off-the-shelf and enhanced, are best suited to the following kinds of tasks:

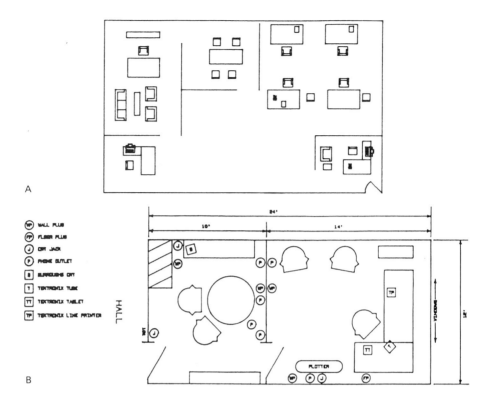

Figure 6.24. Single-line office layout. These are easily accomplished with personal-computer-based CAD. (A, courtesy of AutoDESK, Mill Valley, California. B, courtesy of United Networking Systems, Inc., Houston, Texas.)

1. *Single-line charts and diagrams*: flow and process charts, small piping and wiring diagrams, organizational charts
2. *Single-line installation drawings*: partitions, foundations, pits, machine tools, electronic equipment, etc.
3. *Small furniture and equipment layouts*: up to 5000 square feet, using standard symbols
4. *Preliminary architectural plans*: for small, rectilinear structures, with minimal dimensioning and line weights required

Much of the drafting work in a typical facilities group falls into these four categories. Personal computers will not match the performance of larger CAD systems, and are not yet productive on working drawings. The chief limitations of these small CAD systems are:

1. *Processing speed*: Larger machines are 30 times faster on some computations.
2. *Displays*: Too small, with too low a resolution for some facilities tasks. This is a result of both hardware (horsepower) and software restrictions.

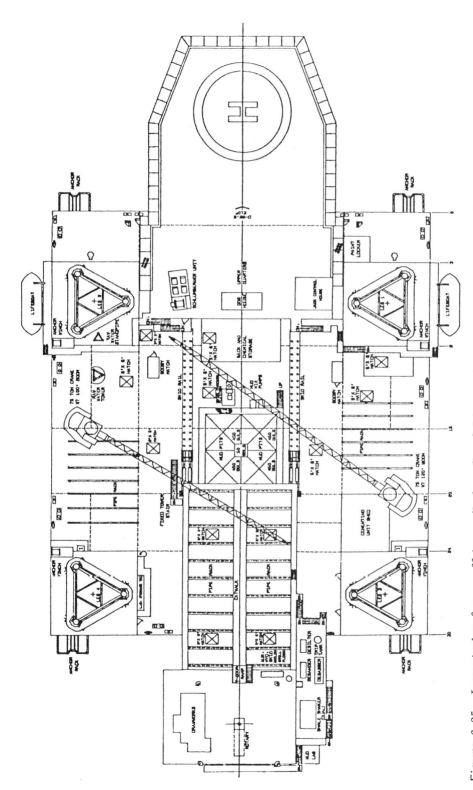

Figure 6.25. Layout plan for an offshore oil rig. Careful inspection reveals that this plan is highly symmetrical about the center line. It also makes repeated use of very few symbols. Such simplifying conditions must be present for a personal-computer-based system to be productive. (Courtesy of United Networking Systems, Inc., Houston, Texas.)

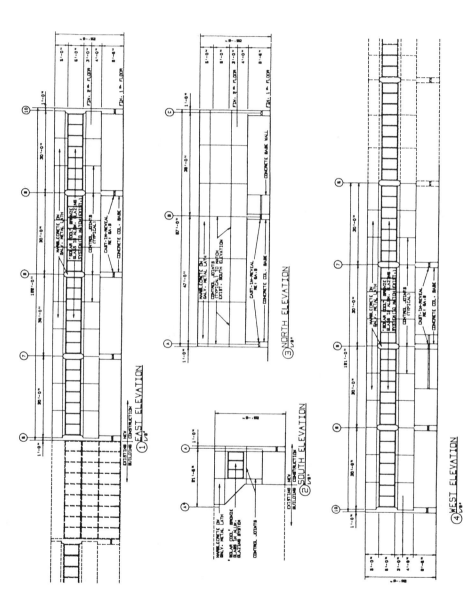

Figure 6.26. Simple architectural elevations. (Courtesy of United Networking Systems, Inc., Houston, Texas.)

3. *Drawing size limits*: From hardware (memory and disk capacity) and often software as well, which may limit the elements in a display or drawing file.
4. *Lack of tailored commands*: For drafting of walls, doors, windows, stairs, and other commonly encountered features of facilities plans. Also commands such as "partial erase" of previously drawn lines, and measurements of distances, angles, and areas.
5. *Lack of reporting capabilities*: For material lists, cost estimates, symbol inventories, etc.
6. *Scaling restrictions*: Some software can only work with integers (whole numbers and fixed decimals). This can limit the user to pre-established drawing scales (1/4 in, 1/8 in, etc.) and may limit the user's ability to fit a drawing or plan to the size of the display or plotter.

As the suppliers of these systems and software begin to focus on architectural and facilities planning applications, these limitations may be corrected. As the power of personal computers increases (and it will) such corrections will be easier to make. At the same time however, personal CAD systems may lose two of their chief current attractions--simplicity and ease of use. They are simple because they are limited. As a result, they are also easy to use and can be mastered in days or weeks of casual use, compared with months of sustained, often full-time use required to master larger systems.

MIS Applications

In this section we will turn our attention to the use of personal computers as management information systems. In this capacity, personal computers are used to perform many tasks that have traditionally been performed on large mainframe or minicomputer systems. We will review several typical uses devoted to facilities and project management. The purpose of each is to provide planners and managers with a greater degree of control. Since Chapter 8 is devoted in full to MIS applications, our discussion here will be limited.

The applications covered in the preceding sections aid the planner in Phases I, II, and III. We will now consider Phases IV and V. As we noted in Chapter 4, these are the two phases supported by MIS technology. Our first example is that of a data base devoted to furniture and equipment. The SPECIFIER is a product of BASICOMP, Inc., and runs on machines supporting the CP/M or MS/DOS operating system. SPECIFIER is shown in Figures 6.27-6.31. At the heart of the data base is a set of vendor catalogs (Figure 6.27). These contain the items which may be used by the planner to satisfy furniture requirements. SPECIFIER can also act as an inventory control device by cataloging existing furniture under a dummy vendor name.

The establishment of space and related furniture standards usually occurs early in a project--in Phases II and III. An example is shown in Figure 6.28. The set of all SPEC CODES contained in these standards constitutes the project data base shown in Figure 6.29. This is in reality the relevant subset of the vendor catalogs already shown. Now, in Phase IV of our project--implementation--we need to arrive at the correct purchase quantities

DATE: 06/30/83

DATA BASE: STEELCASE

SPEC CODE	MANUFACTURER	PRODUCT #	PRODUCT	DESCRIPTION	WIDTH	DEP/LENG	HEIGHT	WEIGHT	COST
SDGTS	STEELCASE	STL.421-311	CHAIR-DESK	TS ARMCHAIR/9201 POL CHROME/5180 IMP. BLU/UPH ARMRESTS	23.5	24.25	34.63	46	327
SDGTS	STEELCASE	STL.421-312	CHAIR-DESK	TILT SWIVEL ARM CHAIR/9201 POL CHROME/5182 IMP. BLU/PED.BASE	23.5	24.25	34.63	46	300
SSWAS	STEELCASE	STL.421-482	CHAIR-SIDE	SIDE CHAIR W-ARMS/CANT.SLED BASE/9201 POL. CHRM/5180 IMP.BLU	23.5	24.25	31.5	36	228
SDPP	STEELCASE	STL.421-520	CHAIR-SECRETARY	SEC POSTURE CHAIR/PED BASE/9201 POL.CHRM/5180 IMP. BLUE	18.5	22	33.25	35	237
FL2M1B36	STEELCASE	STL.836-200/HF	FILE-LATERAL	2 12IN. LTR.SIZE SHELVES/DOORS/LT.OAK TOP/835 BLACK/9201 PC	36	18	29	136	326
DDPWD3060	STEELCASE	STL.W-3026030F	DESK-DOUBLE PED	24-3/4KNEESPACE/FLUSH BACK PANEL/#3963LT. OAK PL TOP/LOCKING	60	30	29	329	951
DDPWD3060	STEELCASE	STL.W-3026030R	DESK-DOUBLE PED	24-3/4KNEESPACE/4IN RECESSED BACK PNL/#3963LT.OAK PL TOP/LCK	60	30	29	313	951
DDPWD3066	STEELCASE	STL.W-3026630F	DESK-DOUBLE PED	30-3/4 KNEESPACE/FLUSH BK PANEL/#3963LT OAK PL TOP/LOCKING	66	30	29	338	976
DDPWD3066	STEELCASE	STL.W-3026630R	DESK-DOUBLE PED	30-3/4 KNEESPACE/RECESSED PANEL/#3963LT OAK PL TOP/LOCKING	66	30	29	322	976
DDPWD3666	STEELCASE	STL.W-3026636F	DESK-DOUBLE PED	30-3/4 KNEESPACE/FLUSH BK PANEL/#3963LT OAK PL TOP/LOCKING	66	36	29	388	1008
DDPWD3666	STEELCASE	STL.W-3026636R	DESK-DOUBLE PED	30-3/4 KNEESPACE/RECESSED PANEL/#3963LT OAK PL TOP/LOCKING	66	36	29	372	1008
C2WKWD2071	STEELCASE	STL.W-30J7120	CREDENZA DB.PED	L-PED BOX FILE/R-PED 3-BOX/3968LT OAK PL TOP/LOCK	71	20	29	252	983
C2WKWD2060	STEELCASE	STL.W-30N6020	CREDENZA DB.PED	L-PED BOX-FILE/R-PED 3-BOX/3968LT OAK PL TOP/LOCK	60	20	29	230	928
CATDS2036	STEELCASE	STL.W-309SC3620	CABINET-TELEPHN	TELEPHONE CAB/2 PEDESTALS BOX-FILE/3963 LT. OAK	36	20	29	155	819
CATHD2036	STEELCASE	STL.W-301TC3620	CABINET-TELEPHN	TELEPHONE CAB/HINGED DBLE DOORS/1-ADJ SHELF/3963 LT. OAK	36	20	29	140	640

DATE: 06/30/83

DATA BASE: HERMAN

SPEC CODE	MANUFACTURER	PRODUCT #	PRODUCT	DESCRIPTION	WIDTH	DEP/LENG	HEIGHT	WEIGHT	COST
SDGTS	HERMAN MILLER	HER.05302/DT/4440	CHAIR-DESK	ERGON/TILT-SWIVEL/ARMS/5-STAR BASE/DARK TONE SHELL/4440-BLUE	26.5	24.75	37.75	46	360
DDPWD3260	HERMAN MILLER	HER.CD101/OK	DESK-DOUBLE PED	L-PED FILE DRWR/R-PED FILE DRWR/WHITE OAK VENEER OK	60	32	28.5	250	1253
DDPWD3260	HERMAN MILLER	HER.CD104/OK	DESK-DOUBLE PED	L-PED 2-BOX DRWR/R-PED FILE DRWR/WHITE OAK VENEER OK	60	32	28.5	250	1307
C2WKDP2370	HERMAN MILLER	HER.CD308/OK	CREDENZA	KNEE SPACE/L&R PED-FILE DRWR/WHITE OAK VENEER OK	70	22.75	26.25	235	1232
S2L3729	HERMAN MILLER	HER.MB101/4440	CHAIR-LOUNGE	FULLY UPHOLSTERED/4440-BLUE-HOPSAK	37	29	24.75	78	593

DATE: 06/30/83

DATA BASE: HAWORTH

SPEC CODE	MANUFACTURER	PRODUCT #	PRODUCT	DESCRIPTION	WIDTH	DEP/LENG	HEIGHT	WEIGHT	COST
MFPFA8C	HAWORTH	HAM.CF-48/N/F-1W	PANEL-ACOUSTICL	2 SIDED FABRIC ACOUSTICAL-CURVED/NEUTRAL TONE TRIM/F-1W WINE	24	2	48	43	374
MFPFA62C	HAWORTH	HAM.CF-62/N/F-1W	PANEL-ACOUSTICL	2 SIDED FABRIC ACOUSTICAL-CURVED/NEUTRAL TONE TRIM/F-1W WINE	24	2	62	56	452
MFPFA66C	HAWORTH	HAM.CF-66/N/F-1W	PANEL-ACOUSTICL	2 SIDED FABRIC ACOUSTICAL-CURVED/NEUTRAL TONE TRIM/F-1W WINE	24	2	66	59	478
MFPFA80C	HAWORTH	HAM.CF-80/N/F-1W	PANEL-ACOUSTICL	2 SIDED FABRIC ACOUSTICAL-CURVED/NEUTRAL TONE TRIM/F-1W WINE	24	2	80	70	520
MFCCPLL01524	HAWORTH	HAM.CTS-215/N/LT.OAK	COUNTER TOP	LIGHT OAK HIGH PRESSURE LAMINATE/NEUTRAL TONE TRIM	24	15	1	14	110
MFCCPLL01536	HAWORTH	HAM.CTS-315/N/LT.OAK	COUNTER TOP	LIGHT OAK HIGH PRESSURE LAMINATE/NEUTRAL TONE TRIM	36	15	1	20	123
MFCCPLL01548	HAWORTH	HAM.CTS-415/N/LT.OAK	COUNTER TOP	LIGHT OAK HIGH PRESSURE LAMINATE/NEUTRAL TONE TRIM	48	15	1	27	130
MFCCPLL01560	HAWORTH	HAM.CTS-515/N/LT.OAK	COUNTER TOP	LIGHT OAK HIGH PRESSURE LAMINATE/NEUTRAL TONE TRIM	60	15	1	33	140
MFDF	HAWORTH	HAM.DFL-11/N	DRAWER-FILE	FILE DRAWER/NEUTRAL TONE/LOCK & KEYS/FILE LETTER OR LEGAL	18.25	15.375	11	11	103
MFDP	HAWORTH	HAM.DS-3/N	DRAWER-SHALLOW	SHALLOW (PENCIL) DRAWER/NEUTRAL TONE/4 COMPARTMENTS	18.25	15.375	3	4	18
MFASD	HAWORTH	HAM.DSD-11/N	STATIONERY DIVD	STATIONERY DIVIDER/NEUTRAL TONE/HAS SIX SECTIONS	16	13	8.5	6	44
MFAFCK	HAWORTH	HAM.FCK-13(LFB)/N	FILE CONV. KIT	FILE CONVERSION KIT FOR LEGAL FILE FOLDERS/3 BARS/NEUTRAL	1	13.125	.5	1.2	14
MFFDPL01436	HAWORTH	HAM.FD-31416/N/LT.OAK	FLIPPER DOOR	NEUTRAL TOP/NEUTRAL TONE TRIP/LIGHT OAK FRONT/LOCK & KEYS	36	14	15.5	22	158
MFFDPL01448	HAWORTH	HAM.FD-41416/N	FLIPPER DOOR	NEUTRAL TOP/NEUTRAL TONE TRIP/LIGHT OAK FRONT/LOCK & KEYS	48	14	15.5	26	171
MFFDPL01460	HAWORTH	HAM.FD-51416/N	FLIPPER DOOR	NEUTRAL TOP/NEUTRAL TONE TRIP/LIGHT OAK FRONT/LOCK & KEYS	60	14	15.5	33	210
MFLFBPL036	HAWORTH	HAM.LFB-3A/N/LT.OAK	LATERAL FILEBIN	LIGHT OAK FRONT/NEUTRAL TONE TRIM/LOCK & KEYS	36	15.5	12.875	62	287
MFLFBPL048	HAWORTH	HAM.LFB-4A/N/LT.OAK	LATERAL FILEBIN	LIGHT OAK FRONT/NEUTRAL TONE TRIM/LOCK & KEYS	48	15.5	12.875	80	318
MFLFBPL060	HAWORTH	HAM.LFB-5B/N/LT.OAK	LATERAL FILEBIN	LIGHT OAK FRONT/NEUTRAL TONE TRIM/LOCK & KEYS	60	15.5	12.875	95	382
MFPSLL	HAWORTH	HAM.LL-17/N	PANEL SUPORTLEG	SUPPORT LEG/LEFT/NEUTRAL TONE	1	17	8	2	32
MFPSLR	HAWORTH	HAM.LR-17/N	PANEL SUPORTLEG	SUPPORT LEG/RIGHT/NEUTRAL TONE	1	17	8	2	32
MFDS	HAWORTH	HAM.MDS-03L/N	DRAWER-SHALLOW	MODULAR DRAWER-CAN STACK/NEUTRAL TONE/LOCK AND KEYS	14.625	20	3	13	104
MFDS	HAWORTH	HAM.MDS-03N/N	DRAWER-SHALLOW	MODULAR DRAWER-CAN STACK/NEUTRAL TONE/NO LOCK	14.625	20	3	13	97

Figure 6.27. Catalog data listed by product number. The Spec Code is mnemonic. "MF" stands for modular furniture, "FT" for file top, "PL" for plastic laminate. Numbers are dimensions. The Spec Code is generic and must be applied to a vendor part number for budgeting and ordering purposes. (Courtesy of BASICOMP, Inc., Mesa, Arizona.)

DATE: 06/30/83

WORK ENVIRONMENT SP-1

SPEC CODE	MANUFACTURER	PRODUCT #	PRODUCT	DESCRIPTION	WIDTH	DEP/LENG	HEIGHT	QTY.	WEIGHT	WEIGHT TOTAL	COST	COST TOTAL
TRDFB36	HAWORTH	HAW.TR-36/N/LT.OAK	TABLE-ROUND	NEUTRAL TONE TRIM & PEDESTAL/LT.OAK HPL TOP	-	36-dia	29	1	45.00	45.00	222.00	222.00
									SUB TOTAL	45.00		222.00
DDFWD3260	HERMAN MILLER	HER.CD104/OK	DESK-DOUBLE PED	L-PED 2-BOX DRWR/R-PED FILE DRWR/WHITE OAK VENEER OK	60	32	28.5	1	250.00	250.00	1307.00	1307.00
C2WKDP2370	HERMAN MILLER	HER.CD308/OK	CREDENZA	KNEE SPACE/L&R PED-FILE DRWR/WHITE OAK VENEER OK	70	22.75	26.25	1	235.00	235.00	1232.00	1232.00
SDBTS	HERMAN MILLER	HER.EA41B/4737	CHAIR-DESK	TILT SWIVEL ARMCHAIR/CASTERS/HOPSAK PLUS 4737-BURGANDY	22.5	24	34	3	32.00	96.00	952.00	2856.00
									SUB TOTAL	581.00		5395.00
FL2M1B36	STEELCASE	STL.B36-200/HF	FILE-LATERAL	2 12IN. LTR.SIZE SHELVES/DOORS/LT.OAK TOP/835 BLACK/9201 PC	36	18	29	1	136.00	136.00	326.00	326.00
									SUB TOTAL	136.00		326.00
									TOTAL	762.00		5943.00

WORK ENVIRONMENT PO-1

SPEC CODE	MANUFACTURER	PRODUCT #	PRODUCT	DESCRIPTION	WIDTH	DEP/LENG	HEIGHT	QTY.	WEIGHT	WEIGHT TOTAL	COST	COST TOTAL
C2WKDP2370	HERMAN MILLER	HER.CD308/OK	CREDENZA	KNEE SPACE/L&R PED-FILE DRWR/WHITE OAK VENEER OK	70	22.75	26.25	1	235.00	235.00	1232.00	1232.00
									SUB TOTAL	235.00		1232.00
SSNAPFU	MUELLER	MUE.2368-M/EK/MU-975	CHAIR-SIDE	LO-BACK/ARMLES/STAT.BASE/MIRROR CHRM BLK EPOXY/LEATH 975BLUE	21.5	26	31.5	1	34.00	34.00	615.00	615.00
TOWD261B	MUELLER	MUE.371B-26/D/MU-11	TABLE-OCCASIONL	OAK VENEERS & SOLIDS/MU-11 LT.OAK/RADIUS SERIES	26	18	22	2	24.00	48.00	375.00	750.00
S2LFUNB	MUELLER	MUE.800-34/MU-875	CHAIR-LOUNGE	FULLY UPHOLSTERED/SUEDE MU-875 BLUE	34	28	26	4	75.00	300.00	1975.00	7900.00
S2S2SFUNR	MUELLER	MUE.800-56/MU-975	SOFA/2-SEATER	FULLY UPHOLSTERED/LEATHER MU-975 BLUE	56	28	26	4	113.00	452.00	2930.00	11720.00
									SUB TOTAL	834.00		20985.00
									TOTAL	1069.00		22217.00

WORK ENVIRONMENT OP-1

SPEC CODE	MANUFACTURER	PRODUCT #	PRODUCT	DESCRIPTION	WIDTH	DEP/LENG	HEIGHT	QTY.	WEIGHT	WEIGHT TOTAL	COST	COST TOTAL
MFPC	HERMAN MILLER	HER.A0212	PANEL CONNECTOR	DARK TONE/CONNECTS TWO PANELS OF THE SAME HEIGHT TOGETHER	3.13	-	80	8	2.00	16.00	8.10	64.80
MFSD12	HERMAN MILLER	HER.A0223	SHELF DIVIDER	METAL/LIGHT NEUTRAL/BOX OF 8		11	5.5	2	14.00	28.00	92.00	184.00
MFFDFA1248	HERMAN MILLER	HER.A0251/4737	FLIPPER DOOR	NEUTRAL LIGHT COVER WITH LOCK/HOPSAK PLUS 4737-BURGANDY	48	12.5	15.5	2	23.00	46.00	158.00	316.00
MFPEC	HERMAN MILLER	HER.A0406LN	PANEL END CAP	NEUTRAL LIGHT	-	-	80	10	7.00	70.00	15.70	157.00
MFPH	HERMAN MILLER	HER.A0409LN	PANEL HINGE	NEUTRAL LIGHT/CONNECTS TWO PANELS OF THE SAME HEIGHT	-	-	80	4	5.00	20.00	9.50	38.00
MFTL48	HERMAN MILLER	HER.A0431LN	TASK LIGHT	NEUTRAL LIGHT/UL LISTED/ONE FLOURESCENT TUBE	48	5.5	1.625	1	9.00	9.00	76.50	76.50
MFSDRM1248	HERMAN MILLER	HER.A0436LN	SHELF-REGULAR	NEUTRAL LIGHT METAL	48	12.5	15.5	2	23.00	46.00	77.50	155.00
MFLFBWD48	HERMAN MILLER	HER.A0470K	LATERAL FILEBIN	NEUTRAL LIGHT TOP/WHITE OAK VENEER FRONT	48	15.5	12.5	2	68.00	136.00	306.00	612.00
MFAFCX	HERMAN MILLER	HER.A0468DT	FILE CONV. KIT	CONVERTS LETTER TO LEGAL SIZE/BOX OF 2-PAIR/DARK TONE	-	-	-	1	1.00	2.00	11.30	22.60
MFTR48	HERMAN MILLER	HER.A0475/4737	TACK BOARD	HOPSAK PLUS 4737-BURGANDY	48	-	48	1	47.00	47.00	155.00	155.00
MFWSPLDK2448	HERMAN MILLER	HER.A0556DK	WORK SURFACE	WHITE OAK VENEER/PANEL HUNG	48	23.75	8.75	2	43.00	86.00	201.00	402.00
MFWSPLDK3048	HERMAN MILLER	HER.A0557DK	WORK SURFACE	WHITE OAK VENEER/PANEL HUNG	48	29.75	11.63	1	53.00	53.00	243.00	243.00
MFSDRM1224	HERMAN MILLER	HER.A0622LN	SHELF-REGULAR	NEUTRAL LIGHT METAL	24	12.5	15.5	1	17.00	17.00	65.00	65.00
MFFDFA1224	HERMAN MILLER	HER.A0623/4737	FLIPPER DOOR	NEUTRAL LIGHT COVER WITH LOCK/HOPSAK PLUS 4737-BURGANDY	24	12.5	15.5	1	12.00	12.00	113.00	113.00
MFTL24	HERMAN MILLER	HER.A0626LN	TASK LIGHT	NEUTRAL LIGHT/UL LISTED/ONE FLOURESCENT TUBE	24	5.5	1.625	1	6.00	6.00	62.00	62.00
MFAC	HERMAN MILLER	HER.A0861LN	ACOUSTIC COND.	TONE & VOLUME CONTROLS/12 FOOT CORD/NEUTRAL LIGHT	8 dia.	-	11	1	6.00	6.00	299.00	299.00
MFPFA4800	HERMAN MILLER	HER.A0878LF/3845	PANEL-ACOUSTICL	2 SIDED FABRIC ACOUSTICAL/NEUTRAL LIGHT TRIM/3845-CHARCOAL	48	2	80	9	89.00	801.00	457.00	4113.00
MFPFA3680	HERMAN MILLER	HER.A0879LF/3845	PANEL-ACOUSTICL	2 SIDED FABRIC ACOUSTICAL/NEUTRAL LIGHT TRIM/3845-CHARCOAL	36	2	80	2	85.00	170.00	394.00	788.00
MFPFA2480	HERMAN MILLER	HER.A0880LF/3845	PANEL-ACOUSTICL	2 SIDED FABRIC ACOUSTICAL/NEUTRAL LIGHT TRIM/3845-CHARCOAL	24	2	80	2	50.00	100.00	326.00	652.00
MFDS	HERMAN MILLER	HER.A0915DE	DRAWER CASE	WHITE OAK VENEER/WITH LOCK	15.5	22.38	3.13	1	13.00	13.00	104.00	104.00
MFTB24	HERMAN MILLER	HER.A0978/4737	TACK BOARD	HOPSAK PLUS 4737-BURGANDY	24	-	15.5	1	10.00	10.00	66.50	66.50
MFTB48	HERMAN MILLER	HER.A0980/4737	TACK BOARD	HOPSAK PLUS 4737-BURGANDY	48	-	15.5	2	17.00	34.00	85.50	171.00
SDBTS	HERMAN MILLER	HER.EA418/4737	CHAIR-DESK	TILT SWIVEL ARMCHAIR/CASTERS/HOPSAK PLUS 4737-BURGANDY	22.5	24	34	4	32.00	128.00	952.00	3808.00
TCODP3660	HERMAN MILLER	HER.ET149/OK	TABLE-OVAL	SEGMENTED BASE/WHITE OAK VENEER	60	36	28.5	1	118.00	118.00	929.00	929.00

Figure 6.28. Furniture standards developed from catalog data. Items are listed alphabetically by vendor. (Courtesy of BASI-COMP, Inc., Mesa, Arizona.)

DATE: 06/30/83

DATA BASE: GENERAL

SPEC CODE	MANUFACTURER	PRODUCT #	PRODUCT	DESCRIPTION	WIDTH	DEP/LENG	HEIGHT	WEIGHT	COST
0	EXISTING	EXT.EXT.0	MISCELLANEOUS	SUPPORT FACILITY/NO ADDITIONAL PRODUCT SPEC REQUIRED	-	-	-	0	0
0	EXISTING	EXT.EXT.1	MISCELLANEOUS	SUPPORT EQUIPMENT & FURN./NO SPEC REQUIRED	-	-	-	0	0
MFDP	HAWORTH	HAW.DS-3/N	DRAWER-SHALLOW	SHALLOW (PENCIL) DRAWER/NEUTRAL TONE/4 COMPARTMENTS	18.25	15.375	3	4	18
MFAFCX	HAWORTH	HAW.FCK-13(LFB)/N	FILE CONV. KIT	FILE CONVERSION KIT FOR LEGAL FILE FOLDERS/3 BARS/NEUTRAL	1	13.125	.5	1.2	14
MFDPLO144B	HAWORTH	HAW.FD-41416/N	FLIPPER DOOR	NEUTRAL TOP/NEUTRAL TONE TRIF/LIGHT OAK FRONT/LOCK & KEYS	48	14	15.5	26	171
MFLFBPLO36	HAWORTH	HAW.LFB-3A/N/LT.OAK	LATERAL FILEBIN	LIGHT OAK FRONT/NEUTRAL TONE TRIM/LOCK & KEYS	36	15.5	12.875	62	287
MFLFRPLO48	HAWORTH	HAW.LFB-4A/N/LT.OAK	LATERAL FILEBIN	LIGHT OAK FRONT/NEUTRAL TONE TRIM/LOCK & KEYS	48	15.5	12.875	80	318
MFDP	HAWORTH	HAW.MDS-336N/N	DRAWER-PEDESTAL	MULTI-DRWR.PED/TWO 3-IN & ONE 6-IN DRWRS/NEUT.TONE/NO LOCK	14.625	20	12	34	309
MFFPA244B	HAWORTH	HAW.P-248F/N/F-1W	PANEL-ACOUSTICL	2 SIDED FABRIC ACOUSTICAL/NEUTRAL TONE TRIM/F-1W WINE	24	2	48	26	238
MFFPA246B	HAWORTH	HAW.P-266F/N/F-1W	PANEL-ACOUSTICL	2 SIDED FABRIC ACOUSTICAL/NEUTRAL TONE TRIM/F-1W WINE	24	2	66	34	280
MFFFA366B	HAWORTH	HAW.P-366F/N/F-1W	PANEL-ACOUSTICL	2 SIDED FABRIC ACOUSTICAL/NEUTRAL TONE TRIM/F-1W WINE	36	2	66	44	364
MFFFA448B	HAWORTH	HAW.P-448F/N/F-1W	PANEL-ACOUSTICL	2 SIDED FABRIC ACOUSTICAL/NEUTRAL TONE TRIM/F-1W WINE	48	2	48	43	336
MFFFA486B	HAWORTH	HAW.P-466F/N/F-1W	PANEL-ACOUSTICL	2 SIDED FABRIC ACOUSTICAL/NEUTRAL TONE TRIM/F-1W WINE	48	2	66	54	408
MFAPO12	HAWORTH	HAW.PO-B/N	PAPER ORGANIZER	NEUTRAL TONE/2 VINYL CONNECTORS W-EVERY ORGANIZER/8 PER BOX	12.375	11	2.5	8	54
MFSOLM1436	HAWORTH	HAW.S-3149/N	SHELF - LOW	NEUTRAL TONE METAL AND END PANELS	36	14	8.875	15	79
MFSORM1144B	HAWORTH	HAW.S-41416/N	SHELF-REGULAR	NEUTRAL TONE METAL AND END PANELS	48	14	15.875	22	87
MFSOLM1144B	HAWORTH	HAW.S-4149/N	SHELF - LOW	NEUTRAL TONE METAL AND END PANELS	48	14	8.875	18	80
MFASD12	HAWORTH	HAW.SD-12	SHELF DIVIDERS	SELF LOCKING/NEUTRAL COLOR/PACKED EIGHT PER BOX	3.25	12.25	6	10	95
MFTB48	HAWORTH	HAW.TB-416/F-1W	TACKBOARD	FABRIC SURFACE/F-1W WINE/	48	1	16	15	97
MFTB48	HAWORTH	HAW.TB-448/F-1W	TACKBOARD	FABRIC SURFACE/F-1W WINE/	48	1	48	46	166
MFTL48	HAWORTH	HAW.TL-4	TASK LIGHT	UNDERSHELF TASK LIGHT/PRISMATIC LENS/FLOURESCENT/NEUTRAL TON	48	4.5	2	8	99
TRDPB36	HAWORTH	HAW.TR-36/N/LT.OAK	TABLE-ROUND	NEUTRAL TONE TRIM & PEDESTAL/LT.OAK HPL TOP	-	36-DIA	29	45	222
TSOPB3O3O	HAWORTH	HAW.TS-3O3O/N/LT.OAK	TABLE-SQUARE	NEUTRAL TONE TRIM & PEDESTAL/LT.OAK HPL TOP	30	30	29	43	218
MFWSPLLO244B	HAWORTH	HAW.WS-424/N/LT.OAK	WORK SURFACE	LIGHT OAK HIGH PRESSURE LAMINATE/NEUTRAL TONE TRIM	48	24	9	49	178
MFWSPLLO3O4B	HAWORTH	HAW.WS-430/N/LT.OAK	WORK SURFACE	LIGHT OAK HIGH PRESSURE LAMINATE/NEUTRAL TONE TRIM	48	30	9	58	199
MFWSPLLO2472	HAWORTH	HAW.WS-624/N/LT.OAK	WORK SURFACE	LIGHT OAK HIGH PRESSURE LAMINATE/NEUTRAL TONE TRIM	72	24	9	70	278
MFPC	HERMAN MILLER	HER.AO212	PANEL CONNECTOR	DARK TONE/CONNECTS TWO PANELS OF THE SAME HEIGHT TOGETHER	-		80	2	8.1
MFSD12	HERMAN MILLER	HER.AO223	SHELF DIVIDER	METAL/LIGHT NEUTRAL/BOX OF 8	3.13	11	5.5	14	92
MFFDOFA124B	HERMAN MILLER	HER.AO251/4737	FLIPPER DOOR	NEUTRAL LIGHT COVER WITH LOCK/HOPSAK PLUS 4737-BURGANDY	48	12.5	15.5	23	158
MFPC	HERMAN MILLER	HER.AO354	PANEL CONNECTOR	DARK TONE/CONNECTS TWO PANELS OF THE SAME HEIGHT TOGETHER	-		48	1	7
MFPEC	HERMAN MILLER	HER.AO4O4LN	PANEL END CAP	NEUTRAL LIGHT	-		80	7	12.6
MFPEC	HERMAN MILLER	HER.AO4O6LN	PANEL END CAP	NEUTRAL LIGHT	-		80	7	15.7
MFPH	HERMAN MILLER	HER.AO4O7LN	PANEL HINGE	NEUTRAL LIGHT/CONNECTS TWO PANELS OF THE SAME HEIGHT	-		48	3	8.1
MFPH	HERMAN MILLER	HER.AO4O9LN	PANEL HINGE	NEUTRAL LIGHT/CONNECTS TWO PANELS OF THE SAME HEIGHT	-		80	5	9.5
MFTL48	HERMAN MILLER	HER.AO431LN	TASK LIGHT	NEUTRAL LIGHT/UL LISTED/ONE FLOURESCENT TUBE	48	5.5	1.625	9	76.5
MFSORM124B	HERMAN MILLER	HER.AO436LN	SHELF-REGULAR	NEUTRAL LIGHT METAL	48	12.5			

DATE: 06/30/83

SPEC CODE SORT DATA BASE: GENERAL

SPEC CODE	MANUFACTURER	PRODUCT #	PRODUCT	DESCRIPTION	WIDTH	DEP/LENG	HEIGHT	WEIGHT	COST
0	EXISTING	EXT.EXT.0	MISCELLANEOUS	SUPPORT FACILITY/NO ADDITIONAL PRODUCT SPEC REQUIRED	-	-	-	0	0
0	EXISTING	EXT.EXT.1	MISCELLANEOUS	SUPPORT EQUIPMENT & FURN./NO SPEC REQUIRED	-	-	-	0	0
0	UNKNOWN	UNK.UNK.0	MISCELLANEOUS	SUPPORT FACILITY/PRODUCT NOT SPECIFIED	-	-	-	0	0
C2NKDP2370	HERMAN MILLER	HER.CD308/OK	CREDENZA	KNEE SPACE/L&R PED-FILE DRWR/WHITE OAK VENEER OK	70	22.75	26.25	235	1232
DDPWD3260	HERMAN MILLER	HER.CD104/OK	DESK-DOUBLE PED	L-PED 2-BOX DRWR/R-PED FILE DRWR/WHITE OAK VENEER OK	60	32	28.5	250	1307
FL2M1B36	STEELCASE	STL.836-200/HF	FILE-LATERAL	2 12IN.LTR.SIZE SHELVES/DOORS/LT.OAK TOP/835 BLACK/9201 PC	36	18	29	136	326
FL4M1B42	STEELCASE	STL.842-450/HF	FILE-LATERAL	4 12IN.LTR.SIZE DRAWERS/LT.OAK TOP/835 BLACK/9201 PC	42	18	-	251	663
MFAC	HERMAN MILLER	HER.A0861LN	ACOUSTIC COND.	TONE & VOLUME CONTROLS/12 FOOT CORD/NEUTRAL LIGHT	8 DIA.	-	11	6	299
MFAFCK	HAWORTH	HAW.FCK-13(LFB)/N	FILE CONV. KIT	FILE CONVERSION KIT FOR LEGAL FILE FOLDERS/3 BARS/NEUTRAL	1	13.125	.5	1.2	14
MFAFCK	HERMAN MILLER	HER.A0468DT	FILE CONV. KIT	CONVERTS LETTER TO LEGAL SIZE/BOX OF 2-PAIR/DARK TONE	-	-	-	1	11.3
MFAPO12	HAWORTH	HAW.PO-8/N	PAPER ORGANIZER	NEUTRAL TONE/2 VINYL CONNECTORS W-EVERY ORGANIZER/8 PER BOX	12.375	11	2.5	8	54
MFASD12	HAWORTH	HAW.SD-12	SHELF DIVIDERS	SELF LOCKING/NEUTRAL COLOR/PACKED EIGHT PER BOX	3.25	12.25	6	10	95
MFDP	HAWORTH	HAW.DS-3/N	DRAWER-SHALLOW	SHALLOW (PENCIL) DRAWER/NEUTRAL TONE/4 COMPARTMENTS	18.25	15.375	3	4	18
MFDP	HAWORTH	HAW.MDS-336N/N	DRAWER-PEDESTAL	MULTI-DRWR.PED/TWO 3-IN & ONE 6-IN DRWRS/NEUT.TONE/NO LOCK	14.625	20	12	34	309
MFDS	HERMAN MILLER.	HER.A0915DE	DRAWER CASE	WHITE OAK VENEER/WITH LOCK	15.5	22.38	3.13	13	104
MFDFA1224	HERMAN MILLER	HER.A0623/4737	FLIPPER DOOR	NEUTRAL LIGHT COVER WITH LOCK/HOPSAK PLUS 4737-BURGANDY	24	12.5	15.5	12	113
MFDFA1248	HERMAN MILLER	HER.A0251/4737	FLIPPER DOOR	NEUTRAL LIGHT COVER WITH LOCK/HOPSAK PLUS 4737-BURGANDY	48	12.5	15.5	23	158
MFFDPLO144B	HAWORTH	HAW.FD-4141&/N	FLIPPER DOOR	NEUTRAL TOP/NEUTRAL TONE TRIP/LIGHT OAK FRONT/LOCK & KEYS	48	14	15.5	26	171
MFLFBPLO36	HAWORTH	HAW.LFB-3A/N/LT.OAK	LATERAL FILEBIN	LIGHT OAK FRONT/NEUTRAL TONE TRIM/LOCK & KEYS	36	15.5	12.875	62	287
MFLFBPLO48	HAWORTH	HAW.LFB-4A/N/LT.OAK	LATERAL FILEBIN	LIGHT OAK FRONT/NEUTRAL TONE TRIM/LOCK & KEYS	48	15.5	12.875	80	318
MFLFBPLO48	HERMAN MILLER	HER.A044/OK	LATERAL FILEBIN	NEUTRAL LIGHT TOP/WHITE OAK VENEER FRONT	48	15.5	12.5	68	306
MFPC	HERMAN MILLER	HER.A0212	PANEL CONNECTOR	DARK TONE/CONNECTS TWO PANELS OF THE SAME HEIGHT TOGETHER	-	-	80	2	8.1
MFPC	HERMAN MILLER	HER.A0354	PANEL CONNECTOR	DARK TONE/CONNECTS TWO PANELS OF THE SAME HEIGHT TOGETHER	-	-	48	1	7
MFPEC	HERMAN MILLER	HER.A0404LN	PANEL END CAP	NEUTRAL LIGHT	-	-	48	4	12.6
MFPEC	HERMAN MILLER	HER.A0406LN	PANEL END CAP	NEUTRAL LIGHT	-	-	80	7	15.7
MFPFA244B	HAWORTH	HAW.P-24BF/N/F-1W	PANEL-ACOUSTICL	2 SIDED FABRIC ACOUSTICAL/NEUTRAL TONE TRIM/F-1W WINE	24	2	48	26	238
MFPFA244B	HERMAN MILLER	HER.A0880LF/3841	PANEL-ACOUSTICL	2 SIDED FABRIC ACOUSTICAL/NEUTRAL LIGHT TRIM/3841-NAVY	24	2	48	30	253
MFPFA2466	HAWORTH	HAW.P-266F/N/F-1W	PANEL-ACOUSTICL	2 SIDED FABRIC ACOUSTICAL/NEUTRAL TONE TRIM/F-1W WINE	24	2	66	34	280
MFPFA2480	HERMAN MILLER	HER.A0880LF/3845	PANEL-ACOUSTICL	2 SIDED FABRIC ACOUSTICAL/NEUTRAL LIGHT TRIM/3845-CHARCOAL	24	2	80	56	326
MFPFA3666	HAWORTH	HAW.P-366F/N/F-1W	PANEL-ACOUSTICL	2 SIDED FABRIC ACOUSTICAL/NEUTRAL TONE TRIM/F-1W WINE	36	2	66	44	364
MFPFA3680	HERMAN MILLER	HER.A097LF/3845	PANEL-ACOUSTICL	2 SIDED FABRIC ACOUSTICAL/NEUTRAL LIGHT TRIM/3845-CHARCOAL	36	2	80	85	394
MFPFA'848	HAWORTH	HAW.P-48BFN/E-1W	PANEL-ACOUSTICL	2 SIDED FABRIC ACOUSTICAL/NEUTRAL TONE TRIM/F-1W WINE	48	2	48	43	336

Figure 6.29. The project data base sorted by manufacturer (top) and by spec code (bottom). The project data base is the set of all items used in the furniture standards. (Courtesy of BASICOMP, Inc., Mesa, Arizona.)

for each piece of equipment. This requires an interface (Figure 6.30) to a workstation or manpower projections program like the one shown earlier in Figure 6.4. Projections are multiplied by standard furniture or equipment elements to find the total requirements for each item (Figure 6.31). These can then be subtracted from quantities on hand to arrive at net quantities to be purchased.

A similar data base from Micro-Vector, Inc., produces the reports shown in Figures 6.32 through 6.35. The furniture allocation is multiplied by departmental headcount projections to arrive at the budget summaries by job title and department. The distribution of the budget can be examined to see if any departments or job titles are receiving unexpectedly or unacceptably large percentages of the total. If the total projected value exceeds the funds available, these two reports can be used to focus on cost reductions. Once an acceptable budget has been developed, the master bidder's list is produced for the current year's requirements. This should, of course, be "netted" against quantities on hand.

These types of calculation programs and furniture data bases are ideal for personal computers. Most planners deal with such a limited subset of any manufacturer's catalog that even small floppy disks will provide enough data storage to support these applications. And, since results will be desired in printout form, a small, character display will be sufficient.

Personal computers can also be used for project planning, scheduling, and control. A good example is COMPUCHART, another product of BASICOMP, Inc., shown here in Figures 6.36 through 6.39. The program can be used on one of two calendar formats: 30 weeks in 5-day breakdown or 36 months in 4-week breakdown. Both formats contain a perpetual calendar through the year 2000. This saves the user from entering calendar reference data. By entering the start date of an activity and its length in days or weeks, the program will automatically plot its bar on the chart. A critical dates listing shown in Figure 6.37 is provided for reference. It is another way of looking at the bar chart and can be used by the planner to write down the completion dates of each activity. Figure 6.38 shows a weekly extraction of critical dates. A separate listing (Figure 6.39) allows the planner to schedule a week's activity by day and by staff or team member. In the example shown, the 15 active tasks during the week of 08/30/82 are broken down into 21 sub-tasks, arranged by start date for each day of the week.

Along with project scheduling, personal computers can also be used to track and control project budgets. A program for this purpose is shown in Figure 6.40. It is called Budget Tracking from Micro-Vector, Inc., and runs on the IBM-PC. The purpose of the program is to subtract commitments (open purchase orders) and expenditures (paid invoices) from a project budget, highlighting the balance remaining, and computing an implicit percentage of project completion. The total project budget can be broken down into categories such as construction, furnishings, and relocation. Numerous projects may be tracked in this manner, with the program producing summaries for all projects in the system.

MIS software of the type we are discussing can be acquired for roughly $2500 per program. While this is not a large sum, the interested planner should check with his data processing department before buying, to see if comparable, easy-to-use software is available, perhaps on a mainframe timesharing system. All data processing groups have some kind of project management system in place to schedule and control their own software development projects. Such a system may be available to the facilities planner. Budget tracking may also be addressed by an existing work-order system in the maintenance department or elsewhere.

DEPARTMENT:CORPORATE ADMINISTRATION
DATE:07/22/82   6-82

STAFF PROJECTION

EXISTING AREA				WORK STATION	AREA PER UNIT	6-82		12-82		6-83		6-84		6-85		6-87	
PER UNIT	QTY.	TOTAL USED				QTY	AREA REQD.	QTY	AREA REQD.	QTY	AREA REQD.	QTY	AREA REQD.	QTY	AREA REQD.	QTY	AREA REQD.
300	1	300	PRESIDENT	PO-1	400	1	400	1	400	1	400	1	400	1	400	1	400
100	1	100	ADMIN. ASSISTANT TO PRESIDENT	SP-1	150	1	150	1	150	1	150	1	150	1	150	1	150
	2	400	SUB TOTAL			2	550	2	550	2	550	2	550	2	550	2	550
		25	AISLE % 40				220		220		220		220		220		220
	2	500	TOTAL			2	770	2	770	2	770	2	770	2	770	2	770

EQUIPMENT & MISCELLANEOUS PROJECTION

EXISTING AREA				WORK STATION	AREA PER UNIT	6-82		12-82		6-83		6-84		6-85		6-87	
PER UNIT	QTY.	TOTAL USED				QTY	AREA REQD.	QTY	AREA REQD.	QTY	AREA REQD.	QTY	AREA REQD.	QTY	AREA REQD.	QTY	AREA REQD.
0	0	0	CONFERENCE ROOM	XSF-1	300	1	300	1	300	1	300	1	300	1	300	1	300
0	0	0	PRESIDENTS PRIVATE WORK AREA	XSF-0	150	1	150	1	150	1	150	1	150	1	150	1	150
0	0	0	PRESIDENTS PRIVATE BATHROOM	XSF-0	75	1	75	1	75	1	75	1	75	1	75	1	75
25	1	25	STORAGE AREA	XSF-0	50	1	50	1	50	1	50	1	50	1	50	1	50
75	1	75	RECEPTION AREA	XSF-2	150	1	150	1	150	1	150	1	150	1	150	1	150
		100	SUB TOTAL				725		725		725		725		725		725
		25	AISLE % 40				290		290		290		290		290		290
		125	TOTAL				1015		1015		1015		1015		1015		1015

SUMMARY FOR CORPORATE ADMINISTRATION

EXISTING AREA			6-82		12-82		6-83		6-84		6-85		6-87	
QTY.	USED		QTY	AREA REQD.	QTY	AREA REQD.	QTY	AREA REQD.	QTY	AREA REQD.	QTY	AREA REQD.	QTY	AREA REQD.
2	500	STAFF	2	770	2	770	2	770	2	770	2	770	2	770
	125	EQUIPMENT & MISCELLANEOUS		1015		1015		1015		1015		1015		1015
2	625	TOTAL	2	1785	2	1785	2	1785	2	1785	2	1785	2	1785

INTERFACE: PROJECTIONS/WORKSTATIONS 6-83

WORK ENVIRONMENT	QUANTITY REQUIRED
PO-1	1
SP-1	2
XSF-1	2
XSF-0	8
XSF-2	3
OP-1	4
OP-2	33
OP-3	18
X	21
XSE-1	14
XSF	22
PO-2	2
XSF-3	1
XSF-4	1

Figure 6.30. Interface between manpower projections and furniture budgeting. (Courtesy of BASICOMP, Inc., Mesa, Arizona.)

Personal-computer-based information systems have their limitations. The programs reviewed here are not adequate for major construction projects, which simply have too many activities. They also require more sophisticated scheduling techniques, labor reports, and accounting controls. But for routine rearrangements and renovations--the majority of a facilities department's projects--personal computer solutions are cost-effective.

CLIENT: XYZ NATIONAL CONSULTING GROUP

SPEC CODE	MANUFACTURER	PRODUCT #	PRODUCT	DESCRIPTION	WIDTH	DEP/LENG	HEIGHT	QTY.	WEIGHT	WEIGHT TOTAL	COST	COST TOTAL
MFDP	HAWORTH	HAW.DS-3/N	DRAWER-SHALLOW	SHALLOW (PENCIL) DRAWER/NEUTRAL TONE/4 COMPARTMENTS	18.25	15.375	3	51	4.00	204.00	18.00	918.00
MFAFCK	HAWORTH	HAW.FCK-13\(LFB)/N	FILE CONV. KIT	FILE CONVERSION KIT FOR LEGAL FILE FOLDERS/3 BARS/NEUTRAL	1	13.125	.5	66	1.20	79.20	14.00	924.00
MFFDPLO1448	HAWORTH	HAW.FD-4141G/N	FLIPPER DOOR	NEUTRAL TOP/NEUTRAL TONE TRIP/LIGHT OAK FRONT/LOCK & KEYS	48	14	15.5	84	26.00	2184.00	171.00	14364.00
MFLFBPLO36	HAWORTH	HAW.LFB-3A/N/LT.OAK	LATERAL FILEBIN	LIGHT OAK FRONT/NEUTRAL TONE TRIM/LOCK & KEYS	36	15.5	12.875	36	62.00	2232.00	287.00	10332.00
MFLFBPLO48	HAWORTH	HAW.LFB-4A/N/LT.OAK	LATERAL FILEBIN	LIGHT OAK FRONT/NEUTRAL TONE TRIM/LOCK & KEYS	48	15.5	12.875	66	80.00	5280.00	318.00	20988.00
MFDP	HAWORTH	HAW.MDS-336N/N	DRAWER-PEDESTAL	MULTI-DRWR.PED/TWO 3-IN & ONE 6-IN DRWRS/NEUT.TONE/NO LOCK	14.625	20	12	18	34.00	612.00	309.00	5562.00
MFFFA2448	HAWORTH	HAW.P-248F/N/F-1W	PANEL-ACOUSTICL	2 SIDED FABRIC ACOUSTICAL/NEUTRAL TONE TRIM/F-1W WINE	24	2	48	36	26.00	936.00	238.00	8568.00
MFFFA2466	HAWORTH	HAW.P-266F/N/F-1W	PANEL-ACOUSTICL	2 SIDED FABRIC ACOUSTICAL/NEUTRAL TONE TRIM/F-1W WINE	24	2	66	117	34.00	3978.00	280.00	32760.00
MFFFA3666	HAWORTH	HAW.P-366F/N/F-1W	PANEL-ACOUSTICL	2 SIDED FABRIC ACOUSTICAL/NEUTRAL TONE TRIM/F-1W WINE	36	2	66	51	44.00	2244.00	364.00	18564.00
MFFFA4848	HAWORTH	HAW.P-448F/N/F-1W	PANEL-ACOUSTICL	2 SIDED FABRIC ACOUSTICAL/NEUTRAL TONE TRIM/F-1W WINE	48	2	48	36	43.00	1548.00	336.00	12096.00
MFFFA4866	HAWORTH	HAW.P-466F/N/F-1W	PANEL-ACOUSTICL	2 SIDED FABRIC ACOUSTICAL/NEUTRAL TONE TRIM/F-1W WINE	48	2	66	285	54.00	15390.00	408.00	116280.00
MFAPO12	HAWORTH	HAW.PO-8/N	PAPER ORGANIZER	NEUTRAL TONE/2 VINYL CONNECTORS W-EVERY ORGANIZER/8 PER BOX	12.375	11	2.5	69	8.00	552.00	54.00	3726.00
MFSDLM1436	HAWORTH	HAW.S-3149/N	SHELF - LOW	NEUTRAL TONE METAL AND END PANELS	36	14	8.875	36	15.00	540.00	79.00	2844.00
MFSDRM1448	HAWORTH	HAW.S-4141G/N	SHELF-REGULAR	NEUTRAL TONE METAL AND END PANELS	48	14	15.875	84	22.00	1848.00	87.00	7308.00
MFSDLM1448	HAWORTH	HAW.S-4149/N	SHELF - LOW	NEUTRAL TONE METAL AND END PANELS	48	14	8.875	102	18.00	1836.00	80.00	8160.00
MFASD12	HAWORTH	HAW.SD-12	SHELF DIVIDERS	SELF LOCKING/NEUTRAL COLOR/PACKED EIGHT PER BOX	3.25	12.25	6	102	10.00	1020.00	95.00	9690.00
MFTB48	HAWORTH	HAW.TB-416/F-1W	TACKBOARD	FABRIC SURFACE/F-1W WINE/	48	1	16	102	15.00	1530.00	97.00	9894.00
MFTB48	HAWORTH	HAW.TB-448/F-1W	TACKBOARD	FABRIC SURFACE/F-1W WINE/	48	1	48	33	46.00	1518.00	166.00	5478.00
									SUB TOTAL	53356.20		335145.00

CLIENT: XYZ NATIONAL CONSULTING GROUP

DATE: 06/30/83

SPEC CODE	MANUFACTURER	PRODUCT	PRODUCT #	DESCRIPTION	WIDTH	DEP/LENG	HEIGHT	QTY.	WEIGHT	WEIGHT TOTAL	COST	COST TOTAL
MFPC	HERMAN MILLER	PANEL CONNECTOR	HER.A0212	DARK TONE/CONNECTS TWO PANELS OF THE SAME HEIGHT TOGETHER	-	-	80	32	2.00	64.00	8.10	259.20
MFSD12	HERMAN MILLER	SHELF DIVIDER	HER.A0223	METAL/LIGHT NEUTRAL/BOX OF 8	3.13	11	5.5	8	14.00	112.00	92.00	736.00
MFFDFA124B	HERMAN MILLER	FLIPPER DOOR	HER.A0251/4737	NEUTRAL LIGHT COVER WITH LOCK/HOPSAK PLUS 4737-BURGANDY	48	12.5	15.5	8	23.00	184.00	158.00	1264.00
MFPC	HERMAN MILLER	PANEL CONNECTOR	HER.A0354	DARK TONE/CONNECTS TWO PANELS OF THE SAME HEIGHT TOGETHER	-	-	48	1	1.00	1.00	7.00	7.00
MFPEC	HERMAN MILLER	PANEL END CAP	HER.A0404LN	NEUTRAL LIGHT	-	-	48	6	4.00	24.00	12.60	75.60
MFPEC	HERMAN MILLER	PANEL END CAP	HER.A0406LN	NEUTRAL LIGHT	-	-	80	40	7.00	280.00	15.70	628.00
MFPH	HERMAN MILLER	PANEL HINGE	HER.A0407LN	NEUTRAL LIGHT/CONNECTS TWO PANELS OF THE SAME HEIGHT	-	-	48	2	3.00	6.00	8.10	16.20
MFTL4B	HERMAN MILLER	PANEL HINGE	HER.A0409LN	NEUTRAL LIGHT/CONNECTS TWO PANELS OF THE SAME HEIGHT	-	-	80	16	5.00	80.00	9.50	152.00
MFSDRM124B	HERMAN MILLER	TASK LIGHT	HER.A0431LN	NEUTRAL LIGHT/UL LISTED/ONE FLUORESCENT TUBE	48	5.5	1.625	4	9.00	36.00	76.50	306.00
MFLFBWO4B	HERMAN MILLER	SHELF-REGULAR	HER.A0436LN	NEUTRAL LIGHT METAL	48	12.5	15.5	8	23.00	184.00	77.50	620.00
MFAFCK	HERMAN MILLER	LATERAL FILEBIN	HER.A0470K	NEUTRAL LIGHT TOP/WHITE OAK VENEER FRONT	48	15.5	12.5	8	68.00	544.00	306.00	2448.00
	HERMAN MILLER	FILE CONV. KIT	HER.A0468OT	CONVERTS LETTER TO LEGAL SIZE/BOX OF 2-FAIR/DARK TONE	-	-	-	8	1.00	8.00	11.30	90.40
MFTB4B	HERMAN MILLER	TACK BOARD	HER.A0475/4737	HOPSAK PLUS 4737-BURGANDY	48	-	48	4	47.00	188.00	155.00	620.00
MFWSPLOK244B	HERMAN MILLER	WORK SURFACE	HER.A0556OK	WHITE OAK VENEER/PANEL HUNG	48	23.75	8.75	10	43.00	430.00	201.00	2010.00
MFWSPLOK304B	HERMAN MILLER	WORK SURFACE	HER.A0557OK	WHITE OAK VENEER/PANEL HUNG	48	29.75	11.63	4	53.00	212.00	243.00	972.00
MFSDRM1224	HERMAN MILLER	SHELF-REGULAR	HER.A0622LN	NEUTRAL LIGHT METAL	24	12.5	15.5	4	17.00	68.00	65.00	260.00
MFFDFA1224	HERMAN MILLER	FLIPPER DOOR	HER.A0623/4737	NEUTRAL LIGHT COVER WITH LOCK/HOPSAK PLUS 4737-BURGANDY	24	12.5	15.5	4	12.00	48.00	113.00	452.00
MFTL24	HERMAN MILLER	TASK LIGHT	HER.A0626LN	NEUTRAL LIGHT/UL LISTED/ONE FLUORESCENT TUBE	24	5.5	1.625	4	6.00	24.00	62.00	248.00
MFAC	HERMAN MILLER	ACOUSTIC COND.	HER.A0861LN	TONE & VOLUME CONTROLS/12 FOOT CORD/NEUTRAL LIGHT	8 dia.	-	11	4	6.00	24.00	299.00	1196.00
MFFFA4B80	HERMAN MILLER	PANEL-ACOUSTICL	HER.A0897BLF/3845	2 SIDED FABRIC ACOUSTICAL/NEUTRAL LIGHT TRIM/3845-CHARCOAL	48	2	80	36	89.00	3204.00	457.00	16452.00
MFFFA3680	HERMAN MILLER	PANEL-ACOUSTICL	HER.A0897LF/3845	2 SIDED FABRIC ACOUSTICAL/NEUTRAL LIGHT TRIM/3845-CHARCOAL	36	2	80	8	85.00	680.00	394.00	3152.00
MFFFA2480	HERMAN MILLER	PANEL-ACOUSTICL	HER.A0880LF/3845	2 SIDED FABRIC ACOUSTICAL/NEUTRAL LIGHT TRIM/3845-CHARCOAL	24	2	80	8	50.00	400.00	326.00	2608.00
MFFFA4B48	HERMAN MILLER	PANEL-ACOUSTICL	HER.A0886LF/3841	2 SIDED FABRIC ACOUSTICAL/NEUTRAL LIGHT TRIM/3841-NAVY	48	2	48	2	65.00	130.00	374.00	748.00
MFFFA2448	HERMAN MILLER	PANEL-ACOUSTICL	HER.A0888BLF/3841	2 SIDED FABRIC ACOUSTICAL/NEUTRAL LIGHT TRIM/3841-NAVY	24	2	48	2	30.00	60.00	253.04	506.09

SUB TOTAL 11574.00

97946.40

Figure 6.31. Total furniture requirements based on a manpower projection. One list is produced for each vendor. (Courtesy of BASICOMP, Inc., Mesa, Arizona.)

```
                    INTERNATIONAL BANK CORP.
                    FURNITURE ALLOCATION BY TITLE                           PAGE 1


                         FURNITURE                        UNIT    EXTENDED
       CODE  TITLE        CODE  DESCRIPTION           NO.  PRICE    PRICE   CROSS REFERENCE #'S

       -CRM  CONFERENCE ROOM    CNCH  KNOLL    ZAPF   8569321  12   88.00   1,056.00
                                CONF  HELIKON          123654   1 1,488.00  1,488.00

       TOTAL FURNITURE ALLOCATION FOR CONFERENCE ROOM             2,544.00


       /CLO  STORAGE CLOSET     SSHL  INMAC  STORSHELVS 7896325   3  180.00   540.00

       TOTAL FURNITURE ALLOCATION FOR STORAGE CLOSET               540.00


       ACCT  ACCOUNTANT         740F  STEVENS  BACKPART 123-SP740F 1 1,100.25 1,100.25
                                ARMC  SMITH             11122   1  300.00    300.00
                                DES2  DUNBAR            4471258  1 1,200.00  1,200.00
                                WRKS  ICF               W444112  1  500.00    500.00

       TOTAL FURNITURE ALLOCATION FOR ACCOUNTANT                  3,100.25


       ARCH  ARCHITECT          740F  STEVENS  BACKPART 123-SP740F 1 1,100.25 1,100.25
                                ARMC  SMITH             11122   2  300.00    600.00  123A      999A
                                F1    KNOLL             SP-720F  1  925.00    925.00
                                TTB   IBM               558421   1  500.00    500.00  PRIME01-A CDC 123-B
                                WRKS  ICF               W444112  1  500.00    500.00

       TOTAL FURNITURE ALLOCATION FOR ARCHITECT                   3,625.25


       CHRP  CHAIRPERSON        ARMC  SMITH    ARMCHAIR 11122   6  300.00   1,800.00
                                DES2  DUNBAR            4471258  2 1,200.00  2,400.00
                                DESK  HAWORTH           11111   2  880.00   1,760.00
                                SHLV  INMAC             22536987 2  250.00    500.00

       TOTAL FURNITURE ALLOCATION FOR CHAIRPERSON                 6,460.00


       CLRK  CLERK              ARMC  SMITH    ARMCHAIR 11122   1  300.00    300.00
                                DES2  DUNBAR            4471258  1 1,200.00  1,200.00
                                FILC  INMAC             4447586  1   85.00     85.00
                                TTB   IBM               558421   1  500.00    500.00

       TOTAL FURNITURE ALLOCATION FOR CLERK                       2,085.00


       FINA  FINANCIAL ANALYST  ARMC  SMITH    ARMCHAIR 11122   1  300.00    300.00
                                CHR2  DESIGN2000        55555   1  500.00    500.00
                                DES2  DUNBAR            4471258  1 1,200.00  1,200.00
                                F1    KNOLL             SP-720F  1  925.00    925.00

       TOTAL FURNITURE ALLOCATION FOR FINANCIAL ANALYST           2,925.00
```

Figure 6.32. Furniture allocation by title. This is a list of furniture stand-
ards--one for each job title. It eliminates the need for a pointer between standard work places and job titles; however, it may require redundant entries
where the furniture list is the same for different job titles. (Courtesy of Micro-Vector, Inc., Armonk, New York.)

PROJECTED FURNITURE BUDGETS BY JOB TITLES   PAGE 1

PROJECTED YEARS

CODE	TITLE	CURRENT	1983	1984	1985	1987	1989
-CRM	CONFERENCE ROOM	2,544.00	2,544.00	2,544.00	2,544.00	2,544.00	2,544.00
ACCT	ACCOUNTANT	24,802.00	27,902.25	34,102.75	37,203.00	40,303.25	46,503.75
ARCH	ARCHITECT	10,125.00	10,125.00	10,125.00	12,150.00	12,150.00	14,175.00
CHRP	CHAIRPERSON	6,460.00	6,460.00	6,460.00	6,460.00	6,460.00	6,460.00
CLRK	CLERK	28,320.00	32,745.00	35,400.00	40,710.00	47,790.00	60,180.00
FINA	FINANCIAL ANALYST	29,250.00	29,250.00	32,175.00	35,100.00	40,950.00	43,875.00
MNGR	MANAGER	18,900.00	18,900.00	18,900.00	20,250.00	20,250.00	20,250.00
PRES	PRESIDENT	4,625.00	4,625.00	4,625.00	4,625.00	4,625.00	4,625.00
PROG	PROGRAMMER	44,485.00	46,655.00	50,995.00	55,335.00	64,015.00	74,865.00
SECY	SECRETARY	37,560.00	40,377.00	42,255.00	45,072.00	48,828.00	50,706.00
SLSP	SALESPERSON	37,020.00	41,339.00	45,658.00	51,211.00	60,466.00	72,189.00
SPVR	SUPERVISOR	11,760.00	11,760.00	11,760.00	11,760.00	11,760.00	11,760.00
SYSA	SYSTEMS ANALYST	1,200.00	1,800.00	2,400.00	3,000.00	3,000.00	3,000.00
VPRE	VICE PRESIDENT	5,500.00	5,500.00	5,500.00	5,500.00	5,500.00	5,500.00
WPSY	WORD PROCESSING SECY	1,500.00	1,500.00	2,000.00	2,000.00	2,500.00	4,000.00

Figure 6.33. The number of employees in each job title is obtained by summing the entries in the departmental space projections (Figure 6.4). Each job title sum is multiplied by the extended price for its furniture allocation (Figure 6.32). The resulting report shows the proportion of the total furniture budget going to each job title. The figures are cumulative. The report is useful when new furniture is contemplated. (Courtesy of Micro-Vector, Inc., Armonk, New York.)

PROJECTED FURNITURE BUDGETS FOR DEPARTMENTS   PAGE 1

HIERARCHICAL UNIT			***PROJECTION YEARS***				
DIVISION	DEPARTMENT	CURRENT	1983	1984	1985	1988	1990
ADMINISTRATION		3,639.00	3,639.00	3,639.00	3,639.00	3,639.00	3,639.00
ADMINISTRATION	LEGAL	19,408.25	27,833.25	36,453.25	41,163.50	42,248.50	51,568.75
ADMINISTRATION	SECRETARIAL	56,423.50	59,593.50	61,178.50	66,402.50	76,106.50	85,031.50
ADMINISTRATION	MICRO COMPUTERS	14,480.00	16,165.00	18,935.00	29,805.00	33,060.00	37,400.00
ADMINISTRATION	MINI COMPUTERS	12,708.00	12,708.00	12,708.00	12,708.00	12,708.00	12,708.00
NEW LEVEL1(W/NO LEV2	MINI COMPUTERS	31,399.50	31,399.50	31,399.50	31,399.50	31,399.50	31,399.50
CORPORATE	ACCOUNTING	48,218.25	52,388.25	59,658.50	67,998.50	83,608.75	105,474.00
CORPORATE	INVESTMENTS	11,864.00	13,949.00	16,034.00	19,058.00	21,143.00	23,228.00
CORPORATE	CEO	19,265.00	19,265.00	19,265.00	19,265.00	19,265.00	19,265.00
CORPORATE	BUDGET	22,918.00	22,918.00	22,918.00	22,918.00	22,918.00	23,418.00
FINANCE	ACCOUNTING	12,955.50	18,140.75	21,241.00	26,426.25	26,426.25	33,696.50
FINANCE	BUDGET	14,900.00	14,900.00	17,825.00	20,750.00	26,600.00	29,525.00
SALES	DOMESTIC	31,136.00	33,604.00	36,072.00	39,157.00	45,944.00	53,965.00
SALES	INTERNATIONAL	16,415.00	18,266.00	20,117.00	22,585.00	25,053.00	28,755.00
SALES	TRAINING	1,258.00	1,258.00	1,258.00	1,258.00	1,258.00	1,258.00
MARKETING	MARKET RESEARCH	3,639.00	3,639.00	3,639.00	3,639.00	3,639.00	3,639.00
MARKETING	SUPPORT SERVICES	2,700.00	2,700.00	2,700.00	2,700.00	2,700.00	2,700.00
COMMUNICATIONS	TELEPHONES	2,700.00	2,700.00	2,700.00	2,700.00	2,700.00	2,700.00
COMMUNICATIONS	TELEX	1,258.00	1,258.00	1,258.00	1,258.00	1,258.00	1,258.00

Figure 6.34. Companion report summing all the job title expenditures for each department. (Courtesy of Micro-Vector, Inc., Armonk, New York.)

INTERNATIONAL BANK CORP.
FURNITURE MASTER BIDDERS LIST FOR CURRENT YEAR

CODE	MANUFACTURER	PART/MODEL#	DESCRIPTION	COLOR	FINISH	FABRIC	LEAD TIME	ORDER QTY.	UNIT PRICE	TOTAL PRICE
740F	STEVENS	123-SP740F	BACKPART	BLACK	MATTE	CUSHY	10	17	$____/____	$____/____/____
A10	ALISON	356	CHAIR	BLACK	FLAT	WOOL	00	22	$____/____	$____/____/____
ARMC	BROWNSON	11122	ARMCHAIR	BROWN			00	110	$____/____	$____/____/____
B15	BROWNSON	2456	DESK	WALNUT	SHINY		00	0	$____/____	$____/____/____
C10	CLEMSON		COUCH	YELLOW	LEATHER	CANE SEAT	00	22	$____/____	$____/____/____
CHR1	KNOLL	22222	BREUER	BEIGE	WALNUT STN		06	230	$____/____	$____/____/____
	Comment: BROWN SEAT STRAW OPTION									
CHR2	DESIGN2000	55555	VISITCHAIR	BRONZE	SMOOTH	DESIGNTEX	11	26	$____/____	$____/____/____
CNCH	KNOLL	8569321	ZAPF	ROSE	OAK SIDES	CANOVAS	00	24	$____/____	$____/____/____
	Comment: USED ONLY IN EXECUTIVE CONFERENCE SEAT-ING									
CONF	HELIKON	123654	CONF TABLE	NATURAL	OAK		00	2	$____/____	$____/____/____
	Comment: 12'X3' OVAL SHAPE TABLE - IDEAL FOR LARGE MEETINGS; 12 MATCHING CHAIRS AVAILABLE (SEE CNCH)									
DES1	KNOLL	10	BIG DESK	RED	PARTICLEBD	NONE	08	0	$____/____	$____/____/____
DES2	DUNBAR	4471258	DESK	DK. BROWN	WALNUT		00	89	$____/____	$____/____/____
	Comment: THIS DESK IS IDEAL FOR NEEDS OF LARGE WORK SPACE; 4-DRAWERS									
DESK	HAWORTH	11111	TRI-MODE	GOLDEN	HIGH GLOSS	PINE	08	17	$____/____	$____/____/____
	Comment: USED BY NORTHEASTERN DIVISION ONLY									
F1	KNOLL	SP-720F	STATION	BLACK	SHINY	WHEAT	06	25	$____/____	$____/____/____
FIL2	IBM	558694	FILECAB	BLACK			00	50	$____/____	$____/____/____
	Comment: LARGEST FILE CABINET AVAILABLE ANYWHERE INCLUDES 300 HANGING FILE HOLDERS - IDEAL FOR MASSIVE AMOUNTS OF RECORDS									
FILC	INMAC	4447586	FILECABINE	BEIGE	SHINY		00	79	$____/____	$____/____/____
	Comment: INCLUDES 100 HANGING FILE HOLDERS PER ONE CABINET									
RECD	IBM	44478963	RECEPTNDSK	BROWN	SHINY		00	28	$____/____	$____/____/____
	Comment: "C" SHAPED DESK INCLUDING THREE 1'X1' DRAWERS AND ONE .5'X4" DRAWER - IDEAL FOR EXECUTIVE SECRETARY									
SHLV	INMAC	22536987	SHELVES	BROWN	STAIN		00	130	$____/____	$____/____/____
SSHL	INMAC	7896325	STORSHELVS	GREY	SHINY		00	6	$____/____	$____/____/____
	Comment: 8'TALL, 5'WIDE, MADE SPECIFICALLY FOR STORAGE MATERIALS									
TAB1	STEELCASE	55555	CONFER.TAB	TAN	GRAINY	OAK	10	3	$____/____	$____/____/____
TAB2	STEELCASE	33333	WORKTABLE	BLACK	FORMICA		02	0	$____/____	$____/____/____
	Comment: DISCONTINUED									
TTB	IBM	558421	TYPE TABLE	BEIGE	SHINY	NONE	00	71	$____/____	$____/____/____
WRK8	ICF	W444112	WRKSTATION	BROWN SURF	BROWN FORMICA		00	119	$____/____	$____/____/____

TOTAL ORDERING PRICE OF FURNITURE FOR CURRENT YEAR $____/____/____

Figure 6.35. Bidder's list reflects the item detail for the current year budget in Figures 6.34 and 6.35. It is an "explosion" of the furniture allocations in alphabetical furniture code sequence. (Courtesy of Micro-Vector, Inc., Armonk, New York.)

COMPUCHART BAR CHART — TIMEFRAME = 30 WEEKS

PROJECT NAME: INTERIOR ARCHITECTURE/DESIGN PROJECT
PROJECT NUMBER: A-073&/882
ORIG ISSUE DATE: 08/09/82
UPDATE: 05/04/83
UPDATE: 3
LAST UPDATE BY: D ERICKSON

ACTIVITY NO. DESCRIPTION	FOOT	TIME START
		NOTE FRAME DATE

Week columns: WK1 WK2 WK3 WK4 WK5 WK6 WK7 WK8 WK9 WK10 WK11 WK12 WK13 WK14 WK15 WK16 WK17 WK18 WK19 WK20 WK21 WK22 WK23 WK24 WK25 WK26 WK27 WK28 WK29 WK30

Frame dates: AUG09 AUG16 AUG23 AUG30 SEP06 SEP13 SEP20 SEP27 OCT04 OCT11 OCT18 OCT25 NOV01 NOV08 NOV15 NOV22 NOV29 DEC06 DEC13 DEC20 DEC27 JAN03 JAN10 JAN17 JAN24 JAN31 FEB07 FEB14 FEB21 FEB28

PHASE: PROJECT RECAP

No.	Description	Foot	Start
1	PROJECT FORMAT	100	080982
2	ANALYSIS	200	081682
3	TECHNICAL SERVICES/FURNISHINGS	900	083082
4	TECH.SERVCS/LEASEHOLD IMPROVEMNTS	700	083082
5	SCHEMATIC PLAN DEVELOPMENT	150	090682
6	FINAL PLAN DEVELOPMENT	200	092082
7	CONSTRUCTION BIDDING PERIOD	200	101882
8	CONSTRUCTION SUPERVISION	400	112282
9	COORDINATION OF PHYSICAL MOVE	450	122082
10	POST OCCUPANCY EVALUATION	100	022183

PHASE: PROJECT FORMAT

No.	Description	Foot	Start
1	SET PROJECT METHODOLOGY & BUDGETS	50	080982
2	PREPARE PROJECT FORMS & FORMATS	50	081182
3	PREPARE BUDGET GUIDELINES	50	081682
4	PREPARE PROJECT GUIDELINES	50	081682
5	PREPARE SPACE PERFORMANCE GDLINE	50	081682
6	PRINT PROJECT FORMS	30	081882

PHASE: ANALYSIS

No.	Description	Foot	Start
1	MANAGERIAL INDOCTRINATION MEETINGS	10	081682
2	CLIENT STAFF COMPLETE FORMS	60	081682
3	EMPLOYEE INDOCTRINATION MEETINGS	20	081782
4	PREPARE EXISTING INVENTORY DATABS	130	081882
5	ANALYZE DATA/VERIFY/UPDATE	90	082482
6	INTERVIEW MANAGEMENT STAFF	50	083082
7	EVALUATE USE OF EXIST'G FURNISHING	50	083082
8	PREPARE SPACE PERFORMANCE SPEC	50	083082
9	PREPARE WORK-ENVIRONMENT TYPICALS	50	083082
10	PREPARE PRELIMINARY BUDGETS/COSTS	50	083082
11	PREPARE SPACE PROJECTIONS	20	090282
12	PREPARE COMMUNICAITN ANALYS MATRIX	20	090282
13	PREPARE DEPT STACKING DIAGRAM	20	090282
14	PRESENT DATA TO CLIENT MANAGEMENT	10	090682
15	UPDATE DATA PER MANAGEMENT INPUT	40	090682
16	PREPARE FINAL SPACE PERFORM. SPEC	20	090882
17	PRESENT FINAL ANALYSIS REPORT	10	091082
18	UPDATE SPACE PROJECTIONS	20	100482
19	PREPARE ONGOING MAINTENANCE MANUL	100	022183
20	UPDATE SPACE PROJECTIONS	20	022881

PHASE: SCHEMATIC PLAN DEVELOPMENT

No.	Description	Foot	Start
1	PREPARE PRELIMINARY SCHEMATICS	80	090682
2	REVIEW PLANS WITH CLIENT	10	091682

PHASE: CONSTRUCTION SUPERVISION

```
1 REVU CONTRACT DOCUMT W/CONTRACTOR   10 112282.          X
2 OVERSEE LEASEHOLD CONSTRUCTION     400 112282.          .XOXOX.XOXOX.XOXOX.XOXOX.XOXOX.XOXOX.
3 INSTALL CARPETING                  100 122782.                                    .XOXOX.XOXOX.
4 ISSUE FINAL DEFICIENCY PUNCHLIST    10 011083.                                                 .X
```

PHASE: COORDINATION OF PHYSICAL MOVE

```
1  PREPARE MOVING BID DOCUMENTS        30 122082.   .XOX
2  PREPARE NEW FURN/INSTALLATION BID    30 122082.   .XOX
3  ISSUE MOVING BID DOCUMENTS           10 122382.      .O
4  ISSUE NEW FURN/INSTALLATION BID      10 122382.      .O
5  OPEN MOVERS BIDS                     10 010683.              O
6  OPEN NEW FURN. INSTALLATION BIDS     10 010683.              O
7  EVALUATE MOVING & INSTALLATIN BIDS   30 010683.              OX.X
8  AWARD MOVING CONTRACT                10 011183.                 .O
9  AWARD NEW FURN INSTALATN CONTRACT    10 011183.                 .O
10 SUPERVISE NEW FURN. INSTALLATION    180 011283.                 .XOXOX.XOXOX.XOXOX.
11 COORDINATE EXISTING FURN TARGING    100 011783.                       .XOXOX.XOXOX.
12 COORDINATE TELEPHONE INSTALLATION   130 011783.                       .XOXOX.XOXOX.XOX
13 SUPERVISION-EXISTG FURNISHGS MOVE   100 013183.                                  .XOXOX.XOXOX.
14 FINE TUNE FAC W/NEW FURN INSTALRS   140 020183.                                  .OXOX.XOXOX.XOXOX.
15 EMPLOYEE INTRO TO NEW PREMISES       10 020383.                                           .O
16 ASSIST EMPLOYEES WITH PHASE MOVE     70 020383.                                           .OX.XOXOX.
17 FINE TUNE FACILITIES WITH MOVERS    100 020783.                                                 .XOXOX.XOXOX.
```

PHASE: POST-OCCUPANCY EVALUATION

```
1 INTERVIEW EMPLOYEES RE FACILITIES   50 022183.                                                               .XOXOX.
2 EVALUATE RESPONSES                  50 022483.                                                               .OX.XOX
3 PREPARE POST-OCCUPANCE REPORT       30 030183.                                                                     .OXO
4 PREPARE STUDY FINDINGS              10 030483.                                                                        .X
```

Figure 6.36. Each line of the project recap represents a phase in the total project. The tasks in each phase are detailed in the balance of the printout. An "X" is used for Mondays, Wednesdays, and Fridays; and "O" for Tuesdays and Thursdays. (Courtesy of BASICOMP, Inc., Mesa, Arizona.)

COMPUCHART CRITICAL DATES LISTING

PROJECT NAME: INTERIOR ARCHITECTURE/DESIGN PROJECT
PROJECT NUMBER: A-036/882 ORIG ISSUE DATE: 08/09/82 UPDATE: 06/28/83 UPDATE: 4

FINISH DATE	FOOT NOTE#	PROJECTED START DATE	TIME FRAME	ACT NO.	ACTIVITY DESCRIPTION
					PHASE: PROJECT RECAP
08-20-82		08-09-82	10D	1	PROJECT FORMAT
09-10-82		08-16-82	20D	2	ANALYSIS
		08-30-82	90D	3	TECHNICAL SERVICES/FURNISHINGS
		08-30-82	70D	4	TECH.SERVCS/LEASEHOLD IMPROVEMNTS
09-24-82		09-06-82	15D	5	SCHEMATIC PLAN DEVELOPMENT
10-15-82		09-20-82	20D	6	FINAL PLAN DEVELOPMENT
		10-18-82	20D	7	CONSTRUCTION BIDDING PERIOD
		11-22-82	40D	8	CONSTRUCTION SUPERVISION
		12-20-82	45D	9	COORDINATION OF PHYSICAL MOVE
		02-21-83	10D	10	POST OCCUPANCY EVALUATION
					PHASE: PROJECT FORMAT
08-13-82	1	08-09-82	5D	1	SET PROJECT METHODOLOGY & BUDGETS
08-17-82		08-11-82	5D	2	PREPARE PROJECT FORMS & FORMATS
08-20-82		08-16-82	5D	3	PREPARE BUDGET GUIDELINES
08-20-82		08-16-82	5D	4	PREPARE PROJECT GIDLINES
08-20-82		08-16-82	5D	5	PREPARE SPACE PERFORMANCE GIDLINE
08-20-82		08-18-82	3D	6	PRINT PROJECT FORMS
					PHASE: ANALYSIS
08-16-82		08-16-82	1D	1	MANAGERIAL INDOCTRINATION MEETINGS
08-23-82		08-16-82	6D	2	CLIENT STAFF COMPLETE FORMS
08-18-82		08-17-82	2D	3	EMPLOYEE INDOCTRINATION MEETINGS
		08-18-82	13D	4	PREPARE EXISTING INVENTORY DATABS
09-03-82		08-24-82	9D	5	ANALYZE DATA/VERIFY/UPDATE
		08-30-82	5D	6	INTERVIEW MANAGEMENT STAFF
08-30-82		08-30-82	5D	7	EVALUATE USE OF EXIST'G FURNISHNG
09-03-82		08-30-82	5D	8	PREPARE SPACE PERFORMANCE SPEC
09-03-82		08-30-82	5D	9	PREPARE WORK-ENVIRONMENT TYPICALS
09-03-82		08-30-82	5D	10	PREPARE PRELIMINARY BUDGETS/COSTS
09-03-82		09-02-82	2D	11	PREPARE SPACE PROJECTIONS
09-03-82		09-02-82	2D	12	PREPARE COMMUNICATN ANALYS MATRIX
09-03-82		09-02-82	2D	13	PREPARE DEPT STACKING DIAGRAM
09-06-82		09-06-82	1D	14	PRESENT DATA TO CLIENT MANAGEMENT
09-09-82		09-06-82	4D	15	UPDATE DATA PER MANAGEMENT INPUT
09-09-82		09-08-82	2D	16	PREPARE FINAL SPACE PERFORM. SPEC
09-10-82		09-10-82	1D	17	PRESENT FINAL ANALYSIS REPORT
		10-04-82	2D	18	UPDATE SPACE PROJECTIONS
10-05-82		02-21-83	10D	19	PREPARE ONGOING MAINTENANCE MANUL
		02-28-83	2D	20	UPDATE SPACE PROJECTIONS

PHASE: SCHEMATIC PLAN DEVELOPMENT

09-15-82	09-06-82	8D	1	PREPARE PRELIMINARY SCHEMATICS
09-16-82	09-16-82	1D	2	REVIEW PLANS WITH CLIENT

PHASE: CONSTRUCTION SUPERVISION

11-22-82	1D	1	REVU CONTRACT DOCUMT W/CONTRACTOR
11-22-82	40D	2	OVERSEE LEASEHOLD CONSTRUCTION
12-27-82	10D	3	INSTALL CARPETING
01-10-83	1D	4	ISSUE FINAL DEFICIENCY PUNCHLIST

PHASE: COORDINATION OF PHYSICAL MOVE

12-20-82	3D	1	PREPARE MOVING BID DOCUMENTS
12-20-82	3D	2	PREPARE NEW FURN/INSTALLATION BID
12-23-82	1D	3	ISSUE MOVING BID DOCUMENTS
12-23-82	1D	4	ISSUE NEW FURN/INSTALLATION BID
01-06-83	1D	5	OPEN MOVERS BIDS
01-06-83	1D	6	OPEN NEW FURN. INSTALLATION BIDS
01-06-83	3D	7	EVALUATE MOVING & INSTALLATN BIDS
01-11-83	1D	8	AWARD MOVING CONTRACT
01-11-83	1D	9	AWARD NEW FURN INSTALAIN CONTRACT
01-12-83	18D	10	SUPERVISE NEW FURN. INSTALLATION
01-17-83	10D	11	COORDINATE EXISTING FURN TAGGING
01-17-83	13D	12	COORDINATE TELEPHONE INSTALLATION
01-31-83	10D	13	SUPERVISION-EXISTG FURNISHGS MOVE
02-01-83	14D	14	FINE TUNE FAC W/NEW FURN INSTALRS
02-03-83	1D	15	EMPLOYEE INTRO TO NEW PREMISES
02-03-83	7D	16	ASSIST EMPLOYEES WITH PHASE MOVE
02-07-83	10D	17	FINE TUNE FACILITIES WITH MOVERS

PHASE: POST-OCCUPANCY EVALUATION

02-21-83	5D	1	INTERVIEW EMPLOYEES RE FACILITIES
02-24-83	5D	2	EVALUATE RESPONSES
03-01-83	3D	3	PREPARE POST-OCCUPANCE REPORT
03-04-83	1D	4	PREPARE STUDY FINDINGS

Figure 6.37. Complete listing of finish and start dates for each line entry on the bar chart. (Courtesy of BASICOMP, Inc., Mesa, Arizona.)

COMPUCHART CRITICAL DATES LISTING

PROJECT NAME: INTERIOR ARCHITECTURE/DESIGN PROJECT
PROJECT NUMBER: A-0736/882
ORIG ISSUE DATE: 08/09/82   UPDATE: 06/28/83   UPDATE: 4

CRITICAL DATES / WEEK OF: 08/30/82

FINISH DATE	FOOT NOTE#	PROJECTED START DATE	TIME FRAME	ACT NO.	ACTIVITY DESCRIPTION
PHASE: PROJECT RECAP					
09-10-82		08-16-82	20D	2	ANALYSIS
		08-30-82	90D	3	TECHNICAL SERVICES/FURNISHINGS
		08-30-82	70D	4	TECH.SRVCS/LEASEHOLD IMPROVEMNTS
PHASE: ANALYSIS					
09-03-82		08-18-82	13D	4	PREPARE EXISTING INVENTORY DATABS
		08-24-82	9D	5	ANALYZE DATA/VERIFY/UPDATE
		08-30-82	5D	6	INTERVIEW MANAGEMENT STAFF
		08-30-82	5D	7	EVALUATE USE OF EXIST'G FURNISHNG
		08-30-82	5D	8	PREPARE SPACE PERFORMANCE SPEC
		08-30-82	5D	9	PREPARE WORK-ENVIRONMENT TYPICALS
		08-30-82	5D	10	PREPARE PRELIMINARY BUDGETS/COSTS
		09-02-82	2D	11	PREPARE SPACE PROJECTIONS
		09-02-82	2D	12	PREPARE COMMUNICATN ANALYS MATRIX
		09-02-82	2D	13	PREPARE DEPT STACKING DIAGRAM
PHASE: TECH.SRVCS/LEASEHOLD IMPRVMNTS					
09-10-82		08-30-82	10D	1	PREPARE BASE BUILDING FLOORPLANS
PHASE: TECHNICAL SERVICES/FURNISHINGS					
		08-30-82	60D	1	RESEARCH FF&E ITEMS-FINISHES-ETC.

Figure 6.38. Data listing for a specific week. (Courtesy of BASICOMP, Inc., Mesa, Arizona.)

COMPUCHART CRITICAL DATES LISTING

PROJECT NAME: INTERIOR ARCHITECTURE/DESIGN PROJECT
PROJECT NUMBER: A-0736/882   ORIG ISSUE DATE: 08/09/82   UPDATE: 06/28/83   UPDATE: 4

DAILY ACTIVITIES / WEEK OF: 08/30/82

FINISH DATE	FOOT NOTE#	PROJECTED START DATE	ACT NO.	ACTIVITY DESCRIPTION	ACCOMPLISHED BY	HOURLY FORECAST M	T	W	T	F	S/S	TOTAL HOURS
		08-30-82	1	INITIAL SCHEDULE COMPLETE/DECIDE PRESENTATION FORMAT 4 CLNT	RME/JT/SW/DL							
		08-30-82	2	REVIEW BASE BLDG. INFO PRIOR TO MTG. W/ENGINEERS	GT/GG							
		08-30-82	3	WRITE PROGRESS REPORT TO CLIENT	RME							
		08-30-82	4	WRITE MINUTES OF AUGUST 25-1982 MEETING WITH CLIENT	RME							
		08-30-82	5	GET CARRY CASE FOR BINDERS AT STATIONERS	DL							
		08-31-82	1	START WORKING DRAWINGS (BASE BLDG. DRWGS)	GG							
		08-31-82	2	RESPONSE DUE BY GT TO SW RE WAY TO PROCEED RE PRESIDENT	GT							
		08-31-82	3	BEGIN CARPET INVESTIGATION RE ITEMS AVAIL.-QUALITY-TIMING	SW							
		08-31-82	4	PREPARE WORKFLOW CHARTS	RME							
		08-31-82	5	PREPARE COMMUNICATION MATRIX-USE C.A.M.	RME							
		09-01-82	1	PREPARE REVISED PROJECT BAR CHART-USE COMPUCHART	GT/SW							
		09-01-82	2	PREPARE CO.ORG CHART-INCORPORATE 5 YEAR SPACE PROJECTIONS	RME							
		09-01-82	3	DETERMINE REQUIRM'S 4 EACH INDIVIDUAL/DO REQUIRED FACILITY	RME							
		09-02-82	1	REQUEST APPROVAL FROM CLIENT TO USE ISOTEC FOR M&E	GT							
		09-02-82	2	ISSUE REVISED PROJECT BAR CHART TO CLIENT	GG							
		09-02-82	3	PREPARE PROJECT BUDGETING PER SIGNED CONTRACT WITH CLIENT	GT							
		09-03-82	1	CK RETURNED FORMS FOR COMPLETNESS RE NAMES-DEPTS-ETC	RME							
		09-03-82	2	PREPARE FUNCTIONAL ORGANIZATION CHART	RME							
		09-03-82	3	PREPARE SPACE PROJECTIONS FOR INPUT	RME							
		09-03-82	4	INPUT SPACE PROJECTIONS INTO COMPUTER	IB							
		09-03-82	5	WRITE LETTER TO BRIAN RE CENTRAL SERVICES CONCEPT	RME							
					TOTAL HOURS							

Figure 6.39. This report allows the planner to expand each week's activities into a list of daily "things to do." If the schedule is tight, each day's assignments can be estimated and balanced against the staff time available, using the form on the right. (Courtesy of BASICOMP, Inc., Mesa, Arizona.)

DETAIL BUDGET TRACKING REPORT AS OF 9/20/83

PROJECT NO.	DESCRIPTION	BUDGET	COMMITMENT	EXPENSE	BALANCE	COMPL
1	ORIGINAL TOTAL:	$ 20,000.00				
	PROJECT 1					
	SUITE 2000					
	100 BROAD STREET					
	NEW YORK NY					
	10021 0					
	Department/Branch Manager:					
	MR. P CAREY					
	Project Manager & Designer					
	Initials: ML NN					
	Approval Date: 12/30/82					
	Completion Date: 0/ 0/ 0					
	Move-in Date: 0/ 0/ 0					
	Estmatd.Close Date: 12/30/83					
	Fait Accompli Date: 0/ 0/ 0					
	Rentable Area: 20,000 Sq.Ft.					
	Usable Area: 19,000 Sq.Ft.					
	Population: 100					
CATEGORY:	CONSTRUCTION	$ 5,000.00				
Commitmt:	1.1.A.CONSTRUCTION COMPANY		$ 500.00			
	Contract Date: 1/ 1/83					
	Invoice: 1.1.A.2					
	Date: 3/ 1/83			$ 65.00		
	Invoice: 1.1.A.3					
	Date: 3/30/83			$ 100.00		
	NEW INVOICE TO SHOW NORA					
Subtotal:	1.1.A.CONSTRUCTION COMPANY		$ 500.00	$ 165.00	$ 335.00	33.0%

CAT.TOTAL: CONSTRUCTION	$ 5,000.00	$ 500.00	$ 165.00	$ 4,500.00	10.0%
CATEGORY: FURNISHINGS/FINISHINGS	$ 6,000.00				
Commitmt: 1.2.A.FURNISHING COMPANY Contract Date: 2/ 2/83		$ 600.00			
Subtotal: 1.2.A.FURNISHING COMPANY	$ 600.00	$ 600.00	$ 0.00	$ 600.00	0.0%
CAT.TOTAL: FURNISHINGS/FINISHINGS	$ 6,000.00	$ 600.00	$ 0.00	$ 5,400.00	10.0%
CATEGORY: TELECOMMUNICATIONS	$ 7,000.00				
Commitmt: 1.3.A.TELEX CORPORATION Work Order: 123 Contract Date: 2/ 2/83		$ 5,000.00			
TOTAL: PROJECT 1	$ 20,000.00	$ 6,622.44	$ 265.00	$ 13,377.56	33.1%
ORIGINAL PROJECT BUDGET TOTAL:	$ 20,000.00				
$ LEFT IN PROJECT BUDGET:	$ 0.00				

Figure 6.40. The total project budget is broken down into cost categories. Within each category, commitments and expenses are recorded as they occur, giving the planner up-to-date information on the current budget balance and the percentage of funds expended. (Courtesy of Micro-Vector, Inc., Armonk, New York.)

Software Sources

We have reviewed the output of some 20 personal computer programs--all aimed directly at the facilities planner. This is but a fraction of what is available. How does one find such software? The information sources in Appendix A are a good place to begin. In general, programs will be obtained from three types of sources, not including in-house development:

1. *General purpose software houses*: These firms write programs for business and professional use. They are not specifically tailored to facilities applications but can be adapted. Spreadsheets, statistical analysis, business graphics, simple drafting, and project scheduling are all available from these sources. Their generality makes many of these programs good investments, even in the presence of more powerful, specialized software.

2. *Facilities-oriented software houses*: There are a growing number of firms serving architectural designers and engineers, interior designers, and facilities planners and managers. Their software is more expensive, but it is typically easier to use and more powerful by virtue of its specialization. It is also likely to be better supported by people knowledgeable in facilities planning. Programs for space projections, calculations, layout routines, drafting, and project management are all available from facilities-oriented sources. These firms may also develop good custom programs using their standard products as a base.

3. *Other facilities professionals*: With the rapid spread of personal computers and the mastery of programming by facilities professionals, there is a growing body of software within the profession, much of which can be had for free. The plant layout program by Khator and Moodie is an example. Its code was published in full by *Industrial Engineering* magazine.

The publishing of personal computer software is a business like any other. However, it is part of an emerging industry in which the distribution channels have not matured. This can lead to some curious and perplexing sourcing arrangements, especially for specialized, facilities-oriented software. The same program may be available from two or three sources, often at different prices. One CAD package is available from a specialized software house which obtains it from a licensee of the authoring firm. If this type of secondary and tertiary distribution has raised the price, the planner should satisfy himself that he is receiving extra support or service or some other value to justify the incremental price. Many of the individual programs reviewed in this chapter can be obtained at a discount when bundled into a larger package of integrated routines. Some of the personal computer programs for facilities planning are also supported by or incorporated into larger CAD systems. The BASICOMP programs are available as part of a larger system from ASID Computer Systems, a division of the American Society of Interior Designers. Drafting software such as T & W Systems' CADAPPLE is sold through regional distributors who may also sell and service CAD equipment such as plotters and digitizers. Distribution arrangements like these are positive when they provide installation, training, and faster response to problems.

If a planner were to acquire the best software in each of a dozen applications, he would no doubt find himself in need of two or three operating systems, language compilers, and possibly two or three different personal computers to make the software run. Remember, good software pays for itself many times over in faster, more effective facilities plans. If the software is as good as it should be, the extra hardware investment will not be wasted. A large facilities group might productively use one personal computer for space projections and block layout, another for CAD, and a third for project management. This approach may be the best, as long as the applications are self-contained, with no need to re-key significant amounts of data.

Capacities and Configurations

At the beginning of this chapter, we deferred a precise distinction between personal and microcomputers. We must now make this distinction before we can shop for equipment. The differences and their implications are summarized in Figure 6.41.

Computer capacities are described in bits. "Bit" is short for binary digit, the basic character of information read by the computer. Bits are joined to make "words" of various lengths. A word is the largest piece of data which the computer can work with at one time. Word length also governs the amount of data that the computer can find in its memory. The longer the word length, the more memory may be addressed. This lessens the time spent fetching data from external disk storage. The rate at which data is moved into and out of memory is measured in bits per input/output (I/O) operation.

For our purposes, we can define personal computers as devices using an 8/8- or an 8/16-bit microprocessor. The first number refers to the bits per I/O operation. The second refers to the internal word length used in computations. Other things being equal, the higher the first number, the faster the I/O; the higher the second number, the faster the internal computation.

Personal computers are currently made by more than 40 companies. Many earlier models used the 8/8-bit, Z-80 microprocessor from Zilog. Newer machines with 16-bit word length typically use the Intel 8088 microprocessor. In most applications the I/O rate has more impact on response than does the internal word length. For this reason, an 8/16-bit device such as the IBM-PC may perform many tasks no faster than an 8/8-bit machine. On the other hand, the 16-bit word length makes use of larger memory, which does have advantages.

Microcomputers are devices that use 16/16-bit and 16/32-bit microprocessors. (32-bit I/O capacities will be available in late 1984.) The greater I/O rate and longer word length offer more speed and make use of more memory than a personal computer. Microcomputers are currently made by more than 60 companies. Both the Intel 8086 (16/16-bit) and the Motorola 68000 (16/32-bit) are popular. Some microcomputers contain auxiliary processors such as the Z-80 used for input/output control, and the Intel 8087, which performs floating point calculations and other mathematical operations. Input/output limitations also apply to microcomputers. Therefore, a 16/32 device may be no faster than a 16/16 in some applications (20). The chief advantage of the 32-bit word is its ability to address more memory.

Memory is measured in kilobytes (K) or megabytes (Mb). One kilobyte is 1024 bytes, where one byte equals 8 bits. One megabyte equals 1000K. In general, 8-bit word length devices are provided with 64K of memory. Those having 16-bit word length come with 64K to 512K, although 128K is most com-

FEATURES	PERSONAL COMPUTERS	MICROCOMPUTERS
Input/Output Capacity Affects speed when data is moved into and out of memory.	8-bit	16-bit
Internal Word Length Affects speed of long calculations. Also determines the amount of memory which may be used.	8-bit 16-bit	16-bit 32-bit
Commonly-supplied Memory Affects speed by reducing the frequency of disk access.	64K - 256K	64K - 512K
Clock Speed Affects speed of long calculations, especially in multi-user, multi-task applications.	4 - 8 MHz	8 - 12 MHz
Operating System Affects speed by taking advantage of hardware features.	MS-DOS, PC-DOS, CP/M Apple DOS, TRS DOS	UNIX, Pick p-System, CP/M-86
Storage Devices Affects speed of input/output. Determines amount of storage on-line.	Dual 5¼" floppy disks	5¼" or 8" hard disk
Auxiliary Processor Affects speed by relieving the main processor of input/output or special math functions such as floating point.	Generally not provided	Available on some machines Often an 8-bit Z-80
Most Suitable Applications	DSS-- calculations, space projections, statistical models, simple plots and graphs, layout scoring. CAE-- simple calculations for energy, lighting, power, etc. CAD-- small installation plans; single-line, highly repetitive. MIS-- simple project management and control. Applications which can be supported on floppy disks.	DSS-- all personal computer applications plus cluster analysis, block layout, and vertical stacking algorithms. CAE-- wide range of analyses, calculations and simulations. CAD (16-bit)-- installation plans, small layouts. CAD (32-bit)-- full range of applications. MIS-- local equipment inventories and data bases. Full range of project management and control. Applications which can be supported by a single hard disk drive.

Figure 6.41.   The difference between personal and microcomputers.

mon. Microprocessors with 32-bit word length may have as much as 1Mb of memory, although 512K is most common. (In theory, a 32-bit device can address up to 16Mb of memory.)

Large memory is important for two reasons. First, it reduces the frequency of slower I/O operations by holding larger amounts of information. More importantly, perhaps, extra memory can be used to support such features as:

1. *On-screen menus and prompts*: reducing the user's need for manuals and memorization of commands
2. *Edit checks*: to catch bad data before it gets in
3. *Integrated software*: (1-2-3, MBA, VisiON) which requires 256K for simultaneous displays and activities

The rate at which a computer operates is governed by its clock. The speed of a clock is measured in megahertz (MHz). The higher the MHz, the faster the computer. Clock speed is geared to the bit capacities of the microprocessor. Those with 8/8- and 8/16-bit capacities operate between 4 and 8 MHz. Those with larger 16/16- and 16/32-bit capacities operate between 8 and 12 MHz. Higher clock speeds amplify the performance differences between personal and microcomputers.

Designers of computer systems have found that 8/8- and 8/16-bit machines running at 4-8 MHz are generally limited to single-task, single-user applications, i.e., "personal computing." Larger 16/16- and 16/32-bit machines, running at 8-12 MHz, are significantly more powerful. They can be used for multiple-task, multiple-user applications--what might be called "departmental computing." The operating systems for these computers have been chosen accordingly. As we discussed earlier in Chapter 4, the CP/M operating system has become the most popular for Z-80-based machines. MS/DOS and PC/DOS are the most popular for Intel 8088 machines. UNIX is the most popular for Motorola 68000-based machines. The Pick system and the USCD p-system are also used with the Motorola 68000.

The differences in speeds and uses of personal and microcomputers lead to some natural differences in data storage devices. Generally, the personal computer is well supported by floppy disk drives. The micro (or departmental) computer often requires a hard disk drive, as do most CAD applications.

Disks contain the information that cannot be held in RAM at any particular moment. Floppy disks (so called because they are flexible) store data in tracks of grooves of magnetically coated Mylar. A read-write head puts information onto the disk in the same way that information is recorded onto magnetic tape. The head touches the disk while it turns. This causes wear and limits drive speed to about 300 revolutions per minute. This, in turn, limits the speed of data storage and retrieval. Since Mylar expands and contracts with changes in temperature, the tracks on the disk cannot be too closely packed, or else the head might read the wrong track.

Floppy disks were originally developed with 8-in. diameters. Today, the 5-1/4-in. size is much more popular. Disk capacity is measured like memory, in kilobytes. It is usually expressed as formatted K, after deducting the amount of space that must be devoted to disk controls and sectoring (blank tracks used to subdivide the disk). Depending upon track densities, the formatted capacities of 5-1/4-in. floppies range from 80K to 360K; 8-in.

floppies range from 300K to 1000K. Formats vary among manufacturers. As a result, floppies can be transferred only between the same model computer and operating system. Still, transferability (and low cost) are the chief advantages of floppy disk drives. If a computer or disk drive malfunctions, the program and data can be run on another machine. If the disk shows wear and begins to yield errors, its contents can be copied onto a new disk and the old one discarded. It is generally a good idea to have two drives. This makes it much easier to copy disks, expands the amount of storage available, and provides redundancy if one unit fails.

The chief limitations of floppy disks are their limited storage capacity and, to a lesser extent, their slow speed. Floppies are adequate for project-oriented, decision support tasks such as space projections, calculations, and simple statistical analyses. Floppies can support clustering analysis and block layout algorithms, but their limitations begin to show. Slowness becomes an impediment to effective CAD applications. Large data bases of corporate facilities and equipment typically exceed the capacity of even several floppies. As a result, they cannot be used effectively by such a system. Floppies are also unacceptable for multi-user systems. For these applications, the planner must use a hard disk drive.

In the early 1970s, IBM developed a 14-in. rigid or hard disk which could hold 30 Mb of data. It was packaged two disks to a unit. The resulting "30/30" configuration came to be known as a Winchester, in remembrance of that company's famous rifle. The term is still used, even though it no longer refers to the 30/30 device.

The metals used in a hard disk are less susceptible than Mylar to temperature. This permits track densities several times greater than those possible on a floppy. The read-write head glides above the disk on an air cushion created by the disk's rotation. While this cushion is only a few millionths of an inch thick, its existence allows the drive to spin at speeds up to 3600 RPM. Consequently, data access if many times faster than a floppy. Hard disk capacities of 20, 30, and even 80 megabytes are now common in 5-1/4-in. drives. With these capacities, even large facilities data bases can be accommodated. With the increase in speed, applications such as drafting are made possible, as well as multiple-task and multiple-user operations.

Hard disks are not removable. This is one of their limitations. Cost and ease of damage are the others. If the disk drive fails, the computer and its software are out of order. The expense of hard disk drives usually precludes a spare. Unlike floppies, hard disks cannot be discarded when they begin to produce errors. Instead, the controller identifies the location of an error and locks out the flawed sector.

Copying and transferring data from a hard disk can be accomplished in three ways. If the drive is a simple fixed disk, its contents can be copied onto multiple floppies for backup or transfer. This is slow and could obviously require many floppies. The best solution is provided by a built-in tape drive. A large-capacity tape cassette is used to extract the contents of the disk. This can be done simply as a back-up measure, or as a way to transfer data between disks. Some drives use a removable disk cartridge. This is expensive and does not address the problem of copying the contents of the disk.

Hard disk drives are delicate devices, sensitive to smoke, dust, and rough handling. They have been made more rugged in recent years through the use of sealed, pressurized cartridges, with automatic head locking when the drive is not in use.

Some users of personal computer systems start with floppy drives and up-grade to a standalone hard disk. This is not as easy as connecting a printer or display. It may be necessary to add a circuit board and other devices to the computer. A new operating system may also be required to format the data, control the disk, and manage the files. Programs developed for use with floppy drives may run no faster on the hard disk. For example, soft-ware written for floppies closes data files after each access. This is impor-tant since floppies are subject to a "head crash" in which data is lost. This approach is not necessary on a hard disk. If the code is not changed, the program will continue to close and reopen files--an operation that may even take longer on the hard disk.

Local area networking (distances up to one mile) is one of the latest ad-vances in personal and microcomputing. Modems and telephone connections, of course, have been available all along. Networking moves data between com-puters, disk drives, and other peripherals at speeds far greater than phone line connections do. This capability allows computers to share a hard disk, or for that matter, to copy files from one floppy drive to another. Several computers can also share specialized printers and plotters. Development of a network may make sense when a large hard disk is required for data stor-age yet several users with personal computers need access to the data.

Two popular network configurations are the star and the ring. In a star, the computers, disk drives, and other peripherals are connected by cable to a central controller. This device contains an interface card or circuit board for each device on the network. These cards and the data transmissions are managed by a microprocessor also contained in the controller. In a ring, the interface card is added to each device--computer or peripheral. These are then connected into a single cable that runs in a loop between all points. With both types of networks, file formats continue to affect the ability to share data.

Networking systems (controllers, interface cards, and cables) are avail-able from a variety of vendors. Those oriented toward personal and micro-computers include: Corvus Systems, San Jose, California; Nestar Systems, Palo Alto, California; and 3 Com Company, Mountain View, California. With or without such networks, most personal and microcomputers can now be equipped to communicate with minicomputers and mainframes, corporate and external data banks. The trend is toward the use of personal computers both as standalone units and as intelligent terminals in larger corporate net-works.

Acquiring a Small Computer

In most larger firms, the decision to purchase a personal computer and the choice of the device itself is a departmental matter. It can often be addressed by the facilities manager, acting on his or her own authority. The decision to purchase a microcomputer and the choice of this device is more often a divi-sional or corporate issue involving the data processing staff. Most firms are attempting to standardize the use of both personal and microcomputers. This will make future networking easier and it may also earn volume discounts or direct, "national account" status with the computer manufacturers. For these reasons, it is wise to consult with the data processing staff, even when the planner has the authority to "go it alone."

Earlier we expressed the philosophy of buying hardware to get software.

This is the correct approach in virtually all cases, even if it puts the planner at odds with a corporate standard. A sensible acquisition process has four steps and it begins with software selection:

1. Find the software that does the best job
2. Identify the machines on which it runs
3. Find suppliers who can demonstrate the machines and will service them locally
4. Choose the best machine

The definition of "best" will vary but should include such factors as price, serviceability, and compatibility with corporate standards, if they exist (16, 17). We will discuss the selection of computer aids more fully in Chapter 9.

7

Computer-Aided Design (CAD)

A Brief History of CAD

Entire books have been written about CAD (29, 31, 40). For our purposes here, a general understanding of hardware and software will be sufficient. Later, in Chapter 10, we will examine the process of selecting and installing a CAD system.

The term "computer-aided design" is something of a misnomer. In fact, as we mentioned in an earlier chapter, CAD systems are really aids for drafting, template manipulation, reproduction, and document storage. Still, because these activities are part of the design process, and because of the engineering and other software that may be provided, CAD systems are cental to computer-aided facilities planning.

The use of computers to create and manipulate "drawings" on a screen dates from the early 1960s. The first successful systems were developed for product design at General Motors and Lockheed. Large, dedicated computers were used, the displays were limited, and the systems cost $4 to $5 million to develop.

By the late 1960s, breakthroughs in electronics made possible the 16-bit minicomputer. This device was powerful enough to support interactive graphics at a fraction of the cost of a dedicated mainframe computer. Simultaneous developments resulted in low-cost displays. Linked with graphics software, it became possible to construct a dedicated CAD system for as little as $250,000

A new industry was born, with a number of companies formed to produce so-called turnkey CAD systems. The typical vendor acquired computers, displays, and related devices from several sources, tied the devices together with proprietary graphics programs, and delivered the resulting system to the customer. The vendor took responsibility for support of both hardware and software, regardless of its original source. As their volumes increased, the largest vendors began to manufacture their own hardware, especially the displays. At the same time, several companies that supplied only hardware decided to enter the turnkey system business.

With one or two exceptions, the dominant suppliers of CAD systems today were all in business by the early 1970s. IBM licensed the CADAM software developed by Lockheed in the 1960s. It was (and still is) offered on mainframe computers. More recently, IBM has introduced minicomputer systems using its own graphics software. Computervision, formed in 1969, based its early systems on a 16-bit minicomputer from Data General. A few years later, Computervision introduced its own minicomputer. It is the only major CAD company today that manufactures its own computer. Intergraph, Applicon, Autotrol, and Calma based their early systems on 16-bit minicomputers from Digital Equipment, Sperry Univac, and Data General. In some cases, CAD vendors replaced or modified the operating systems for these computers to make them more efficient in handling graphics tasks. Most early systems used storage tube displays from Tektronix.

The first widespread uses of turnkey systems turned out to be within the computer industry itself, for the design of integrated and printed circuits. Increasing sales were also made for mechanical part design--the original application of CAD nearly 10 years before.

Early uses of CAD for architectural design and facilities planning date from roughly 1970 (25). The interior design firm of Saphier, Lerner and Schindler used punched card inputs and computer-driven plotters to produce floor plans and furniture layouts. By the early 1970s, the firm had developed its own interactive system using a Hewlett-Packard computer and Hughes graphics displays. Known today as Environetics International, the firm has its CAD system installed in three U.S. offices. It is used for all aspects of interior design.

In Boston, also about 1970, the architectural firm of Perry, Dean and Stewart installed a CAD system using a Digital Equipment minicomputer. This system was used for a full range of architectural applications, including relationship analysis, bubble diagrams, and block studies. A much-refined successor to this system, called PEAC, is offered today by Decision Graphics, Inc., in Southboro, Massachusetts. Architect Clifford Stewart is still innovating. His current firm, the Stewart Design Group, uses a microcomputer-based CAD system from Graphic Horizons of Boston (22).

By the mid 1970s, other progressive architectural firms were also developing CAD systems in-house. Among them were Skidmore, Owings and Merrill in Chicago and Albert C. Martin & Associates in Los Angeles. The software developed by Albert C. Martin has since been licensed to ARCAD of Los Angeles, California.

British groups, such as Applied Research of Cambridge and Compeda, have been a source of many CAD innovations over the years. They were also active in the early 1970s, developing software for such diverse applications as process plant and hospital design. Refined successors to these early creations are currently marketed in the United States by PRIME and McAuto, among others.

Among the early turnkey vendors, Autotrol was perhaps the first, in about 1972, to install a system for process plant design. By the mid-1970s, Autotrol had become a leading supplier for such applications as piping, wiring, and instrumentation diagrams. From these early uses, Autotrol systems were extended to such tasks as building design and store, office, and plant layout.

Throughout the early and mid-1970s, software improved steadily. The capability to relate graphic symbols to non-graphic data became standard. This made possible improved bill of material reports and calculation and costing routines. Vendors expanded their standard symbol libraries, saving the purchaser much of the time spent drafting them into the system. Graphic data bases for 3-dimensional drafting became widely available, as well as color displays. However, both 3-D and color taxed the capacity of 16-bit minicomputers and severely degraded performance on multi-station systems.

Fortunately, by the late 1970s, additional advances were made in electronics. As a result, computer manufacturers were able to provide 32-bit minicomputers at reasonably low prices. The additional power of these machines allowed CAD vendors to provide color and 3-D with less degradation, and without resorting to expensive mainframe links. These 32-bit minis were also well suited to the heavy computational tasks of computer-aided engineering.

Also during the late 1970s, the concept of the intelligent graphics workstation emerged, taking advantage of the dropping costs for 16-bit microprocessors. These workstations, pioneered by Tektronix among others, combined displays and processors in a single desktop computer. When connected to a disk drive and plotter, the result was a computer-aided design system for well under $100,000. Such systems lacked the data storage capacity and computational speeds of larger mini-based systems. However, they made CAD available to a much wider range of users. More importantly perhaps, the lower cost of such systems allowed new vendors such as Arrigoni Computer Graphics and Iconica to open up the architectural and facilities planning markets. By 1982, over 30 vendors offered design and drafting systems for under $100,000, with some as low as $35,000. Quite a few of these systems are oriented toward architecture and facilities planning. Personal CAD software discussed in Chapter 6 first appeared in 1981 and proliferated throughout 1982 and 1983. Today, there are over a dozen software products with more appearing each month. Many can be bought in retail stores for prices between $250 and $2000.

The 32-bit microcomputer discussed in the previous chapter is the latest advance to affect CAD technology. The power of 32-bit internal capacity closes the gap between intelligent workstations and the traditional mini-based CAD systems. Today it is possible to have 3-D and color capability in a complete desktop system for well under $100,000. And such systems can perform most of the heavy computational tasks traditionally reserved for minicomputers. Workstations can be linked in a network to share peripherals for input, output, and storage. With this approach there is no loss of performance as additional stations are added. And, if one station fails, the others are still in operation.

The intelligent workstation with networking also lets the first-time user start with big system performance at a reasonable price. The overhead of the minicomputer and its controlled environment (electrical, air-conditioning, filtering, etc.) is avoided. On the other hand, additional intelligent stations are much more expensive than the simple display terminals that would be added to a central minicomputer. So, at a certain number of stations, the capital investment is comparable and then becomes lower with the central approach.

Sigma Design and Arrigoni Computer Graphics were the first to apply the 32-bit workstation approach to architectural CAD systems. Among the large vendors, Autotrol was the first to offer a 32-bit intelligent station with networking. Computervision, Calma, and others have announced similar offerings, and most vendors can be expected to offer this technology in the next year or two.

In 1983 there were probably 1500 CAD systems in use among architectural firms and corporate facilities departments. This is a small number when compared to the total number of planning and design groups which might potentially use a CAD system. Still, it gives us plenty of experience from which to generalize about the current uses of CAD in the facilities planning process. The most common uses are:

1. *Documentation of existing facilities*: redrawing or tracing of as-built drawings and overlays (for the subsequent uses below)

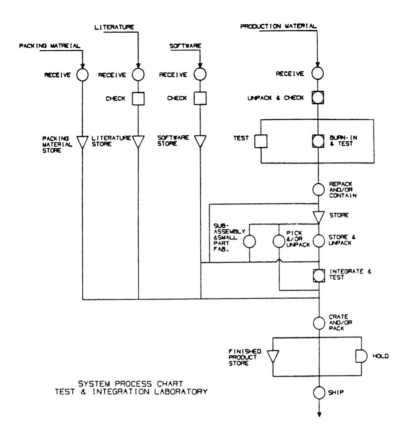

SYSTEM PROCESS CHART
TEST & INTEGRATION LABORATORY

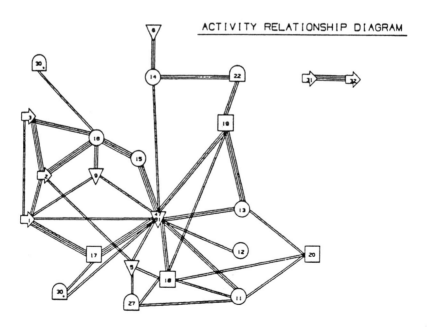

ACTIVITY RELATIONSHIP DIAGRAM

Figure 7.1. CAD can be used for simple diagrams in the early phases of planning. These were produced on the Unigraphics system. (Courtesy of McAuto, St. Louis, Missouri.)

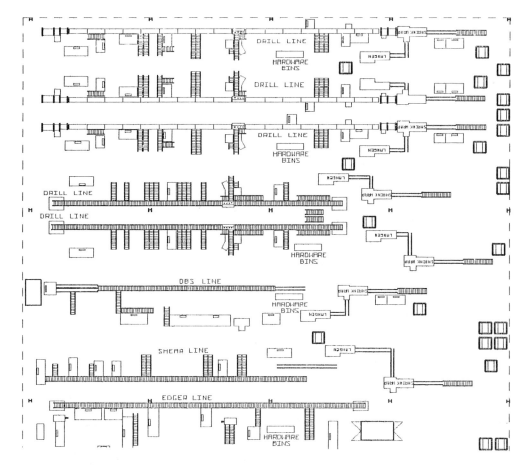

Figure 7.2. Factory layout with the PEAC system. This layout is made up of the symbols shown in Figure 7.10. (Courtesy of Decision Graphics, Inc., Southborough, Massachusetts.)

2. *Area measurements*: calculation of spaces occupied by tenants or various activities within the facility
3. *Planning and designing*: drafting changes to existing facilities and developing the drawings for new ones. Typical applications include: building and interior design, layout of furniture or equipment, and interference checks for process-oriented plants
4. *Graphic inventories*: keeping track of people, space and assets with annotated drawings (as opposed to lists)
5. *Visualizations*: dynamic three-dimensional displays used in lieu of physical models, isometric drawings and renderings (13, 14).

In addition to these basic uses, a growing number of systems provide additional facilities planning or facilities management software. This allows them to be used for many of the functions described elsewhere in this book--relationship diagramming, generation of block layouts, vertical stacking plans, nongraphic inventories and even project management. Figures 7.1 through 7.5

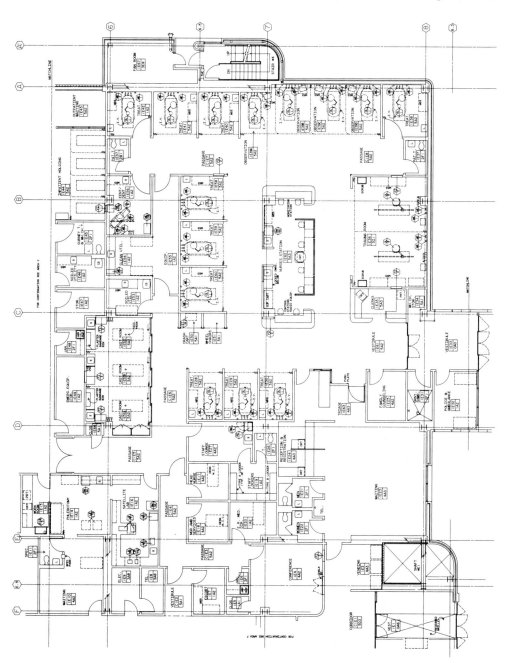

GRAPHIC DATA BASE MANAGEMENT

EQUIPMENT TRACKING

SORT BY ROOM

RM.#	RM. NAME	ID	QT	DESCRIPTION
IE4	CONFERENCE	FI2		FILM ILLUMINATOR
IE14	PHLEBOTOMY	RMAA		REFRIGERATOR
		LB175		BLOOD DRAWING CHAIR
IE15	WORK ROOM	PTS3		PNEUMATIC TUBE SYSTEM
IE16	SATELLITE LAB	RMA2		REFRIGERATOR-FLOOR STDG.
		LBI72		CENTRIFUGE
		LBI70		SLIDE STAINER
		LBI74		T.S. METER
		LBI73		BINOCULAR MICROSCOPE
		LBI71	2	BLOOD CELL COUNTER
		RMA		REFRIGERATOR
		RE1		MICROWAVE OVEN
		OL1		EXAM. LIGHT
IE27	STAFF LOUNGE			
IE28	TREATMENT	ME1		SPHYMOMANOMETER
		ME2		OTHAL-OTOSCOPE
IE29	TREATMENT	OL1		EXAM. LIGHT
		ME1		SPHYMOMANOMETER
IE30	TREATMENT	ME2		OTHAL-OTOSCOPE
		ME1		SPHYMOMANOMETER
IE36	TRAUMA ROOM	ME2		OTHAL-OTOSCOPE
		FI2	2	FILM ILLUMINATOR-DOUBLE
		ME1	2	SPHYMOMANOMETER
		OL8	2	MINOR SURGICAL LIGHT
IE38	TREATMENT	OL1		EXAM. LIGHT
		ME1		SPHYMOMANOMETER
IE39	OBSERVATION	OL1	4	OTHAL-OTOSCOPE
		ME1	4	SPHYMOMANOMETER
IE40	TREATMENT	OL1		OTHAL-OTOSCOPE
		ME1		SPHYMOMANOMETER
IE41	TREATMENT	OL1		OTHAL-OTOSCOPE
		ME1		SPHYMOMANOMETER
IE42	TREATMENT	OL1		OTHAL-OTOSCOPE
		ME1		SPHYMOMANOMETER
IE50	CRASH CART	ST18		OTHAL-OTOSCOPE
IE52	CASTROOM	FI2		FILM ILLUMINATOR-DOUBLE
IE54	CLEAN UTILITIES	FT1		ICE MAKER DISPENSER
IE56	EENT/DENT	ME1		FILM ILLUMINATOR-SINGLE
		ME06		SPHYMOMANOMETER
		ME17A		COAGULATOR
		ME19		ENT CHAIR

SORT BY EQUIP.

ID	DESCRIPTION	QT	RM.#	RM. NAME
FI1	FILM ILLUMINATOR-SINGLE		IE56	EENT/DENT
FI2	FILM ILLUMINATOR-DOUBLE	2	IE4	CONFERENCE
			IE36	TRAUMA ROOM
		3	IE52	CASTROOM
			IE62	NURSES STATION
			CIE7	PASSAGE
LBI70	SLIDE STAINER		IE16	SATELLITE LAB
LBI71	BLOOD CELL COUNTER	2	IE16	SATELLITE LAB
LBI72	CENTRIFUGE		IE16	SATELLITE LAB
LBI73	BINOCULAR MICROSCOPE		IE16	SATELLITE LAB
LBI74	T.S. METER		IE16	SATELLITE LAB
LBI75	BLOOD DRAWING CHAIR	2	IE14	PHLEBOTOMY (2)
ME1	SPHYMOMANOMETER		IE28	TREATMENT
			IE29	TREATMENT
			IE30	TREATMENT
		2	IE36	TRAUMA ROOM
			IE38	TREATMENT
		4	IE39	OBSERVATION
			IE40	TREATMENT
			IE41	TREATMENT
			IE56	TREATMENT
			IE58	TREATMENT
			IE60	TREATMENT
ME2	OTHAL-OTOSCOPE		IE28	TREATMENT
			IE29	TREATMENT
			IE30	TREATMENT
		2	IE36	TRAUMA ROOM
			IE39	OBSERVATION
		4	IE40	TREATMENT
			IE41	TREATMENT
			IE58	TREATMENT
			IE59	TREATMENT
			IE60	TREATMENT
ME10	ICE MAKER DISPENSER		IE54	CLEAN UTILITIES
ME16	COAGULATOR		IE56	EENT/DENT
ME17A	SLIT LAMP		IE56	EENT/DENT
ME19	ENT CHAIR		IE56	EENT/DENT
ME06	TREATMENT CAB.		IE56	EENT/DENT
OL1	EXAM. LIGHT		IE28	TREATMENT
			IE29	TREATMENT

CONSTRUCTION COST ESTIMATE

SORT BY WALL MAT'L A

RM.#	RM. NAME	FIN. CODE	UNIT COST	CEIL HT.	CIRC.	FLR. AREA	WALL AREA	COST EXT.
IE3	VENDING	4A3	.45	0	26	131	209	94
IE5	CONFERENCE	4A2	.45	0	66	251	531	239
IE7	CLOSET	IA1	.45	0	18	20	145	65
IE7	ESCORT	IA2	.45	0	34	72	275	124
IE9	TELEPHONE	3A8	.45	0	33	51	265	120
IE9	ELECTRIC	3A8	.45	0	64	65	509	229
IE10	WAITING	4A2	.45	0	54	211	433	159
IE14	PHLEBOTOMY	1A2	.45	0	35	183	279	126
IE15	WORK ROOM	1A2	.45	0	61	68	491	221
IE16	SATELLITE LAB	1A2	.45	0	34	213	299	135
IE17	HEAD NURSE	4A2	.45	0	73	81	287	129
IE18	MGR. AMB. SERVICES	1A2	.45	0	36	86	299	135
IE19	STAFF LOCKERS	1A2	.45	0	37	88	311	140
IE20	E.R. MED. DIR.	IA2	.45	0	39	584	795	358
IE23	WAITING	4A3	.45	0	99	124	358	161
IE24	RECEPT. AND REGISTRATION	4A3	.45	0	45	94	312	169
IE26	FAMILY COUNCILING	4A2	.45	0	39	129	376	169
IE27	STAFF LOUNGE	4A2	.45	0	30	86	241	109
IE28	TREATMENT	5A2	.45	0	30	86	241	109
IE29	TREATMENT	5A2	.45	0	30	86	241	109
IE30	TREATMENT	1A2	.45	0	44	124	359	162
IE32	POLICE AND AMBULANCE	OAO	.45	0	57	186	452	203
IE33	VESTIBULE	5A1	.45	0	26	21	183	82
IE34	WHCH. STORAGE	5A1	.45	0	11	92	88	40
IE35	CLERKS	5A3	.45	0	43	115	347	156
IE38	OBSERVATION	5A2	.45	0	57	369	456	205
IE40	TREATMENT	5A2	.45	0	30	93	236	106
IE41	TREATMENT	5A2	.45	0	30	93	236	106
IE42	TREATMENT	1A2	.45	0	40	100	320	140
IE47	SOILED UTIL.	1A2	.45	0	77	77	290	131
IE50	STORAGE	1A2	.45	0	39	113	381	172
IE51	CRASH, EQUIP. STORAGE	5A1	.45	0	48	91	91	41
IE52	CASTROOM	5A2	.45	0	82	16	653	294
IE54	WHEELCHAIRS	1A2	.45	0	79	366	292	132
IE55	EQUIPMENT STO.	IA2	.45	0	37	92	308	202
IE58	CLEAN UTILITIES	5A2	.45	0	56	157	448	202
IE59	TREATMENT	5A2	.45	0	31	102	248	170
IE60	TREATMENT	5A2	.45	0	31	102	244	110
IE61	VESTIBULE	5A2	.45	0	31	102	244	110
IE62	NURSES STATION	5A3	.45	0	68	380	544	245
IE63	TRIAGE	4A3	.45	0	55	157	440	198

SUBTOTAL WALL FINISH A   14973   6334

Figure 7.3. Office layout and associated data base management. Equipment symbols are given an ID and description. Their placement associates them with a room number for subsequent inventory listing. Each room is given a wall finish code for subsequent cost estimating. This example produced on a Calcomp system. (Courtesy of Design Logic, Inc., Oakland, California.)

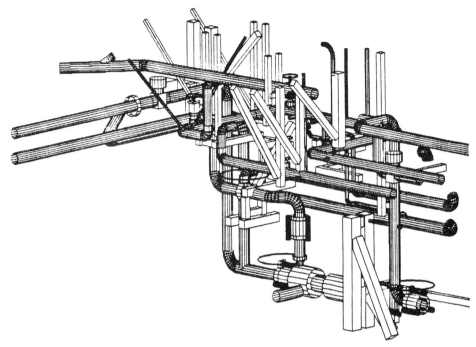

Figure 7.4. Process plant visualization using a three-dimensional data base and shading. This example by Construction Systems Associates, Inc., uses the firm's Space Modeling and Interference Detection System.

show several uses of CAD systems by architects and facilities planners. Additional uses will appear throughout this chapter.

   With the exception of visualization, and the facilities management software just described, virtually any of the 40 to 50 CAD systems on the market today could be applied to the uses listed above. This poses a serious challenge to the interested planner or manager--how to decide which system is best? We can start by understanding the three price-performance categories which exist in the market today. These are defined in Figure 7.6, using six key features:

   1. Internal capacity (word length in bits)
   2. Clock speed (MHz)
   3. Method of floating-point calculation (software or hardware)
   4. Memory (in kilobytes)
   5. Storage capacity (in kilobytes)
   6. Storage devices (type and size)

   As noted in the previous chapter, smaller internal capacities (8- and 16-bit word length) limit the speed of processors in performing graphics tasks. While this may not be disturbing to the novice or occasional user, it becomes tedious to the experienced, full-time user. Noticeable delays occur in executing commands and generating display images. These inhibit the user's train of thought and productivity, especially on large, complex drawings.

Figure 7.5. CAD rendering using standard textures.

The value of large memory and its relationship to word length were discussed in the previous chapter. Memory is an important price-performance variable in CAD. Small memory means more frequent disk access, slowing the system to a noticeable degree.

The approach to floating-point calculations is another important factor in system performance. In floating-point calculations, the computer decides for itself where to put the decimal point. This allows a variable to take on the widest possible range of values. A floating-point number can be thought of in scientific notation: 3.12345×10^6, with the computer keeping track of the power of 10. Floating-point calculations determine the resolution and accuracy of the CAD system. They also make it possible to rescale a drawing to virtually any proportions (display size, paper size, etc.) and to "zoom in" and "zoom out" of a very large plan. Floating-point calculations can be accomplished with special program code, or with a microprocessor which is used whenever such calculations are required. The hardware (microprocessor) approach is much faster than software. The net result is a faster system. (As we noted earlier, in Chapter 6, some personal-computer-based systems are restricted to integer calculations, with attendant restrictions on resolution,

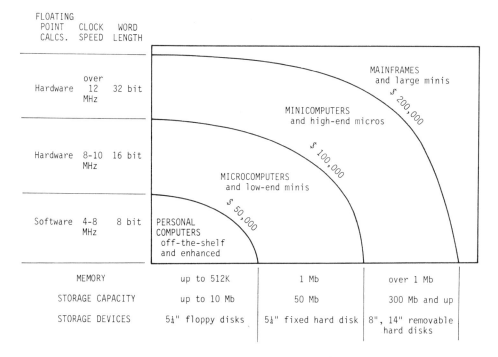

Figure 7.6. CAD price-performance categories as of 1983. Curves represent first station prices including all necessary peripherals and software.

scaling, and zooming.) Today, some smaller CAD systems use the software approach to floating-point. The trend, however, is toward the inclusion of a special microprocessor. So, this advantage of larger systems may disappear in the near future.

Small storage limits the amount of data available to the user. In facilities planning, this may mean an inconvenient or unacceptable limit to the square footage that can be reached. Drafting may have to be segmented into bays or some arbitrary split. This can preclude the planner from seeing and working on the whole facility. It also causes tedious problems and delays when drafting along the boundaries. Delays will occur every time a disk must be changed or a new tape loaded.

In the face of limited storage, good archives become very important. The planner may have to use several disks or tapes to store a complete floorplan, building, or project. On the other hand, keeping track of a few disks and tapes is still easier than rummaging through the typical planner's flat and roll files to fetch overlays and drawings.

Unfortunately there are no standards relating square footage to disk storage for different types of facilities. In fact, it is almost impossible to generalize. Storage requirements vary with the graphic detail and density of plans and drawings, and the amount of annotation they contain. The number of colors and overlays used by the planner is another influence. Finally, the structure of the data base and the disk format used by the vendor lead to varying storage efficiencies.

Personal computer CAD systems and their uses were discussed in the previous chapter. These systems are slow when compared to the larger micros, minis, and mainframes. On the other hand, personal computer systems are low-cost ($10,000 to $50,000) and easy to learn. They offer a good way to get familiar with CAD. Unfortunately, most offer little or no upgrade path if demands grow in the future.

Micro and low-end minicomputer systems are the most common form of CAD installation today. Most have 16- or 32-bit capacities. Many of the microcomputer systems use 5-1/4-in. fixed, hard disks for storage. Data is archived onto cartridge tapes. Minicomputer-based systems offer larger disks--both fixed and removable. Proprietary and UNIX operating systems are common. Software for architectural and facilities applications is available. CAD systems in this category have good response time. The minicomputer versions are typically two- or three-station installations and there is some degradation in response over a single-station. Color displays up to 19 in. are becoming standard. Most vendors offer software compatibility in the event that the system is outgrown. This means that the planner's data base (on tape and disk) can be copied onto the larger, often 32-bit system. The displays and peripherals can be retained, but the processor is replaced. Micro and low-end minicomputer systems sell from $50,000 to $100,000.

High-end microcomputers and large minis have 32-bit capacities. These are state-of-the-art CAD systems with virtual memories as well as large on-line storage devices. The result is a high-performance system that can readily access a large number of drawings. Here too, proprietary and UNIX operating systems are common. Architectural and facilities-oriented software, a wide range of color, 3-dimensional visualization, and large, high-resolution displays are all fairly standard offerings in this category.

The minicomputer systems can support 4-6 terminals before degradation becomes a problem. The micro-based systems avoid this problem with networking. A few vendors are solving the degradation problem by linking a 32-bit intelligent display to their central minicomputer. This is an expensive approach, but it does give the best of both worlds for the large capacity and performance-oriented user. The issue of upgrading is moot, since both the micro- and mini-based systems offer state-of-the-art performance.

Mainframe computers can be used for CAD. In fact, the largest systems today are large mainframes driving dozens of time-sharing displays. These large systems, however, are used in mechanical engineering and are rare in facilities planning. The high expense of the central computer is spread across a great many displays, making it an economical approach for very large installations. In fact, such systems show degradation in response, unpredictable delays and, on rare occasions, system-wide down time. However, until recently the only alternative has been a very expensive flock of dispersed minicomputers. With networking, up to 100 or more intelligent 32-bit stations can be connected. It remains to be seen what impact this approach will have on the future of mainframe installations.

Regardless of price-performance category, all CAD systems use essentially the same hardware devices for input, display, and output. These devices take the place of the tools used in manual drafting.

Input Techniques and Devices

CAD systems operate on graphic and non-graphic information supplied by the planner. Typically, the graphic input data include:

1. *Schematic sketches*: to be converted into diagrams of material flow, production process, and duct, pipe, and wire routings
2. *Dimensioned sketches*: to be converted into plans and drawings of buildings, equipment, duct work, and piping and wiring runs
3. *Existing drawings*: floor plans, overlays, and elevation drawings to be entered for subsequent recall and modification
4. *Symbols and templates*: to be entered for subsequent recall and positioning in future designs and plans
5. *Text or notes*: to be displayed as graphic elements on diagrams, plans and drawings, symbols or templates

The non-graphic input to a CAD system consists of:

1. *Text, notes, charts, tables and lists*: to be held in memory and reported as non-graphic reference data about a diagram, plan, drawing, symbol, or template
2. *Operating instructions*: to govern the operation of the CAD system

For graphic inputs, two practical techniques can be used: on-screen drafting or the digitizing table. Non-graphic input is achieved with a variety of devices--keyboards, touch-sensitive and cursor-controlled "menu" panels, and even voice for operating instructions. The choice of technique and input device is often a matter of personal preference. In other cases, the equipment or system supplier may provide only one means of input. The most common approaches are summarized in Figure 7.7.

Sketches are generally drafted into the CAD system at the display terminal. A cursor is moved about the display to create the desired line work. Text and notes are added with a keyboard. Dimensioned plans and drawings are produced by keying-in the necessary dimensions as the cursor is moved about. There are a variety of cursor control devices available to the planner. These include:

1. Directional keys
2. Thumbwheels
3. Joystick
4. Track ball
5. Light pen on screen
6. Stylus on tablet
7. Puck on tablet
8. Mouse
9. Touch-sensitive panel

Each device has its proponents, and again the choice is one of preference, if not dictated by the supplier.

There are two schools of thought regarding the entry of existing drawings. One favors on-screen drafting as just described. The other favors the use of a digitizing table. With this technique the drawing is placed on a table containing a grid of closely spaced electrical wires. Other types of tables use solid,

INPUT DATA	PRIMARY INPUT TECHNIQUE
schematic sketches	on-screen drafting
dimensioned sketches	on-screen drafting
existing plans and drawings	on-screen drafting or digitizing, correcting for stretch in both directions
symbols and templates	on-screen drafting
text, notations, charts and tables	keyboard
operating instructions	keyboard, menu, or voice

Figure 7.7. Input to CAD.

electromagnetic surfaces or acoustical sensing, but the wire grid is most common. The planner holds a stylus or cursor "puck" which contains a transducer. When the cursor is activated, its position above the grid is recorded as an X-Y coordinate. This information is then sent to the CAD graphics processor for storage, analysis and display. By "touching" the endpoints of each line in the drawing, it is effectively "traced" into memory. A variety of cursor and stylus designs are available. The most useful cursors have multiple control buttons that can be used to perform certain editing and data processing functions as the line work is entered.

The planner needs a relatively high resolution display within view while digitizing. As the endpoints of each line are entered, the line should appear on the screen. In this way, the planner keeps track of progress and visually checks for errors and omissions.

The point of digitizing an existing drawing is to avoid figuring and keying in the dimensions of each line segment. This can save a great deal of time under certain circumstances; however, it does sacrifice some accuracy when compared to on-screen drafting.

The accuracy of the digitizing process depends in part on the equipment being used, especially the resolution of the wire grid, and the design of the stylus or cursor. Cross-hair cursors, for example, are influenced by parallax, in which case the planner may slightly misplace the cross hairs with respect to the point being entered. The accuracy of digitizing is influenced most by the stretch of the drawing. A good CAD system will provide the capability to correct or normalize for stretch in both X and Y axes.

Another key issue in digitizing is economics. Large digitizing tables are expensive, costing between $12,000 and $20,000. The planner should study the data entry process before buying, to insure that the time savings is great enough to justify the investment. Will there be a continuing need to enter existing drawings, or is this need limited to a start-up period? Among the installations known to your author, about two-thirds use digitizers for entry of existing drawings.

With either input technique, the facilities planner should give serious thought to editing and improving existing drawings as they are entered. It may not be necessary to enter all of the drawing's details. And it may also make sense to separate the drawing's contents onto separate layers or overlays.

Interactive Graphic Displays

The display terminal is the most prominent feature of a CAD system. Three
types are in widespread use:

1. Direct-view storage tube
2. Vector-refresh
3. Raster-scan

The direct-view storage tube was the earliest display device. Its image is
created by an electron beam which penetrates a monochromatic phosphor coating (usually green) on the inside of the screen. The image remains on the
screen, flicker-free, as long as power is supplied to the device. Unfortunately,
this also makes it impossible to erase any part of the image. Erasure is achieved
only by "repainting" the entire image, with offending lines removed. While the
repaint is extremely fast, even on a dense image, it can be distracting during
highly interactive tasks such as layout planning. On the positive side, storage tubes give the highest possible resolution--limited only by the grain of the
phosphor coating itself. This limit is generally expressed as 4096 by 4096 equivalent points per screen. Storage tubes are relatively inexpensive. They are
well suited to tasks such as digitizing, where they can be used to verify the
input of an existing drawing. They are also good for on-screen drafting of
symbols and standard details which are created once and then stored away.

Vector-refresh devices overcome the erasure limitation of storage tubes by
using a short-lived phosphor and repainting the image on a continuous basis.
This approach also provides dynamic movement or "dragging" of symbols on the
screen. However, the display will flicker whenever the repaint time exceeds
1/30th of a second. Repaint time is generally a function of the number of lines
displayed.

Vector-refresh devices can also provide limited color in one of two ways.
Beam penetration displays have two layers of phosphor, one green and one red.
By varying the beam's strength, this type of display can produce green, red,
or a mixture of the two. Shadow-mask displays use a screen coating of red,
green, and blue phosphor dots. Three electron beams, one for each color,
pass through a sieve-like mask which directs them to the correct dot. A variety of colors can be created by varying the mixture of activated dots. This
approach, however, sacrifices some of the resolution of beam penetration. The
terms random-scan, stroke-vector, stroke-refresh, calligraphic, and direct-
beam refreshed technology all refer to what we have just described as vector-
refresh.

Both storage tube and vector-refresh devices are driven by graphic data
that is stored in a vector format. Each element in an image (or drawing) is
described by the X-Y coordinates of its points. Blank areas are not described.
The resulting data base lends itself well to line work and economizes on memory requirements. However, this data format and the vector display make it
practically impossible to fill or shade solid areas of an image. This limits the
utility of vector-refresh (and storage tubes) for certain types of visualization
tasks.

Raster-scan is the third and perhaps most popular type of display. Raster-
scan (also called raster-refresh and video) is essentially the same technology as
home television. Electron beams continuously scan the entire display from top to

bottom, line by line. Monochromatic (black and white) displays use a single beam. Color displays use the three-beam, shadow-mask approach just described.

The electron beams cover the screen from top to bottom, 60 times per second. However, interlaced displays scan only every other line. As a result, the image is totally re-scanned at a rate of only 30 times per second. This is the minimum rate required to avoid flicker. Interlaced displays are less expensive than non-interlaced--those which scan every line 60 times per second. But they sacrifice some brightness, contrast, and stability when viewing dense images.

Raster-scan devices can be driven by data in a vector format. However, certain conversions or "rasterizations" are required and this reduces performance. Ideally, raster displays should be driven by graphic data stored in raster or "pixel" (picture element) format. This approach divides an image (or drawing) into pixels, each with its own memory address. Attributes like color, brightness, and data class are assigned to each address. Blank areas of a drawing are described by "empty" addresses. A drawing recorded in this format needs more data storage than one recorded in vector format. However, the additional storage cost is not a significant portion of total system cost.

The primary benefit of raster-scan devices is their ability to handle a wide range of colors, including solid fill and shading. The number of colors actually available to the user varies by vendor and model.

Naturally, the more active colors, the higher the cost of the display. Raster-scan also provides selective erasure and such interactive features as continuous panning (horizontal scan), scrolling (vertical scan), and zooming (scaling "up" or "down"). All of these capabilities are provided with a flicker-free image. The chief limitation is low resolution.

Raster data can be stored at a density of 4096 by 4096 pixels per image--equivalent to storage tube or vector-refresh resolution. However, due to limitations in software and control devices, it is not yet possible to display 4096 by 4096 pixels in a refresh mode. With raster displays, a large amount of data must be thrown onto the screen every 30th or 60th of a second. At present, a few systems can achieve 2000 x 2000 pixels per image. Generally, though, 1024 x 1024 is considered high resolution. Medium resolution is generally considered to be 512 x 512. Low resolution is 256 x 256. At lower resolutions, images become "blocky" around the edges. Diagonal lines exhibit a "stairstep" effect known as the "jaggies." This condition exists only on the display and direct screen copies. It does not appear on pen plots.

In facilities planning, raster resolutions can be applied as follows:

1. *High (1024 x 1024)*: acceptable for virtually all detailed layout planning and drafting tasks.
2. *Medium (512 x 512)*: adequate for block layout and site plans. Can be disturbing in detailed work. Good for charts, graphs, and tables.
3. *Low (256 x 256)*: adequate for simple block layouts, charts, graphs, and tables. Serious limitations for detailed layout and drafting.

Proponents of raster displays claim that they are bright enough to be used in normal office light. By contrast, storage tube and vector-refresh do require dim light. Many users of raster, especially interlaced displays, still choose dim light to improve their view of the screen.

The cost of color raster displays has dropped significantly since the mid-1970s. The premium for color over black and white is not significant when compared to the total investment in a system. Nor is the premium that great for raster over vector-refresh and storage tube displays. Still, a system's cost can be held down using the minimum display capabilities for each application, including vector or storage where color or selective erase are not important. The display required for digitizing is a good example.

Screen size is also a major influence on cost, regardless of display type. Displays suitable for facilities applications range from a low of 13 inches to a high of 19 inches in diagonal width. The ability to work with the smaller screen is a matter of preference and budget.

Output Devices

When it comes to output, the planner can choose from a sometimes bewildering array of devices, as shown in Figure 7.8. Generally, a CAD system produces four types of output.

1. *Finished plans and drawings*: primary outputs of the system at various desired scales.
2. *Working copies*: quick, small reproductions of the display image, scaled by default to fit the paper size of the output device.
3. *Reports*: lists, notes, tables, and charts. May include business graphics from a data base of non-graphic information.
4. *Presentation aids*: typically slides, photographs or transparencies of color, and 3-D screen images. Especially useful in interior and architectural design.

The pen plotter is the most common output device in computer-aided design. There are several types: flatbed, beltbed, drum, and grit wheel. Each produces plans and drawings on a variety of media, including paper, vellum, polyester and acetate. Certain types of plotters do better on some media than others. All use multiple pens to produce different colors and line widths.

Electrostatic and photographic plotters offer alternatives to pen wear and maintenance problems. Electrostatic plotting fuses toner to the output media. It is much faster than pen plotting but it is also expensive, and its line quality is not as fine as pen plotting. Electrostatic plotting can be justified for high-volume production of large, dense drawings. In this setting, the slower speed of pens and the downtime due to pen wear and ink consumption can be significant. Photographic plotting is slow and expensive, but it produces very high quality line work. It can save a production step where blueprints are used.

Facilities planners need not be overly concerned with plotting accuracy and line quality. A plotter accurate to the width of a human hair is not really necessary in a profession which works to the width of a sharp pencil. And, if an electrostatic plot, for example, is properly dimensioned, the slight fuzziness of its finer lines are unlikely to impede the construction crew.

In the early phases of facilities planning, outputs such as land-use plans, site plans, and block layouts are largely conceptual documents which can be interpreted even at a very small scale. While these outputs can be plotted,

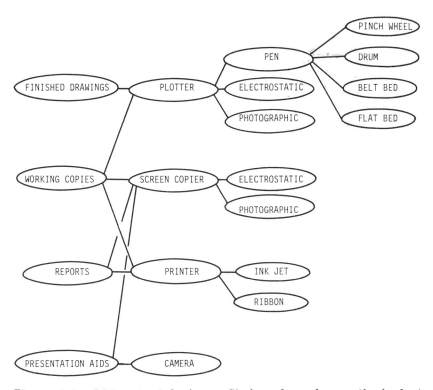

Figure 7.8. CAD output devices. Choices depend upon the budget, volume of output, and quality requirements.

a copying device or printer operating directly from the screen is generally sufficient.

As detailed plans and drawings are developed, the planner will need to make frequent, quick copies of the display screen. These are used to check and verify the design and to record its development. Two devices are commonly used to produce these so-called "hard copies." One is a photographic copier that makes a high quality black and white or "gray shade" image on photographic paper. The other device is the familiar electrostatic copier which uses a toner--either black and white or color. Ink jet and ribbon printers can also be used for color hard copies. These devices are often referred to as printer/plotters. The ink jet devices operate at a faster rate but may need extra maintenance. Ribbon printers are more limited in their color quality and range, but they are also less expensive.

All types of hard copy units can be used to report non-graphic data stored in the system. However, if these reports are frequent or lengthy, a common dot-matrix printer should be used.

Today's most powerful CAD systems can produce very exciting color and 3-D images on the screen. However, these lose a great deal of their impact when reduced to plots and hard copies. To capture the impact of these images, and to recreate the rotation of objects on the screen, cameras can be attached to the display. These produce color slides, photographs and transparencies. Color photocopiers can be used for much the same purpose. Such output is

very useful in presenting and gaining approval of facilities plans and designs.

Software and Operations

Software is the least visible yet most important part of any CAD system. Powerful hardware can often be outperformed in use by good software on a modest
piece of equipment. It is not appropriate or even possible, since products
are always changing, to evaluate specific vendor systems in a book of this nature. We can, however, identify those aspects of software and operation which
are of major importance to facilities planners. In this way, we can help the interested planner do a better job of comparing the systems available to him.

If we were to list all the tasks performed in making a layout plan or drawing, we would have a good list of necessary software and operational features.
Grouping these tasks, we would find that they fall into three types of activity:
drafting, reproduction, and non-graphic tasks.

Drafting includes all line work produced with pencil, pen, tape, films, and
transfers. This includes the creation of symbols for layout planning and other
types of drawings. It also includes tracings of all types, lettering, and texturing. Reproduction includes all cutting and pasting, overlay and sepia work,
and the production of prints and templates. Our non-graphic tasks include:
notation, measuring lengths and areas, counting objects, and the use of these
measurements and counts in subsequent calculations. CAD assists in all of
these tasks. To distinguish among different CAD systems, it is valuable to
compare their performance and capability in the following respects:

1. Input and creation of lines and text
2. Creation of symbols with attributes
3. Dimensioning and scaling
4. Measuring areas
5. Constructing lists or take-offs
6. Manipulation of symbols and line work
7. Handling of screen images
8. Three-dimensional visualizations
9. Operational commands and menus
10. Support capabilities

CAD software must allow for easy input and creation of basic line work and
geometries. Multiple line types and weights must be provided as well as the
ability to place them in parallel, perpendicular, tangent, and angled positions.
Circles, ellipses, arcs, splines, and fillets should be easily created. Several
text fonts are desirable. Multiple fonts are a matter of taste and style rather
than necessity. They may be more important perhaps to architects than to facilities planners. Several types of textures should also be provided.

When creating symbols, the software must allow for non-graphic attributes
describing the symbol to be entered and retrieved for various uses. It should
be possible in layout planning, for example, to create machinery symbols or
templates and to enter all relevant data about them. Such data might include:
make and model, asset number, age, electrical requirements, etc. An example

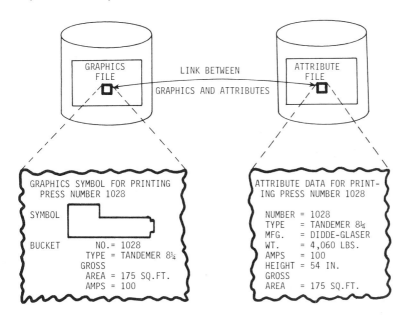

Figure 7.9. Illustration of dual CAD data bases--one for graphic symbols, the other for related non-graphic attributes. Symbol dimensions and attributes are obtained from the survey form shown earlier in Figure 2.11. In this scheme, describing the Intergraph CAD system, the graphics file contains a display-able "bucket" or subset of the attribute file. The bucket is user-defined and can be varied for different planning tasks.

of desirable capabilities appears in Figure 7.9. Typical symbols for a factory layout are pictured in Figure 7.10.

Many systems claim to have automatic dimensioning. But in most CAD software it is not truly automatic. Instead, once a line's endpoints are re-corded in the data base, its length to scale can be displayed, usually in Im-perial or metric. Some software may automatically generate arrowheads, ex-tension lines, and dimension lines. The more sophisticated software available lets the user define different types of arrowheads. It also displays lengths in either decimals or feet and inches, and can automatically convert from one to the other, as well as to metric.

Dimensions should always be displayed from the data base, i.e., from the true endpoints of the line. Some systems display the distance between two points as identified by the cursor. This is equivalent to scaling and is sub-ject to error, especially if there is no ability to "snap" the cursor to the end-point of the line in question.

Area measurement capabilities are related to dimensioning. The software should be capable of calculating and displaying the area of any enclosed space. The calculation should be based on the coordinates held in the data base, not on perimeter points touched by a cursor. The latter is analogous to a planime-ter and may be subject to error.

The software should produce formatted lists, bills of material, or take-offs from finished plans and drawings. At a minimum, these should be com-plete symbol counts. More sophisticated software may produce reports using

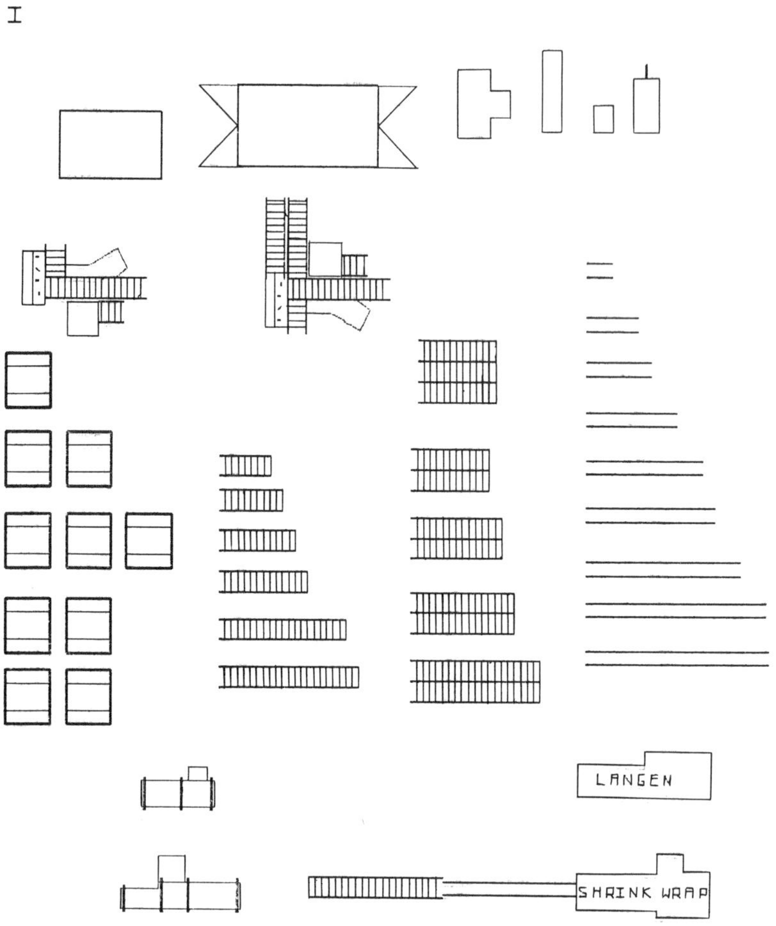

I

Figure 7.10. Machinery and equipment symbols used to produce the factory layout shown in Figure 7.2. These correspond to Mylar, acetate, or vinyl templates used in manual layout planning. (Courtesy of Decision Graphics, Inc., Southborough, Massachusetts.)

the symbol attributes on file, or even calculations and extensions of attribute data to arrive at cost estimates. Automating these activities is CAD's greatest contribution to productivity improvement. A good example of a material listing capability is shown in Figure 7.11.

Several types of manipulations of symbols and line work are useful. The software should allow objects to be defined from the combining or layering of previously entered symbols. These objects are, in effect, "higher order" symbols. Rotation and mirroring of symbols, objects, and line work in general is essential. The software should also provide "snapping." This is a feature that automatically moves the cursor to a line or a point, once it is placed within a specified distance from the line or point. This feature overcomes many problems imposed by screen resolution and manual cursor positioning.

Several image handling capabilities are essential. The first is layering. This corresponds to the function of a tracing or acetate overlay in conventional drafting. Often, it is not practical or desirable to put all a drawing's information onto a single layer. The user must be able to define multiple layers and put certain categories of information into each. The software should then permit these layers to be displayed individually or together, at the user's will. When two or more layers are displayed, the user must be able to choose the layer(s) upon which he wishes to work, and the others should not be affected by any actions taken.

Examples of typical overlays are shown in Figures 7.12 and 7.13. Successful use of CAD requires the definition of a reasonably complex drawing data base. (See Figure 7.14.) A certain amount of procedural "overhead" and control is necessary to insure that each drawing is treated consistently and that information is in fact stored in the correct location. Forms and indexes like those shown in Figures 7.15 and 7.16 are essential.

Windowing is another essential image handling capability. A window is a bounded area within a display image. The software must permit such areas to be easily defined, named as objects, moved about and enlarged. This function is analogous to that of a pair of scissors or a knife in conventional drafting.

Often, when working at a large scale, the display image can show only a portion of the entire drawing or plan. Under these circumstances the capability to "zoom, pan, and scroll" becomes very important. Zooming refers to the scaling of display elements making the image larger or smaller. This is usually done in discrete fashion by repainting. Panning refers to horizontal "scanning" to reveal different portions of the drawing on the display image. Scrolling refers to vertical scanning. These are features of raster technology. Moving about a drawing with storage tube or vector-refresh requires the use of windowing to reach the area of interest.

Three-dimensional visualizations are an often confusing aspect of CAD software. Most facilities planning work is done with two dimensions. Three-dimensional drawings are used occasionally for illustrations, renderings, and presentation aids. The notable exceptions are process plant design and conveyor layouts in materials handling systems. In these tasks, three-dimensional visualizations are essential to detect interferences and inadequate clearances.

Not surprisingly, most CAD software is two-dimensional. In some cases, a capability called "2-1/2-D" is provided. With it, the user can "extrude" two-dimensional drawing features such as walls or partitions to a specified height. These can then be viewed from any angle and displayed in isometric or perspective form. Many planners will find this sufficient for their occasional visualization tasks.

Full, three-dimensional software carries X, Y, and Z coordinates for every point in the data base, although the Z coordinate may remain unused. Such software exhibits the property of "coherence." This means that action taken in any view will be immediately reflected in all views. Three-dimensional images are displayed either as so-called wire-frames or as solids models. Wire-frame models suffer from the display of "hidden" lines, which must be removed manually from the image. This is time-consuming and impractical, especially if the image contains a great amount of detail. Representative uses of wire-frame, three-dimensional images are presented in Figures 7.17 through 7.21.

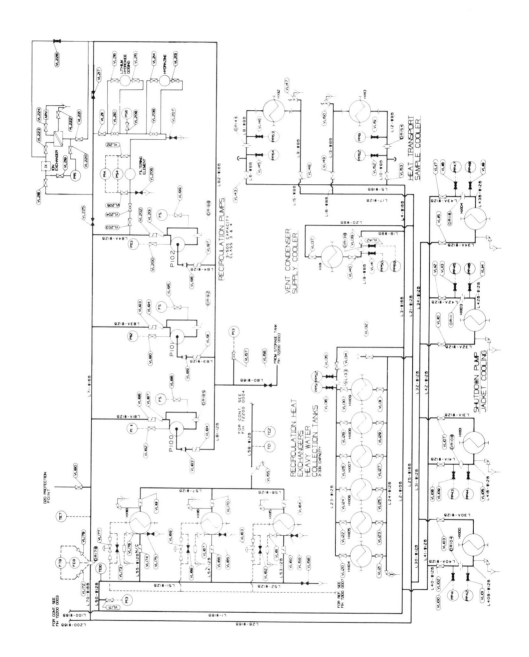

POWER SYSTEM SERVICES						
DESIGN AND CONSTRUCTION DIVISION		VALVE LIST				
MECHANICAL DEPARTMENT					SHEET	1 of 3

:PROJECT	RECIRCULATED COOLING	:REV :	DATE	:APPROVED
:P&ID DIAGRAM	650-0019	: :	:	:
:SYSTEM:	WATER SYSTEM	: :	:	:

+++++ VALVE LIST +++++

VALVE NO	FUNCTION	SIZE	MATERIAL CODE NO	TYPE	:RAT:END:OPR :CLS:PRP:TYP	:BD: DESIGN :MT:TEMP: PRESS	DESIGN TEMP :NO	:LINE:SUPPLIER: NO :MANUFACT:	VALVE DWG NO	:REQN-NO :OR PO NO
VL100		:12		GL VLV	: B :			:L40A:		
VL101		:12		GL VLV	: B :			:L40B:		
VL102		:12		VDIAPH	: B :			:L40A:		
VL103		:12		GL VLV	: B :			:L30A:		
VL105		:12		VDIAPH	: B :			:L41B:		
VL106		:12		VDIAPH	: B :			:L41A:		
VL107		:12		GL VLV	: B :			:L31A:		
VL109		:12		VDIAPH	: B :			:L41A:		
VL110		:12		VDIAPH	: B :			:L32A:		
VL112		:12		V CL	: B :			:L42A:		
VL113		:12		V CL	: B :			:L42B:		
VL114		:12		VDIAPH	: B :			:L34A:		
VL115		:12		VDIAPH	: B :			:L43A:		
VL117		:12		V CL	: B :			:L43B:		
VL118		:12		V CL	: B :			:L23F:		
VL119		:12		VDIAPH	: B :			:L24F:		
VL120		:8		VDIAPH	: B :			:L23E:		
VL121		:8		VDIAPH	: B :			:L24E:		
VL122		:8		VDIAPH	: B :			:L23D:		
VL123		:8		VDIAPH	: B :			:L24D:		
VL124		:8		VDIAPH	: B :			:L23C:		
VL125		:8		VDIAPH	: B :			:L24C:		
VL126		:8		VDIAPH	: B :					
VL127		:8		VDIAPH	: B :					

Figure 7.11. Piping plan and corresponding value list. (Courtesy of Calma Company, Santa Clara, California.)

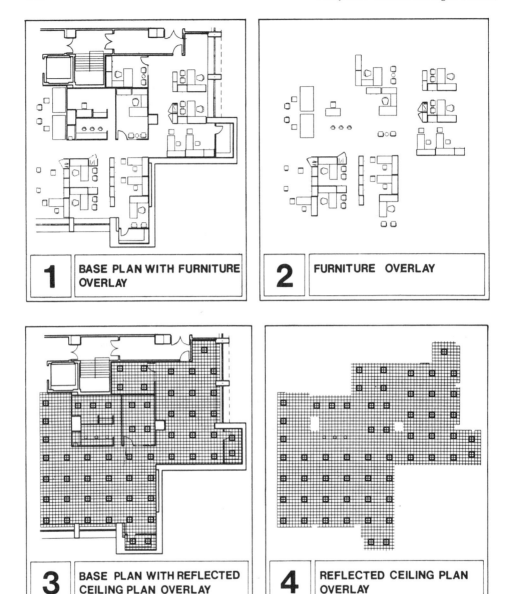

Figure 7.12. Typical CAD overlays. (Courtesy of Decision Graphics, Inc., Southborough, Massachusetts.)

The latest approach to three-dimensional solids modeling is the volumetric approach in which the data base is composed of solid primitives such as blocks, wedges, cones, cylinders, and spheres. This is in contrast to the X, Y, and Z connected-line approach of most available software. All three-dimensional software is time-consuming to load. It is also more expensive than two-dimensional software and requires more computer capacity. The facilities planner should be sure that the value received is worth the extra effort and expense. For most, it is probably not.

As noted earlier, the design of process plants and conveyor handling systems are two special cases where three-dimensional modeling is essential. Here,

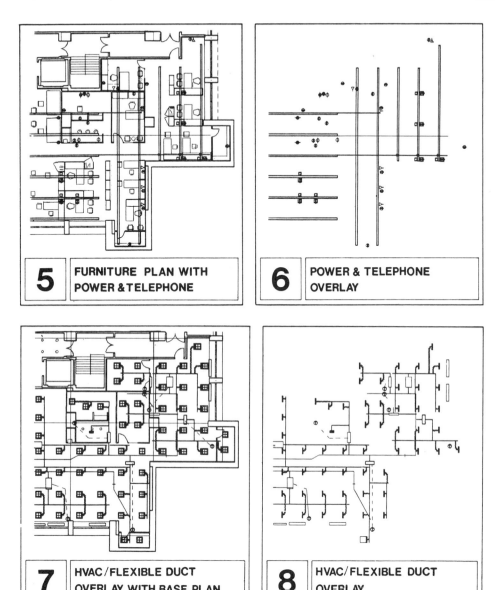

Figure 7.13. More CAD overlays. Routings such as these are best viewed on a color CAD system. Eight colors are usually sufficient for most overlay tasks. (Courtesy of Decision Graphics, Inc., Southborough, Massachusetts.)

CAD offers capabilities that are difficult or impossible to achieve with traditional physical models (11). The most valuable capabilities are summarized in Figure 7.22.

CAD software that provides the features we have discussed requires the user to master a great many operations, so many, in fact, that the system may be clumsy and difficult to use. The planner may have to spend several months of regular and even continuous use to become proficient. This is almost always the case with large and general purpose systems. It is less true of those that are specifically oriented toward architecture and facilities planning.

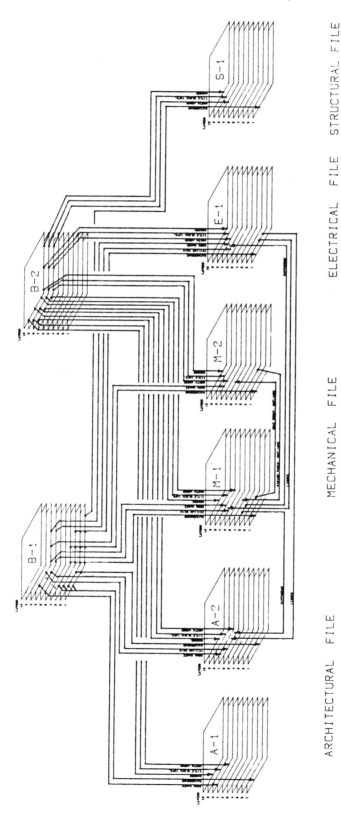

Figure 7.14. Visualization of a drawing data base. File B-1 contains the background (building, walls, doors, windows, columns, etc.), the ceiling grid, and room names. File B-2 contains the north arrow, keyplan, scale, title block, and border. The contents of base files B-1 and B-2 are "copied" into the various discipline sheets, using a standard layering convention. Discipline-specific data also can be copied into several sheets. The lighting layout, for example, may be shared between the electrical and architectural files. (Courtesy of O'Connor Consulting, Inc., Union Lake, Michigan.)

GRAPHIC DESCRIPTION FORM

CLIENT # 1000	PROJECT		HBA#

SET UP			DESCRIPTION		BORROWED DRWG.	
LAYER	PEN	TEXT			TYPE	LIBRARY #
1	V		BACKGROUND LAYERS 1-4		S	B-1
2	N		DUCTWORK LAYOUT			
3	B		EQUIP. OUT-LINES & V.A.V. REHEAT BOXES			
4	N		CEILING DIFFUSERS			
5	D / N / D		EQUIP. INSIDE LINES DETAIL / GRILLES & REGISTERS / OUTLINE OF PUMPS, BOILERS, ETC. LAYER 3		S	M-2
6	D	3/32	NOTES			
7	D		CEILING GRID LAYER 7 / LIGHTING LAYOUT LAYER 2		S / S	B-1 / E-1
8			ROOM NAMES & SHEET TITLE FREEZE LAYER 8 / NORTH ARROW, KEYPLAN,& SCALE FREEZE LAYER 8		S / S	B-1 / B-2
9	N	1/8"	TITLE BLOCK INFORMATION STANDARD LAYER 9 / TITLE BLOCK INFO.		S	B-2
10			TITLE BLOCK & BORDER LAYER 10		S	B-2

PENS
V- VERY DIM
D- DIM
N- NORMAL
B- BRIGHT

DRWG. TYPES
C- COMBINE
S- SHAPED
D- DIGITIZED

PLOTTING INSTRUCTIONS:

WORKING DRAWINGS - ALL LAYERS,BALL POINT

FINAL PLOT - LAYERS 1-6,8-9 INK ON OWNERS MYLAR

Figure 7.15. Control form for sheet M-1. This form shows the origins of borrowed information and the layer into which it is placed. Within each layer, several pen types may be used to distinguish between graphic elements or achieve certain display and plotting effects. Note the instruction to omit the ceiling grid (Layer 7) on final plots. (Courtesy of O'Connor Consulting, Inc., Union Lake, Michigan.)

SHEET INDEX

SHEET	TITLE	ACTIVE LAYERS
A- 1	COVER/TITLE SHEET	1,49
A- 2	SITE PLAN	49,104
A- 3	BLDG. EXISTINGAND DEMOLITION PLAN	2,3,49
A- 4		
A- 5		
A- 6		
A- 7	BLDG. FLOOR PLAN WITH WALL AND FLOOR SCHEDULE	5,6,11,16,17,18,26,49
A- 8	BLDG. FLOOR PLAN WITH CONSTRUCTION DIMENSIONS	5,6,9,11,18,19,49,200
A- 9	BLDG. REFLECTED CEILING PLAN	5,6,14,15,18,19,49
A-10	BLDG. FLOOR PLAN WITH DOOR AND WINDOW LAYOUT	5,6,18,19,30,49
A-11	DOOR SCHEDULE AND SPECIFICATIONS	28,49
A-12	WINDOW SCHEDULE AND SPECIFICATIONS	29,49
A-13	BLDG. FLOOR PLAN WITH FINISH SCHEDULE	27,49

LAYER INDEX

LAYER	LAYER NAME	USED ON SHEETS
1	TITLE SHEET	1
2	EXISTING BLDG. PLAN WITH COLUMNS LOCATION	3
3	EXISTING BLDG. OFFICE PLAN	3
4	BLDG. REFFRENCE BAY LINES	7,8,9,10
5	CONSTRUCTION BLDG. PLAN	7,8,9,10
6	CONSTRUCTION BLDG. OFFICE PLAN	
7	DOCK CONSTRUCTION PLAN	
8		
9	DIMENSIONS	
10	FURNITURE	7,8,
11	REVISIONS NOTES	7,8
12	LUNCH ROOM FURNITURE	
13	COMPUTER ROOM HARDWARE	
14	FIRE EXIT SIGNS	9
15	REFLECTIVE CEILING	9
16	RAISED FLOOR	7
17	RMA SPECS.	7
18	DEPARTMENT NAMES	7,8,9,10
19	ROOM NUMBERS	8,9,10
20	TEST & INTEGRATION RACKS	
21	WAREHOUSE RACKS	
22	SMALL PARTS STORAGE RACKS	
23	PICK, PACK & COLLATE	
24	TEST & BURN-IN RACKS	
25	FORKLIFTS (4)	
26	WALL SCHEDULE	7
27	FINISH SCHEDULE	13
28	DOOR SCHEDULE	11
29	WINDOW SCHEDULE	12
30	DOORS & WINDOWS LAYOUT	10
31	REPAIR AND RETURNS	
32	PREPRODUCTION	
49	1901 TITLE BLOCK	1,2,3,7,8,9,10,11,12,13
50	TITLE BLOCK	
51	BAY LINE DESIGNATIONS	
52	AREA BY DEPARTMENTS	
75	COLUMN CENTER LINE REFERENCE (EAST TO WEST)	
100	SITE PLAN WITH PARKING SPACES	
101	SITE PLAN DIMENSIONS (INCHES)	
102	SITE MCAUTO SIGN	
103	K19-1 BLDG. SITE PLAN	
104	(PROPOSED SITE PLAN)	2
200	SHORT CENTER LINE FOR COLUMNS	3
240	WALL SPECS.	
250	BAY LINE SPACING DIMENSIONS (INCHES)	
255	SHEET INDEX	
256	LAYER INDEX	

Figure 7.16. Examples of sheet and layer index for a plant layout project.
(Courtesy of McAuto, St. Louis, Missouri.)

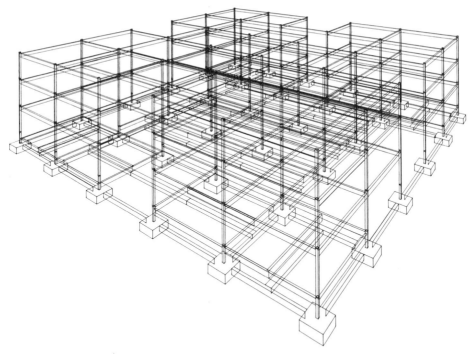

Figure 7.17. Visualization of steel framing. (Courtesy of Applied Research of Cambridge, Ltd., England.)

Virtually all CAD software is "menu-driven." At any point, the user has displayed in front of him a variety of system commands, graphic functions, and symbols. These are typically displayed in one of several ways:

1. *Taped-down paper or paperboard*: A printed display is placed in a known location on a digitizing tablet. By touching a stylus or cursor to the desired portion of the display, an action takes place.
2. *On the graphics screen*: A few permissible commands or functions may appear on the graphics display screen to the side or underneath the drawing image. By "touching" these with a cursor, action takes place. One action is typically to display other additional menu selections since only a few can be shown at any one time.
3. *On a special menu device*: Special function keys, dedicated display screen or other input device. By touching with a finger or light-pen, actions take place.

System commands initiate basic activities such as plotting, saving a drawing to disk, deleting a file, naming a file, etc. Function commands control the images on the display. These commands include such actions as creating, moving, and deleting a line or point; changing scales, displaying a grid, zooming, panning, and scrolling. Finally, symbol menus retrieve pre-entered symbols from storage and display them on the screen.

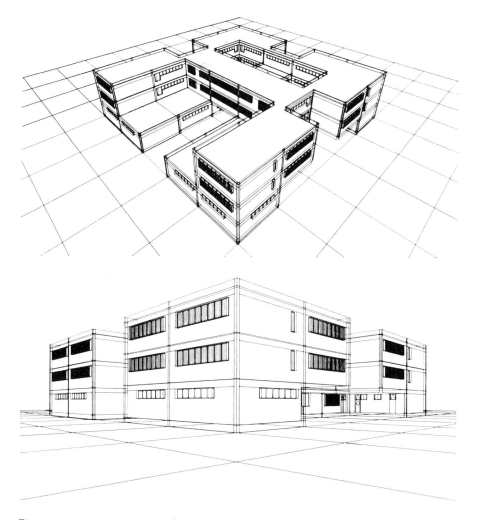

Figure 7.18. Two views of a building with most hidden lines removed. (Courtesy of Applied Research of Cambridge, Ltd., England.)

Outwardly, many menus look alike. And upon first inspection, the operation of many CAD systems appears the same. The viewer should remember, however, that the system is typically being shown by a trained, experienced user. The prospective purchaser should set aside time to examine the structure of commands and functions, and the resulting complexity or simplicity of the menus themselves. Four examples of menus are shown in Figures 7.23 through 7.26. Note the extensive vocabulary which must be mastered to use a CAD system. Each figure represents just one of many menus that are necessary to make full use of the system.

Large and general purpose systems typically have a set of very basic function commands. Often, to perform a specific drafting task, several of these commands must be used in sequence before any action takes place on

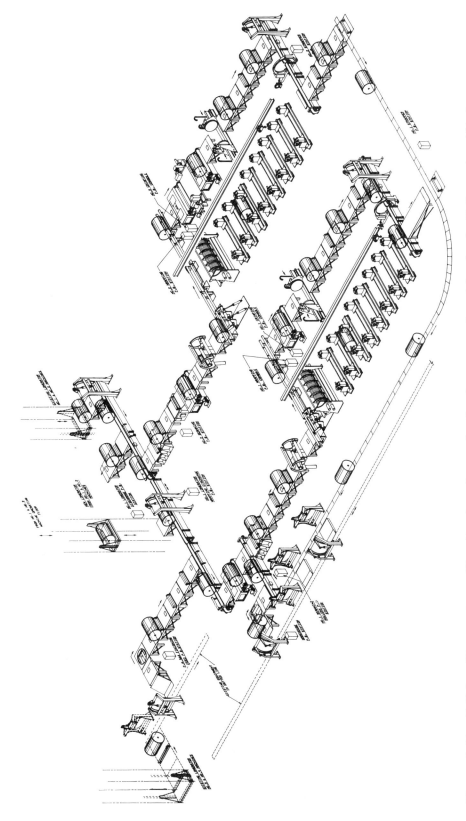

Figure 7.19. Roll-handling system. Visualization of a production layout, produced by Sandwell Computer-Aided Drafting, Atlanta, Georgia.

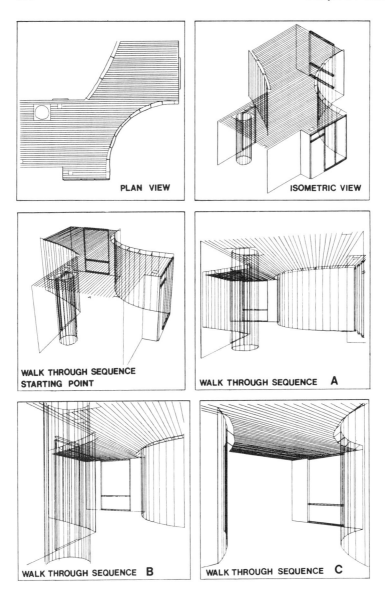

Figure 7.20. Two views and a walk through sequence. (Courtesy of Decision Graphics, Inc., Southborough, Massachusetts.)

the display. If such tasks and sequences are repetitive, operation of the system becomes tedious and prone to error. In such cases, it is desirable to create what is known as a "macro." This is a single command defined by the user that executes automatically a series of more basic commands provided by the system. "Macro-ing" is the capability of the user to string together any available commands, dynamically, while working on a drawing. If the software and its function commands are not architecturally or facilities-oriented,

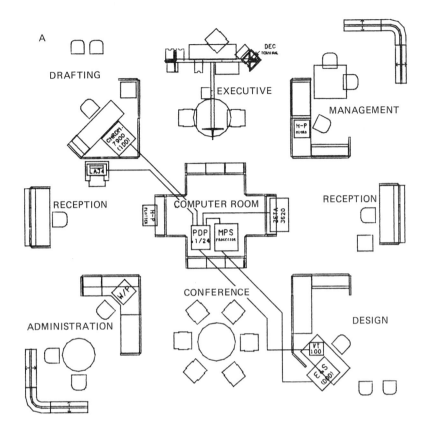

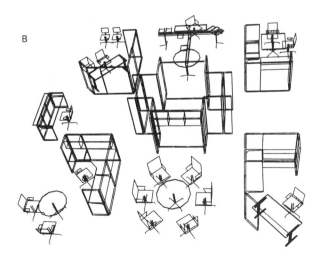

Figure 7.21. Plan (A) and perspective (B) of a furniture layout. (Courtesy of CORE, Herman Miller, Inc., Grandville, Michigan.)

1. Hard interferences -- detection of conflicts in the
    placement of pipes, ducts, trays, conduits, and
    structural members. "Un-buildable" conditions.

2. Soft interferences -- detection of inadequate spacing
    and clearance for: equipment installation and removal,
    operator and maintenance access, doorways, vehicles, etc.

3. Space allocation -- definition of zones reserved or
    occupied by a specific discipline: structural, elec-
    trical, piping of various types, etc.

4. Special studies -- production of composite drawings
    to show progress, scheduled completions for individual
    components and structures. Also sketches and views to
    aid in field engineering and construction.

5. Separation requirements and tolerances -- to verify that
    proper distances have been achieved between critical
    systems and components. Also that necessary separating
    structures such as concrete walls are present.

6. Temperature and operational displacements -- modeling
    extreme positions of components under "hot and cold,"
    "on and off" conditions.

7. Alignment and conductivity checks -- insuring proper
    placement of pipes and pipe hangers, anchor plates,
    equipment and its foundations, pipe-to-pipe connections,
    and in general, all interfaces between disciplines.

8. Erection and equipment rigging -- modeling the rigging
    paths for major equipment to insure that proper clearances
    have been provided.

9. Jet impingement analysis -- modeling the fluid jet and
    resulting impact that could occur if pipes break or
    rupture.

Figure 7.22. Uses of CAD in process plant design. (From Ref. 11.)

the ability to create macros is essential. Facilities-oriented commands should
directly address the creation and manipulation of such features as walls, partitions, doors, windows, stairs, and the like.
    Some useful support capabilities in CAD software include:

1. *Plotting controls:* which allow the planner to plot only the latest re-
    visions or a windowed portion of an entire drawing. Also the ability
    to plot in "background," i.e., while the display is in simultaneous
    use for drafting.
2. *Job accounting:* to keep track of the time spent by each user on each
    drawing and by all users on a given project.
3. *Protection and recovery from mistakes:* special pauses to keep the in-
    experienced user from making a serious mistake, especially inadvertent
    erasures. Some ability to recover if a mistake is made.

EJECT	NO	OPERATION COMPLETE	REGENERATE DISPLAY	RETURN TO TOP LEVEL	WIDER SELECTION	PLOT							
3	4	5	6	7	8	9	10	11	12	13	14	15	16
C	D	E	F	G	H	I	J	K	L	M		SPACE	O
P	Q	R	S	T	U	V	W	X	Y	Z		RETURN	5
5/b	%	.	,	∇	&	<	>	(	)	#		BACK SPACE	.

BLANK NBLANK	DELETE	FILE/ TERMINATE	TABLET MANAGEMENT	DATA BASE MANAGEMENT	DISPLAY CONTROL	DATA VERIFY/ VAR MGMT
LINE			**CIRCLE**		**CURVES**	
NTER OORDS	SCREEN POSITION	ENTER COORDS	SCREEN POSITION	ENTER CENTER	CONICS	TRIANGLE
ELTA	JOIN	TANGENT TO	CENTER POINT RADIUS	CENTER POINT TANGENT CURVE	SPLINE	RECTANGLE
ARC ENTER	HORIZ/VERT OR AXIS	POLAR	CENTER POINT EDGE POINT	CENTER POINT 2 EDGE POINTS	STRING	N-GON
AT A RAMETER	PARALLEL THRU POINT	PARALLEL DISTANCE	THROUGH 3 EDGE POINTS	EDGE POINT TANGENT CURVE	OFFSET CURVE	TAPERED OFFSET SPLINE
ITERSECT CURVES	PERP TO LINE	CHAMFER	FILLET	INSCRIBED IN THREE LINES	CONVERT POLYGON TO LINES	CONVERT CURVE TO POLYGON
ULTIPLE POINTS	N-SEG	INFINITE STATUS	NO CENTER INDICATION	WITH CENTER INDICATION	GROUP	MIRROR
			FILLET			
IK CHAIN ROM ALL	**DELETE** SINGLE FROM ALL	CHAIN FROM ALL	MULTIPLE CHAIN SELECT	EASY	TRANSLATE	ROTATE
OM ALL OUT REGION	FROM ALL IN REGION	FROM ALL OUT OF REGION	**TRIM/EXTEND** ONE END	TWO ENDS	DUPLICATE AND TRANSLATE	DUPLICATE AND ROTATE
LEVELS	ALL DISPLAYED POINTS	ALL NOT DISPLAYED POINTS	MIDDLE	2 CURVES AT INTERSECT	STRETCH	DI HORIZONTAL
NK LEVELS	ALL POINTS	LAST ENTITY	EASY TRIM	EASY EXTEND	RETRIEVE A PATTERN	CHAINED HORIZONTAL
ET FONT SOLID	RENAME CURRENT PART	LIST ON-LINE PART FILE	DELETE A PART	PART MERGE	CREATE A PATTERN	ANGULAR
	VIEW CONSTRUCTION				**DRAFTING**	
DASHED	SIMPLE PROJECTION	ISOMETRIC PROJECTION	SHAFT END VIEW	SECTION	TEXT/ ARROW	TEXT SIZE
PHANTOM	ENTER SCALE	**ZOOM** NEW CENTER	NEW ORIGIN	RETURN TO BASE SCALE	JUSTIFY MODE	DECIMAL PLACES
		DIAGONAL	AUTO			

Figure 7.23. Close-up of a drafting menu. (From *Computers for Design and Construction*, Vol. 1, No. 1, 1982.)

4. *Quick and easy back-up:* so that inordinate amounts of time are not spent archiving and creating back-up copies.
5. *Interfaces to other systems:* to bring in or send out data, and to use other computer resources through the medium of the CAD display.

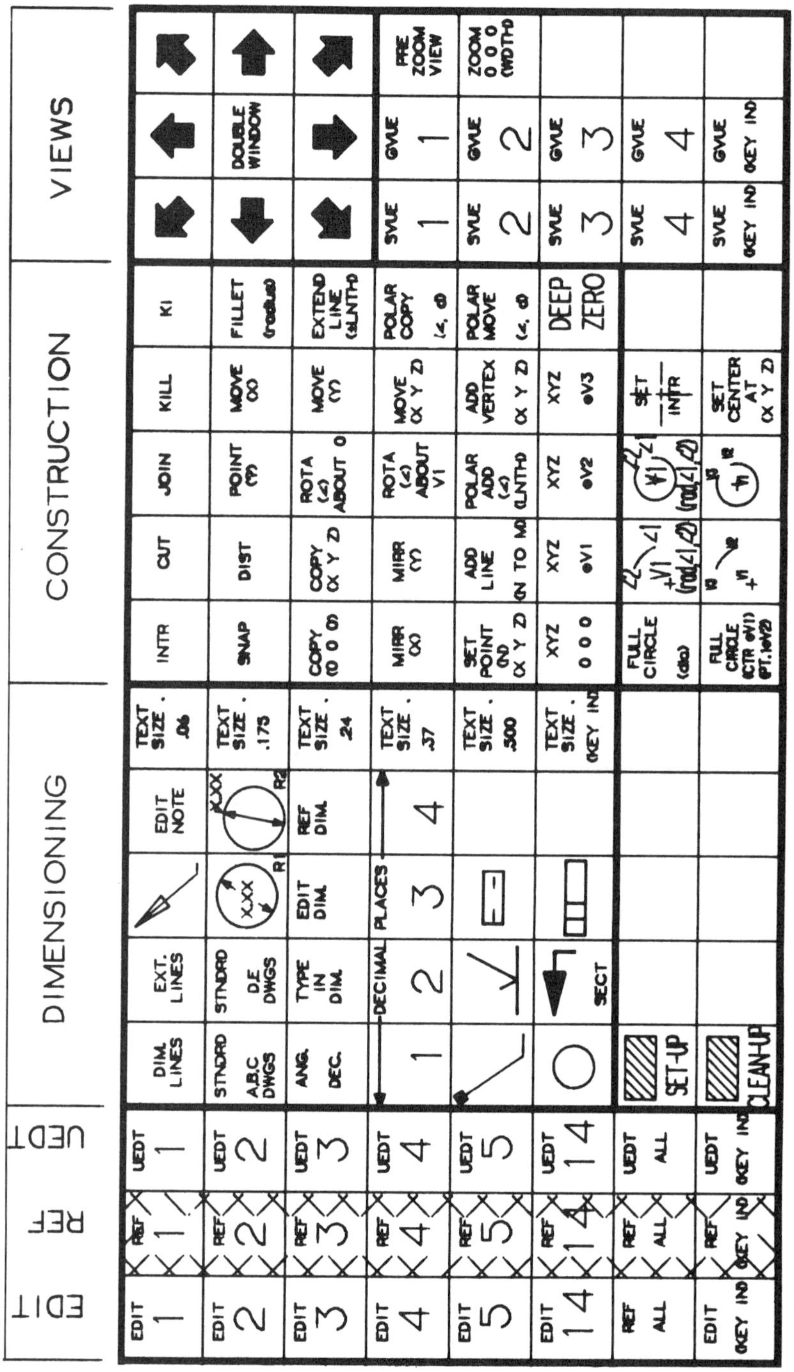

Figure 7.24. User-defined menu of dimensioning and drafting commands. (Courtesy of Applicon, Inc., Burlington, Massachusetts.)

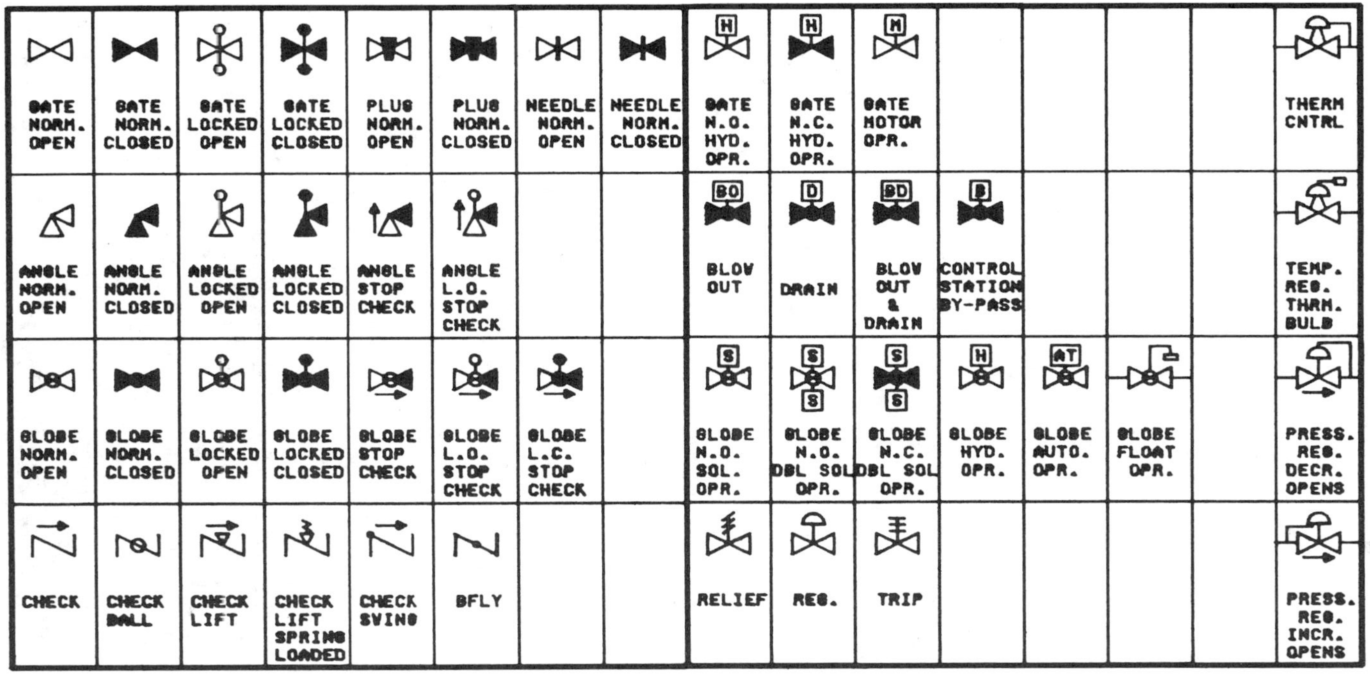

Figure 7.25. Menu of ANSI standard symbols. (Courtesy of Applicon, Inc., Burlington, Massachusetts.)

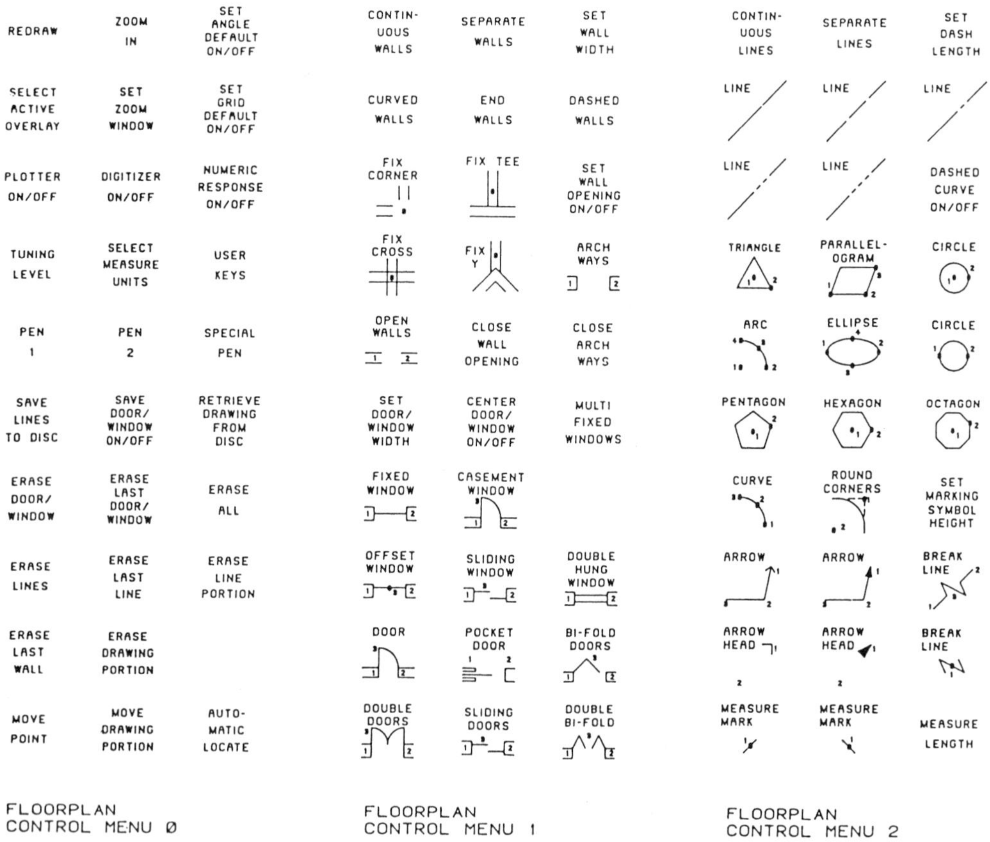

REDRAW   ZOOM IN   SET ANGLE DEFAULT ON/OFF   CONTIN-UOUS WALLS   SEPARATE WALLS   SET WALL WIDTH   CONTIN-UOUS LINES   SEPARATE LINES   SET DASH LENGTH

SELECT ACTIVE OVERLAY   SET ZOOM WINDOW   SET GRID DEFAULT ON/OFF   CURVED WALLS   END WALLS   DASHED WALLS   LINE   LINE   LINE

PLOTTER ON/OFF   DIGITIZER ON/OFF   NUMERIC RESPONSE ON/OFF   FIX CORNER   FIX TEE   SET WALL OPENING ON/OFF   LINE   LINE   DASHED CURVE ON/OFF

TUNING LEVEL   SELECT MEASURE UNITS   USER KEYS   FIX CROSS   FIX Y   ARCH WAYS   TRIANGLE   PARALLEL-OGRAM   CIRCLE

PEN 1   PEN 2   SPECIAL PEN   OPEN WALLS   CLOSE WALL OPENING   CLOSE ARCH WAYS   ARC   ELLIPSE   CIRCLE

SAVE LINES TO DISC   SAVE DOOR/WINDOW ON/OFF   RETRIEVE DRAWING FROM DISC   SET DOOR/WINDOW WIDTH   CENTER DOOR/WINDOW ON/OFF   MULTI FIXED WINDOWS   PENTAGON   HEXAGON   OCTAGON

ERASE DOOR/WINDOW   ERASE LAST DOOR/WINDOW   ERASE ALL   FIXED WINDOW   CASEMENT WINDOW   CURVE   ROUND CORNERS   SET MARKING SYMBOL HEIGHT

ERASE LINES   ERASE LAST LINE   ERASE LINE PORTION   OFFSET WINDOW   SLIDING WINDOW   DOUBLE HUNG WINDOW   ARROW   ARROW   BREAK LINE

ERASE LAST WALL   ERASE DRAWING PORTION   DOOR   POCKET DOOR   BI-FOLD DOORS   ARROW HEAD   ARROW HEAD   BREAK LINE

MOVE POINT   MOVE DRAWING PORTION   AUTO-MATIC LOCATE   DOUBLE DOORS   SLIDING DOORS   DOUBLE BI-FOLD   MEASURE MARK   MEASURE MARK   MEASURE LENGTH

FLOORPLAN CONTROL MENU 0   FLOORPLAN CONTROL MENU 1   FLOORPLAN CONTROL MENU 2

Figure 7.26. Command menus for floorplan drafting. Note the tailored commands on Menu 1. Such commands are of great value to facilities planners. Creation of floor plans using general purpose commands (Figure 7.23) is tedious at best. (Courtesy of Arrigoni Computer Graphics, Los Gatos, California.)

6. *Non-CAD software*: for such tasks as forecasting space requirements, managing leases, project scheduling, maintaining equipment inventories. (If the CAD system is fully used for drafting, running such software may have no value and may even cause problems.)

While these features are not necessarily essential, many good CAD systems do include them.

Common CAD Configurations

In spite of the great variety of devices for input, output, and display, CAD systems are configured in a limited number of ways. These are pictured in Figures 7.27 through 7.29.

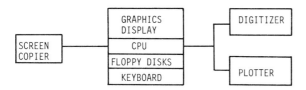

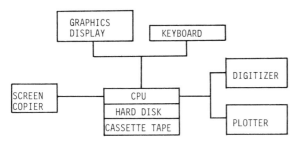

Figure 7.27. Single-station systems may have integral displays and keyboards, or they may consist of separate components.

1. *Single-station, standalone systems*: These are duplicated rather than expanded.
2. *Two-station, shared-resource systems*: These share some combination of storage, peripherals, and processing devices. They may be expanded to a modest degree.
3. *Large-scale systems*: Using dedicated minicomputers, large mainframes, or local area networks.

Each turnkey vendor, and any in-house development effort, will deliver some variation of these basic configurations. Each has its place, which is a function of budget, desired performance, and the required number of initial and future display stations.

Sourcing CAD Systems

Turnkey vendors are the most common source of CAD systems. Over 40 companies offer systems to architects and facilities planners. Choosing the best one is difficult. We will review the selection and installation process thoroughly in Chapters 9 and 10.

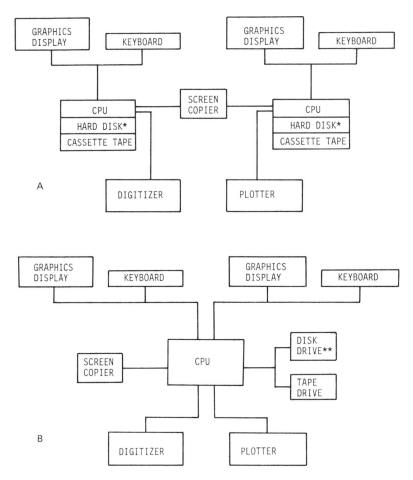

Figure 7.28. Two-station systems may be constructed with separate personal or microcomputers sharing peripheral devices (A). Or, they may consist of a central minicomputer supporting both stations and all peripherals (B). In the former approach, it is common to dedicate one station to input (digitizing) and one to output (plotting). Data is shared by passing tape cassettes (or floppy disks) between the two stations. *5-1/4 in. fixed. **8 in., 14 in. removable.

With the growing use of CAD systems, an increasing number of experienced operators, systems programmers, and consultants are available. More firms are hiring such experienced specialists to put together CAD systems in-house. By acquiring software and a display, and connecting it to an existing mainframe or minicomputer, it is possible to get started with one or two stations at a very low cost. However, such moves must be planned carefully with future growth in mind, and with the active participation of the data processing staff. The ultimate costs of in-house development may be quite high when compared to the turnkey approach.

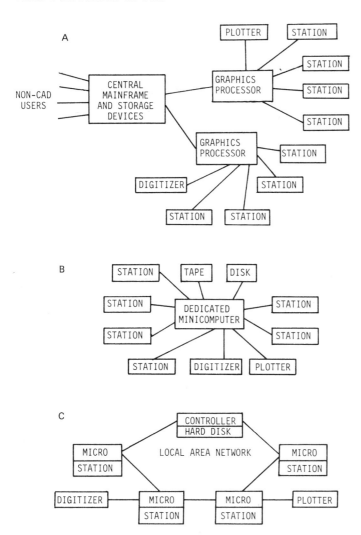

Figure 7.29. Three approaches to large, multi-station systems. The local area network, C, is emerging as the preferred approach since it avoids degradation. In the other approaches, A and B, response deteriorates as stations contend for access to the CPU. A large (300 Mb or more) hard disk is placed in the network to contain the central data base. Portions are sent as needed to the disk at each requesting station. Data may also be passed from one station to any others.

Future Directions in CAD

While CAD is a maturing technology, leading suppliers of hardware and software will continue to announce major improvements. The most significant developments in the next 5 years are likely to include:

1. *Easier input and operation*: Using higher-level command languages, voice, and touch-sensitive controls. The dropping costs of microprocessors and internal memories will permit a significant investment of system resources to ease-of-use.
2. *More sophisticated color imagery*: At ever lower cost. Solid fill and shading and three-dimensional solids visualizations will become common. Planners will learn to work and possibly become dependent upon the use of color.
3. *Flat display devices*: Using plasma discharge technology. Together with increasing miniaturization and compactness of processing units, the flat display will result in much smaller, less obtrusive systems.
4. *Improved electrostatic plotting*: In high-resolution color, at a reasonable price. Once planners and designers come to accept the appearance of electrostatic output, this could become the most popular form of plotting.
5. *A virtual end to storage limitations*: The plummeting cost of hard disks, the perfection of video disks, and breakthroughs in memory technology will greatly simplify or eliminate problems of archiving and data storage.
6. *More powerful applications software*: Three-dimensional data bases with solids modeling will become standard. So too will the integration of computer-aided engineering (calculations) with CAD graphics for input and output. Vendors will offer a steady stream of new capabilities, such as freehand sketching now available from Formtek, and 3-D robot simulation from McAuto.
7. *Networks will become standard*: For new CAD installations. Operating systems and networking software will be enhanced. Extensive sharing of data files and peripherals--even computing capacity--between intelligent stations will be common and less expensive.
8. *Less use of turnkey vendors*: And more in-house development will occur as the number of experienced CAD professionals grows. The purchase and in-house installation of third-party CAD software will be common. In response, turnkey vendors will "unbundle" and sell their software separately.
9. *Less expensive digitizers and plotters*: Will help reduce overall system cost. At present many large digitizers and plotters give unnecessary accuracy at unnecessary expense for facilities planning applications. Peripherals are now 50% of some systems' prices.
10. *Better, faster training*: Will reduce learning time from three to six months to as little as one or two. On-site, programmed learning (possibly using videodisks) or other "online" approaches will take the place of two- and three-week schools at vendor locations.

The trade shows, directories, and consultants listed in the Appendixes will help the interested planner stay abreast of these and other developments as they occur in the years ahead.

8

Management Information Systems (MIS) Applications

When this book was first conceived in 1980, this chapter was to cover main-frame and large minicomputer applications. The information systems to be reviewed are all multi-user systems. Several require 10 Mb or more of online data storage and 500K of RAM for data manipulation. At that time, such capabilities were generally limited to mainframes and minis. As we have discussed in Chapters 4 and 6, such capabilities can be had today on small microcomputers. Most commercially available MIS software is still written for larger computers--at least this is true of the facilities-oriented systems described here. In the next few years however, software suppliers will convert virtually all of their systems to run on microcomputers.

The real purpose of this chapter is to establish the role of large information systems--not large computers--in facilities planning and management. The characteristics of a large information system include:

1. *Data base management*: at least some of the features of a DBMS as described earlier in Chapter 4
2. *Standard reports*: a wide variety of regular reports, with multiple uses and wide dissemination
3. *Clerical users*: typically performing updates and queries, and running reports
4. *Ad hoc reports*: the ability to produce nonstandard reports on a one-time, special purpose basis
5. *Operational uses*: controlling labor and funds, paying rent, cost accounting
6. *Links to non-facilities groups*: and their data bases, typically accounting and personnel

Large information systems are used to manage and control large amounts of manpower, financial, and physical resources. For this reason, development and acquisition costs can range from $50,000 to $200,000 or more, and still be justified if the system improves the organization's performance. This is most definitely *not* personal computing within the facilities planning organization. Large systems are truly part of corporate data processing.

We will examine momentarily several large information systems, each devoted to a different aspect of facilities planning and management. Yet the goal of each is the improved utilization of resources. To achieve this goal may require significant organizational changes and major commitments to standard terminologies and procedures. It may take a year or more to fully install such systems and get meaningful results. But once installed, a large information system should remain in use for years, until major changes in organization or facilities render it obsolete. In keeping with our practice in previous chapters, the software here will be reviewed in order of increasing sophistication and complexity.

Project Management

In its broadest definition, project management includes the following activities:

1. Budgeting
2. Scheduling of events, manpower, and equipment
3. Bidding and purchasing
4. Order and invoice processing
5. Payroll
6. Accounting
7. Progress reporting

On major facilities projects such as construction of a power plant, a mass transit system, or a dam, the project management system is *the* data processing resource for the contractors and the project's sponsors. If the planner's corporation acts as its own general contractor or construction manager, then a full-scale system with the features above is probably in order. But the needs of most corporate facilities groups are less sophisticated, since the projects are smaller and much of the necessary data processing will be performed by contractors or an outside construction management firm.

Most facilities projects are quite small in scope--a departmental rearrangement or move, the remodeling of a floor, or perhaps a maintenance effort. The work may be done largely by company personnel. There may be no bidding on outside work, no special payroll, and few, if any, invoices. In these routine projects, the facilities planner or manager wants to schedule the work, report progress, account for the manhours of company personnel, and record a few expense items. The planner's organization may already have a computer-based system for processing work orders. Existing purchasing and accounts payable systems may take care of the outside expense items. In this situation, the only remaining need for computer aids is in scheduling and progress reporting. In fact, the COMPUCHART system, reviewed in Chapter 6, may be all that is required. We mentioned in passing, back in Chapter 6, that facilities planners may be able to adapt the project scheduling and reporting systems used by their data processing groups. These are typically mainframe or mini-computer-based, and could be made available on a time-sharing basis. These systems are manpower- and data-processing-oriented, but for the planner who is short on funds for new software they represent an inexpensive way to get started. An example of such a system appears in Figures 8.1 and 8.2. The

```
WEDNESDAY, MAY 19, 1974 .RUN 212                    QUICK-TROL                          07:42 AM  PAGE  16
REPORT PROJECT STATUS ID(AR201)              PROJECT GENERAL STATUS REPORT
PROJECT TITLE: NEW ACCOUNTS RECEIVABLE       CUSTOMER ID: ACCT DEPT        ORGANIZATION: DATA PROCESSING SYSTEMS
PROJECT ID:    AR-01                         PROJECT MODE: IN PROGRESS     MANAGER:      JACK MAPLETON
                                                                                         QUALITY DATA PRODUCTS, INC.
```

				PERSONNEL *EQUIPMENT*			*PLANNED* *ACTUAL*		*TIME SCHEDULE*				*PLANNED/ACTUAL CALENDAR DATES*					
PHASE	TASK	NODE	TITLE	HRS PLAN	HRS USED	%	MISCEL-LANEOUS	TEST RUNS	DAYS PLAN	DAYS PAST	DAYS LEFT	%	START	COMPLETE	EST % COMP	PLANNED BUDGET	ACTUAL EXPENSE	%
001		(1)	NEW ACCOUNTS RECEIVABLE															
			PROGRAM DESIGN	400 / 25	493 / 24	123 / 96	350 / 650						10/07/73	12/18/73		3275	4615	141
	D100	(2)	GENERAL FLOW	160 / 40	171 / 35	106	700 / 1012		46				10/01/73 10/07/73	11/15/73 11/17/73		2140	2547	119
	D200	(3)	MODULE DESIGN	120	118	97	700 / 577		42				10/21/73 10/21/73	12/01/73 11/28/73		1720	1576	92
	D300		INTERFACE DESIGN	120	140	116	550 / 1171		45				11/01/73 11/04/73	12/15/73 12/16/73		1630	2426	149
			* PHASE TOTAL *	800 / 25	921 / 24	115 / 96	2300 / 3410		76				10/01/73 10/07/73	12/15/73 12/18/73		8765	11164	127
002		(4)	PROGRAM DEVELOPMENT	320 / 200	301 / 207	96 / 103	500 / 462		74	82		110	12/18/73 12/23/73	03/01/74		3900	3800	97
	C100		DATA BASE MANAGEMENT	600 / 40	616 / 35	102 / 86	200 / 157	254	65				12/18/73 12/23/73	02/20/74 02/22/74		16550	15,026	91
	C200		TRANSACTION EXECUTORS	580 / 50	585 / 49	100 / 98	200 / 252	316	118	82	36	69	12/18/73 12/23/73	04/14/74	65	19260	19078	99
	C300		REPORT GENERATION	340 / 40	342 / 44	100 / 109	200 / 213	208	63	68		113	01/01/74 12/23/73	03/01/74	90	14410	15559	108
			* PHASE TOTAL *	1840 / 330	1843 / 334	100 / 101	1100 / 1084	778	113	82	36	69	12/18/73 12/23/73	04/14/74	62	54120	53463	99
003			IMPLEMENTATION	400					45				04/01/74	05/15/74		2880		
			* PROJECT TOTAL *	3040 / 355	2764 / 358	90 / 100	3400 / 4494	778	227	160	67	70	10/01/73 10/07/73	05/15/74	71	65765	64627	98

```
* * * * * *FOOTNOTES* * * * * *FOOTNOTES* * * * * *FOOTNOTES* * * * * *FOOTNOTES* * * * * *FOOTNOTES* * * * * *FOOTNOTES* * * * * *
*   (1)  SYSTEM UNDER DEVELOPMENT        (2)  SYSTEM AND DETAIL FLOWCHARTS NOW AVAILABLE.                                          *
*   (3)  MODULE DESIGN AND SPECIFICATIONS COMPLETE.   (4)  SLIPPAGE FROM 4/1 to 4/14 (SEE REVISION REPORT)                         *
* * * * * *FOOTNOTES* * * * * *FOOTNOTES* * * * * *FOOTNOTES* * * * * *FOOTNOTES* * * * * *FOOTNOTES* * * * * *FOOTNOTES* * * * * *
```

Figure 8.1. Data processing project management report, showing man-hours, expenses, and dates for each phase and task.

MONDAY..APRIL 18. 1974..RUN 197   QUICK-TROL   05:47 AM PAGE 35

REPORT PROJECT REVIEW

TO DATE PROJECT REVIEW   QUALITY DATA PRODUCTS, INC.

PROJECT ID	NOTE	PROJECT TITLE	*PERSONNEL* *EQUIPMENT*				*MISCELLANEOUS*			*PLANNED* *ACTUAL*		*PROJECT TOTAL*		
			PLAN'D HOURS	EXP'ND HOURS	RECORDED EXPND'URE	%	PLANNED BUDGET	RECORDED EXPND'URE	%	TEST RUNS	TEST RUNS	PLANNED BUDGET	RECORDED EXPND'URE	%
AR201	(1)	NEW ACCOUNTS RECEIVABLE	3040	2764	20718.50	90	3400.00	4493.72	132		778	65765.00	64627.01	98
			355	358	39414.79	100								
AR202		RJE TRANSIENT CONVERSION	1340	922	6876.88	68	100.00	1250.32	125	120	109	25400.00	20923.10	82
			50	43	12795.90	85								
AR203		PAYROLL PGM CONVERSION FOR PEP	1740	722	5273.03	41	1500.00	313.52	21	180	34	27300.00	8937.55	33
			45	11	3351.00	24								
AR204	(2)	GENERAL SYSTEM MAINTENANCE	1075	432	3240.00	40	2000.00	717.65	36	200	67	27462.50	13426.85	49
			60	32	9469.20	52								
AR205		FUTURE SYSTEMS STUDY	1786	1138	8532.75	63	2200.00	1753.65	80	125	150	28495.00	28286.40	99
			45	60	18000.00	133								

* * * * * * * * *FOOTNOTES* * * * * * * * * * * *FOOTNOTES* * * * * * * * * *FOOTNOTES* * * * * * * * * * *

(1) SYSTEM UNDER DEVELOPMENT   (2) PROJECT WAS CANCELLED BY APARS ON 10/12/73

* *

Figure 8.2. Data processing project management report, showing status of active projects in a given user department.

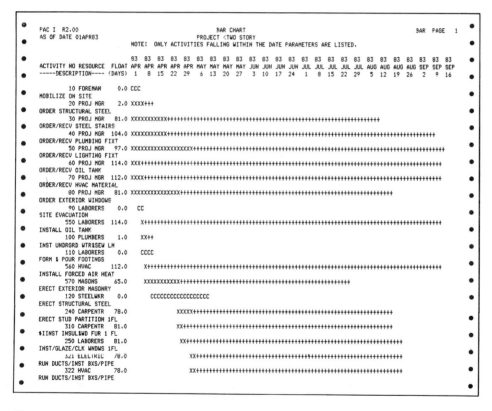

Figure 8.3. Critical path report and bar chart. Activities identified by a "C" are on the critical path. (Copyright 1983, AGS Management Systems, Inc. Further reproduction prohibited.)

reports shown are two of several produced by the Quick-Trol project management and billing system from Quality Data Products, Inc., of Spokane, Washington. The system tracks man-hours, expenses, and calendar progress by task, on multiple projects. Summaries can be made for all projects in the system (Figure 8.1) or for a department (Figure 8.2). Most of the routine space

```
PAC I  R2.00              ACTIVITY COST REPORT                      COST PAGE   1
AS OF DATE 01APR83        PROJECT <TWO STORY

ACTIVITY NO RESOURCE                                      ACTUAL                  %
-----DESCRIPTION------    ESTIMATED   PROJECTED           TO DATE    REMAINING  CPL

**PROJECT** PROJ MGR
TWO STORY STEEL & CONCRE       0.00   113208.00             0.00     113208.00    0

    10 FOREMAN
MOBILIZE ON SITE
    20 PROJ MGR                0.00      300.00             0.00        300.00    0
ORDER STRUCTURAL STEEL
    30 PROJ MGR                0.00      800.00             0.00        800.00    0
ORDER/RECV STEEL STAIRS
    40 PROJ MGR.               0.00     2080.00             0.00       2080.00    0
ORDER/RECV PLUMBING FIXT
    50 PROJ MGR .              0.00     1920.00             0.00       1920.00    0
ORDER/RECV LIGHTING FIXT
    60 PROJ MGR                0.00     3680.00             0.00       3680.00    0
ORDER/RECV OIL TANK
                               0.00      480.00             0.00        480.00    0
```

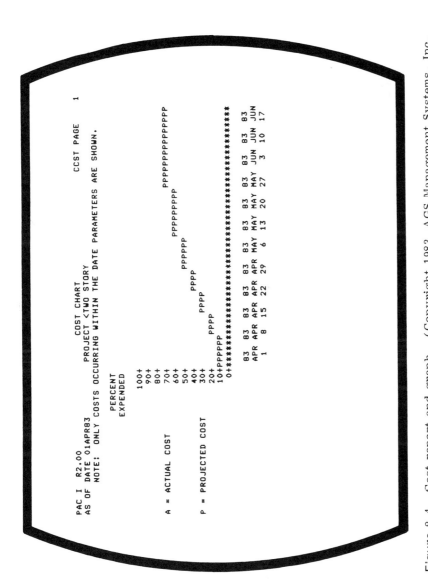

Figure 8.4. Cost report and graph. (Copyright 1983, AGS Management Systems, Inc. Further reproduction prohibited.)

```
PAC I  R2.00                          MASTER MENU                           MENU

        COMP (COMPUTE)    CNTS (COUNTS)      EJEC (EJECT)     EXIT (EXIT)
        HELP (HELP)       OPTS (OPTIONS)     PREV (PREVIOUS)  PRIN (PRINT)
        GLOS (GLOSSARY)   SAVE (SAVE)        SELE (SELECT)    SORT (SORT)
        TOTS (TOTALS)     TOTB (BREAK)       WRIT (WRITE)     EXAM (EXAMINE)


ENTER FUNCTION: PRIN

PAC I  R2.00                     SELECT FIELDS TO PRINT                     PRIN

ENTER FIELDS BY NUMBER, CODE, OR DESCRIPTION (N, C, OR D): D
ENTER FIRST      DESCRIPTION: PROJECT NAME
ENTER SECOND     DESCRIPTION: PROJECTED START DATE
ENTER THIRD      DESCRIPTION: PROJECTED FINISH DATE
ENTER FOURTH     DESCRIPTION: ESTIMATED TIME
ENTER FIFTH      DESCRIPTION: ACTUAL TIME TO DATE
ENTER SIXTH      DESCRIPTION: END
WIDTH OF REPORT IS NOW  62 CHARACTERS


ENTER FUNCTION: WRIT

PAC I  R2.00                      WRITE THE REPORT                          WRIT

ENTER OUTPUT FILE NAME: RW1.OUT
IS THIS A NEW FILE (Y/N)? Y

ENTER NAME OF NEXT PROJECT FILE: ENGINR.MAS

ENTER NAME OF NEXT PROJECT FILE: CLINIC.MAS

ENTER NAME OF NEXT PROJECT FILE: SAMPLE.MAS

ENTER NAME OF NEXT PROJECT FILE: END
```

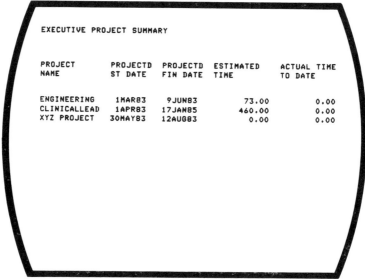

```
EXECUTIVE PROJECT SUMMARY


PROJECT        PROJECTD   PROJECTD   ESTIMATED    ACTUAL TIME
NAME           ST DATE    FIN DATE   TIME         TO DATE

ENGINEERING    1MAR83     9JUN83        73.00         0.00
CLINICALLEAD   1APR83     17JAN85      460.00         0.00
XYZ PROJECT    30MAY83    12AUG83        0.00         0.00
```

Figure 8.5. Interactive inquiry to determine the starting and finish dates, estimated and actual times for three active projects. (Copyright 1983, AGS Management Systems, Inc. Further reproduction prohibited.)

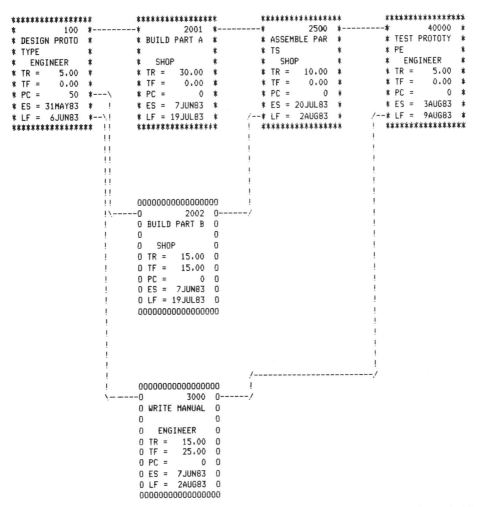

Figure 8.6. Critical path network using printer graphics. (Copyright 1983, AGS Management Systems, Inc. Further reproduction prohibited.)

planning provided to corporate departments could be tracked with such a system. Critical path calculations, however, are not provided. On larger facilities projects these are of great value.

Critical path calculations and diagrams are typically found only in architectural-, engineering- or construction-oriented systems. A good example is PAC I from AGS Management Systems, Inc., of King of Prussia, Pennsylvania. The major features of PAC I include:

1. Critical path calculation and bar charting, showing early and late start and finish, and float in days (Figure 8.3)
2. Cost and expenditure reporting with graphics as well as tabular reporting (Figure 8.4)
3. Interactive inquiry and report generation (Figure 8.5)
4. Printer graphics for critical path network diagramming (Figure 8.6)

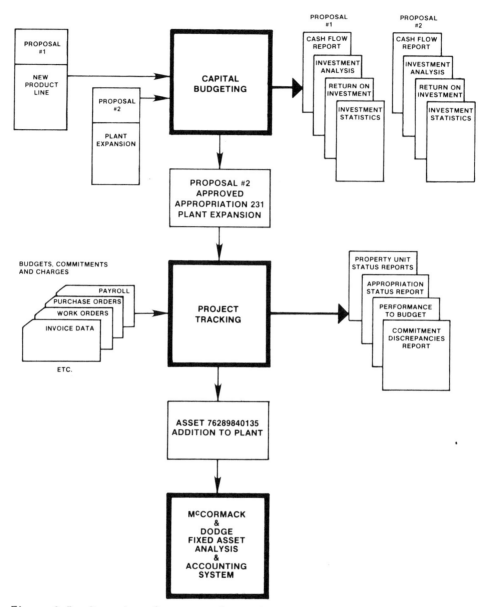

Figure 8.7. Overview of a system for capital budgeting and project tracking. (Copyright 1979, McCormack and Dodge Corporation.)

The PAC I system and others like it have the power to handle all but the largest construction projects. They are "planning oriented" with emphasis on scheduling and project reporting--features of most importance to facilities planners.

Some organizations with large capital budgets need information systems that keep track of capital appropriations. CPA Plus, from McCormack and Dodge Corporation, is an example of a system that addresses these accounting and financial needs. As shown in Figure 8.7, CPA Plus is a computer aid to the entire capital budgeting process. The project-tracking software relates to a

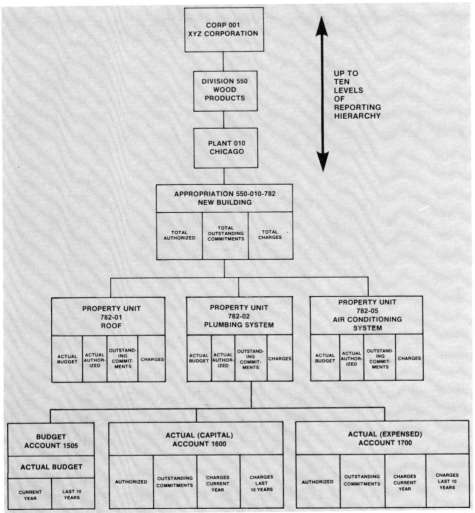

Figure 8.8. A view of the CPA Plus data base for project tracking. (Copyright 1979, McCormack and Dodge Corporation.)

companion system, FA Plus, for fixed asset management. The capital budgeting portion of the system provides discounted cash flow comparisons and project rankings valuable in choosing among competing facilities investments. Our interest here, however, is in project tracking--particularly the record-keeping and reporting structure shown in Figure 8.8.

Approved projects are entered into a master file. Each project is divided into a set of property units to simplify tracking and control. These units include such elements of the facility as: roof, plumbing system, air-conditioning system, etc. Budget amounts for each unit are entered by quarter or year. Up to 10 years of data may be entered. As the project progresses, spending commitments (purchase orders and work orders) are compared to actual expenses (paid invoices, payroll records) and to the latest project budget. Discrepancies are reported for management attention. Sample reports appear in Figure 8.9. A number of software suppliers offer systems similar to CPA Plus.

CORP--001 MCCORMACK-DODGE
CIX040

COMMITMENT DISCREPANCIES
RUN DATE 02/01/78

APPROPRIATION	PROPERTY UNIT	ACCOUNT	COMMITMENT	THIS CHARGE	COMMITTED AMOUNT	CHARGED AMOUNT
REQ-100	016	15C4-001	PC-663912334	INV-31234836	55,000.00	56,200.00
REQ-210	011	1510-010	PO-394877186	INV-17683297	700.00	1,115.00
REQ-217	013	1515-030	PO-487683219	INV-87781912	1,250.00	1,375.00
REQ-217	014	1515-030	PO-487683220	INV-79321685	1,250.00	1,375.00

EXHIBIT 10

APPROPRIATION STATUS VARIANCE REPORT   02/01/78

APPROPRIATION	PROJECT	ACCOUNT	AUTH. AMOUNT	EST. TO COMPLETE	OUTSTANDING COMMITMENTS	SPENT TO DATE	INDICATED TOTAL COST	VAR. FROM ORIGINAL
001-DIV03	AB-011	1503	400,000	400,000	200,000	200,000	800,000	400,000-
001-DIV03	AB-012	1500	150,000	25,000	75,000	50,000	150,000	0
APPROPRIATION TOTAL			550,000	425,000	275,000	250,000	950,000	400,000-
001-DIV05	CI-020		60,500	10,000	15,275	30,500	55,775	4,725-
001-DIV05	CI-021		140,000	113,000	40,100	54,600	167,000	27,000-
001-DIV05	CI-022		55,010	2,000		55,600	47,700	7,310
APPROPRIATION TOTAL			255,510	125,000	55,375	90,100	270,475	14,965-
002-PLTA1234567	123456	0123456789	100,000	20,000	15,000	65,000	100,000	0

VARIANCES ARE LISTED BY SUB-UNIT AND SUMMARIZED FOR TOTAL APPROPRIATION

EXHIBIT 11

CORP	APPROPRIATION	PROPERTY UNIT	AUTH. AMOUNT	OUTSTANDING COMMITMENTS	BUDGET TO DATE	SPENT TO DATE	VAR FROM BUDGET	SCHED. COMPLETION
MAD	010-12345	C0144	25,000	1,000	15,000	13,400	600	12-31-78
MAD	010-12345	00145	12,500	1,100	6,000	3,445	1,455	10-25-77
	TOTALS		37,500	2,100	21,000	16,845	2,055	
MAD	010-12346	00184	7,500	25	7,500	7,676	201-	05-01-76
MAD	010-12346	00186	8,100	2,343	8,100	5,750	7-	05-01-76
MAD	010-12346	00187	4,000	400	4,000	3,852	252-	10-11-77
	TOTALS		19,600	2,768	19,600	17,278	446-	
TOTAL			57,100	4,868	40,600	34,123	1,609	

CORP-001 McCORMACK-DODGE   PERFORMANCE TO BUDGET   02/01/78

EXHIBIT 12

Figure 8.9. Project-tracking reports showing commitments, charges, discrepancies, and budget variance. (Copyright 1979, McCormack and Dodge Corporation.)

Systems of this type can be installed on any popular mainframe or minicomputer. The amount of memory and the disk storage requirements need to be planned in advance by data processing professionals working with the facilities planner.

The three systems just reviewed are far superior to manual methods, which are slow, tedious, and much more prone to error. However, to use them the planner and his organization may have to institute a number of procedural controls, such as the routing of all documents through key entry. It may be necessary to develop a more formal approach to budgeting. And, more rigorous scheduling techniques such as critical path may also be required to get the full benefits of computer-based project management.

Fixed Asset Management

As suggested in Figure 8.7, there is a natural progression from project tracking to fixed asset management, once the newly constructed or acquired facilities are placed in service. At this point, the financial and accounting groups assume a variety of reporting requirements. These include:

1. *Additions to book value*: which include initial date of service, service life, and depreciation method. These are used for subsequent depreciation calculations during the life of the asset. An example of such a report is shown in Figure 8.10, taken from the PIMS Property Information and Management System by University Computing Corporation, Dallas, Texas.
2. *Replacement cost*: a mandatory reporting requirement in many organizations. This type of report incorporates appraisals, book value, and accumulated depreciation from various system files. Index calculations are provided to account for inflation. An example from PIMS is shown in Figure 8.11.
3. *Insurance planning*: can also be accomplished with book and replacement cost data contained in a fixed asset management system. The use of PIMS for this purpose is shown in Figure 8.12.
4. *Tax reporting*: required of all corporations to substantiate depreciation expenses and investment tax credits. Tax reports draw on the same data used elsewhere in the system for depreciation and replacement cost accounting. The example in Figure 8.13 is from the FA Plus system by McCormack and Dodge Corporation.
5. *Lease accounting*: for organizations that must report the capitalized value of leases. An example is shown in Figure 8.14.

While these reports are of primary value to the financial and accounting arms of the organization, the facilities planner can help to insure their accuracy and completeness. Coordination between facilities planners and accountants can eliminate redundant data entry and duplicate files of asset information. Also, the planner may find he is able to use some of the fixed asset system reports in lieu of creating or maintaining his own. A good example is the use of PIMS shown in Figure 8.15 for constructing preventive maintenance reports. With a fixed asset system installed on a mainframe, time-sharing system, this report could be produced in the facilities department while all others remain in accounting.

REQUEST NO. 104
REPORT NO.   110
RUN DATE 02/12/79
RUN TIME 10.08.40

COMPANY 100 NAME   100

BOOK ADDITIONS REPORT
JAN, 1977 THRU DEC, 1977

PAGE   1

ITEM NUMBER	UNIT NUMB	DESCRIPTION	ENT SERV	DEPR METH	EST LIFE	M T Y P C	I Y P G	A U C	N / U	BASIS CODES	ADDITION DATE	QUANT	BASIS AMOUNT	ACCUM DEPR THRU YEAR END	PRIOR DEPR
DIVISION	A001														
HEAVY EQUIPMENT	1000040														
100000043		HEAVY EQUIPMENT	3/77	SL	120	D	2	U	N	1101111	12/77	1	7,500.00	625.00	562.50
100000044		COMPUTER EQUIPT	4/77	SL	120	D	2	U	N	1101111	12/77	1	35,000.00	2,625.00	2,333.33
100000045		COPY MACHINE	4/77	SL	120	D	2	U	N	1101111	12/77	1	10,400.00	780.00	693.33
100000046		AIR CONDITIONER	4/77	SL	120	D	2	U	N	1101111	12/77	1	1,867.20	140.04	124.48
100000047		AUTO	5/77	SL	36	D	2	U	N	1101111	12/77	1	6,123.59	1,360.80	1,190.70
HEAVY EQUIPMENT	1000040 TOTALS												60,890.79	5,530.84	4,904.34
LIGHT EQUIPMENT	1000050														
100000048		TYPEWRITER	6/77	SL	96	D	2	U	N	1101111	12/77	1	250.00	18.23	15.63
100000049		CALCULATOR	7/77	SL	96	D	2	U	N	1101111	12/77	1	250.00	15.63	13.02
100000050		RECORDER	8/77	SL	96	D	2	U	N	1101111	12/77	1	250.00	13.02	10.42
LIGHT EQUIPMENT	1000050 TOTALS												750.00	46.88	39.07
FURNITURE	1000060														
100000051		FURNITURE	9/77	SL	120	D	2	U	N	1101111	12/77	1	3,568.00	118.93	89.20
100000052		DESK	9/77	SL	120	D	2	U	N	1101111	12/77	1	458.00	15.27	11.45
100000053		CHAIR	10/77	SL	120	D	2	U	N	1101111	12/77	1	126.00	3.15	2.10
100000054		TABLE	11/77	SL	120	D	2	U	N	1101111	12/77	1	525.00	8.75	4.38
100000055		BOOKCASE	12/77	SL	120	D	2	U	N	1101111	12/77	1	418.00	3.48	
FURNITURE	1000060 TOTALS												5,095.00	149.58	107.13
FURNISHINGS	1000070														
100000056		FURNISHINGS	12/77	SL	120	D	2	U	N	1101111	12/77	1	3,982.50	33.19	33.19
FURNISHINGS	1000070 TOTALS												3,982.50	33.19	33.19
DIVISION	A001 TOTALS												70,718.29	5,760.49	5,050.54
COMPANY 100 NAME	100 TOTALS												70,718.29	5,760.49	5,050.54

Figure 8.10. Book additions report. (Courtesy of University Computing Company, Dallas, Texas.)

REQUEST NO. 103
REPORT NO. 620
RUN DATE 02/12/79
RUN TIME 10.08.40

COMPANY 100 NAME 100

CURRENT VALUE ESTIMATES REPORT
JAN, 1977 THRU DEC, 1977

PAGE 1

ITEM NO	UNIT NO	BASIS COST	PURCH YR	BK LIFE	EST VALUE	EST YR	CLASS CODE	BASE YEAR FACTOR	CURRENT VALUE	CURRENT DEPR EXPENSE	CURRENT VALUE ESTIMATES ACCUMULATED DEPR EXPENSE	NET VALUE
HEADQUARTERS PROPERTY	1010 400											
100000012		300.00	75	066			00000000000001	1.0881355	326.44	59.36	168.17	158.27
100000013		300.00	75	066			00000000000001	1.0881355	326.44	59.36	168.17	158.27
100000015		300.00	75	066			00000000000001	1.0881355	326.44	59.36	168.17	158.27
100000016		300.00	75	066			00000000000001	1.0881355	326.44	59.36	168.17	158.27
100000018		1,000.00	71	120			00000000000001	1.3601694	1,360.16	136.02	929.45	430.71
100000019		1,000.00	71	120			00000000000001	1.3601694	1,360.16	136.02	929.45	430.71
100000020		1,000.00	71	120			00000000000001	1.3601694	1,360.16	136.02	861.44	498.72
100000021		1,000.00	71	120			00000000000001	1.3601694	1,360.16	136.02	861.44	498.72
100000022		1,000.00	71	120			00000000000001	1.3601694	1,360.16	136.02	861.44	498.72
100000023		1,000.00	71	120			00000000000001	1.3601694	1,360.16	136.02	861.44	498.72
100000030		1,000.00	71	120			00000000000001	1.3601694	1,360.16	136.02	861.44	498.72
100000031		1,000.00	71	120			00000000000001	1.3601694	1,360.16	136.02	861.44	498.72
100000032		1,000.00	71	120			00000000000001	1.3601694	1,360.16	136.02	861.44	498.72
100000033		1,000.00	71	120			00000000000001	1.3601694	1,360.16	136.02	861.44	498.72
100000034		1,000.00	71	120			00000000000001	1.3601694	1,360.16	136.02	861.44	498.72
100000035		1,000.00	71	120			00000000000001	1.3601694	1,360.16	136.02	861.44	498.72
100000036		1,000.00	71	120			00000000000001	1.3601694	1,360.16	136.02	861.44	498.72
100000037		1,000.00	71	120			00000000000001	1.3601694	1,360.16	136.02	861.44	498.72
100000038		400.00	74	120			00000000000001	1.1464285	458.57	45.86	160.50	298.07
100000039		600.00	74	120			00000000000001	1.1464285	687.85	68.79	240.75	447.10
100000043		7,500.00	77	120			00000000000001	1.0000000	7,500.00	625.00	625.00	6,875.00
PROPERTY 400 TOTALS		23,700.00							28,994.42	2,881.37	13,895.11	15,099.31
PROPERTY 600												
100000051		3,568.00	77	120			00000000000001	1.0000000	3,568.00	118.93	118.93	3,449.07
PROPERTY 600 TOTALS		3,568.00							3,568.00	118.93	118.93	3,449.07
PROPERTY 601												
100000052		458.00	77	120			00000000000001	1.0000000	458.00	15.27	15.27	442.73
PROPERTY 601 TOTALS		458.00							458.00	15.27	15.27	442.73
PROPERTY 602												
100000053		126.00	77	120			00000000000001	1.0000000	126.00	3.15	3.15	122.85

Figure 8.11. Current value estimates report to satisfy requirements of the Securities and Exchange Commission. (Courtesy of University Computing Company, Dallas, Texas.)

REQUEST NO. 101
REPORT NO. 950
RUN DATE 02/12/79
RUN TIME 13.08.40

COMPANY 100 NAME 100
ACTUAL CASH VALUE - ASSET VALUATION REPORT
AS OF DEC, 1977

PAGE 1

HEADQUARTERS
SECTION/LOCATION 1010 A001

ITEM NUMBER	LOC/SECT CENT	COST CENT	PROP CATG	ACCOUNT NUMBER	DATE ENTERED SVC	LIFE IN MO.S	INSURANCE PROPERTY GROUP	REMNG LIFE MO.S	APPRAISAL OR BASIS COST	APPREC FACTOR	REPLACEMENT INSURABLE VALUE	DEPR PRCNT	ACTUAL CASH VALUE
100000007	A001	1010	300	10000030	7/70	240	1	150	10,000.00	1.04	10,400.00	5.0	9,880.00
100000012	A001	1010	400	10000040	3/75	66	2	32	300.00	1.01	303.00	1.0	299.97
100000013	A001	1010	400	10000040	3/75	66	2	32	300.00	1.01	303.00	1.0	299.97
100000015	A001	1010	400	10000040	3/75	66	2	32	300.00	1.01	303.00	1.0	299.97
100000016	A001	1010	400	10000040	3/75	66	2	32	300.00	1.01	303.00	1.0	299.97
100000018	A001	1010	400	10000040	3/71	120	2	38	1,000.00	1.03	1,030.00	5.0	978.50
100000019	A001	1010	400	10000040	3/71	120	2	38	1,000.00	1.03	1,030.00	5.0	978.50
100000020	A001	1010	400	10000040	9/71	120	2	44	1,000.00	1.03	1,030.00	5.0	978.50
100000021	A001	1010	400	10000040	9/71	120	2	44	1,000.00	1.03	1,030.00	5.0	978.50
100000022	A001	1010	400	10000040	9/71	120	2	44	1,000.00	1.03	1,030.00	5.0	978.50
100000023	A001	1010	400	10000040	9/71	120	2	44	1,000.00	1.03	1,030.00	5.0	978.50
100000030	A001	1010	400	10000040	9/71	120	2	44	1,000.00	1.03	1,030.00	5.0	978.50
100000031	A001	1010	400	10000040	9/71	120	2	44	1,000.00	1.03	1,030.00	5.0	978.50
100000032	A001	1010	400	10000040	9/71	120	2	44	1,000.00	1.03	1,030.00	5.0	978.50
100000033	A001	1010	400	10000040	9/71	120	2	44	1,000.00	1.03	1,030.00	5.0	978.50
100000034	A001	1010	400	10000040	9/71	120	2	44	1,000.00	1.03	1,030.00	5.0	978.50
100000035	A001	1010	400	10000040	9/71	120	2	44	1,000.00	1.03	1,030.00	5.0	978.50
100000036	A001	1010	400	10000040	9/71	120	2	78	1,000.00	1.03	1,030.00	5.0	978.50
100000037	A001	1010	400	10000040	9/71	120	2	78	1,000.00	1.03	1,030.00	5.0	978.50
100000038	A001	1010	400	10000040	7/74	120	2	78	400.00	1.02	408.00	2.0	399.84
100000039	A001	1010	400	10000040	2/77	120	2	78	600.00	1.02	612.00	2.0	599.76
100000041	A001	1010	200	10000020	2/77	600	1	589	256,845.00	1.01	259,413.45	1.0	256,819.31
100000042	A001	1010	301	10000030	3/77	240	2	230	2,955.00	1.01	2,984.55	1.0	2,954.70
100000043	A001	1010	400	10000040	3/77	120	2	110	7,500.00	1.01	7,575.00	1.0	7,499.25
100000044	A001	1010	401	10000040	4/77	120	2	111	35,000.00	1.01	35,350.00	1.0	34,996.50
100000045	A001	1010	405	10000040	4/77	120	2	111	10,400.00	1.01	10,504.00	1.0	10,398.96
100000046	A001	1010	409	10000040	4/77	120	2	111	1,867.20	1.01	1,085.87	1.0	1,867.01
100000047	A001	1010	410	10000040	5/77	36	2	28	6,123.59	1.01	6,184.82	1.0	6,122.97
100000048	A001	1010	501	10000050	6/77	96	2	89	250.00	1.01	252.50	1.0	249.97
100000049	A001	1010	502	10000050	7/77	96	2	90	250.00	1.01	252.50	1.0	249.97
100000050	A001	1010	503	10000050	8/77	96	2	91	250.00	1.01	252.50	1.0	249.97
100000051	A001	1010	600	10000060	9/77	120	2	116	3,568.00	1.01	3,603.68	1.0	3,567.64
100000052	A001	1010	601	10000060	9/77	120	2	116	458.00	1.01	462.58	1.0	457.95
100000053	A001	1010	602	10000060	10/77	120	2	117	126.00	1.01	127.26	1.0	125.98
100000054	A001	1010	603	10000060	11/77	120	2	118	525.00	1.01	530.25	1.0	524.94
100000055	A001	1010	606	10000060	12/77	120	2	119	418.00	1.01	422.18	1.0	417.95
100000056	A001	1010	700	10000070	12/77	120	2	119	3,982.50	1.01	4,022.32	1.0	3,982.09
SECTION/LOCATION A001 TOTALS									356,718.29		360,875.46		356,263.64
HEADQUARTERS 1010 TOTALS									356,718.29		360,875.46		356,263.64

Figure 8.12. Asset valuation report for planning insurance levels. (Courtesy of University Computing Company, Dallas, Texas.)

PAGE 2

CORP 001   ABEL CORPORATION

TAX DEPRECIATION AND ITC REPORT
ACRS ASSETS
FISCAL YEAR ENDED 12-31-81

CLAS CODE	ITEM NUMBER	TAX ACQ DATE	TAX-RFG DATE	TAX MTH	TAX LIFE GENERAL	TAX COST	TAX DEPR ANNUAL	TAX RESERVE	TAX ITC AMOUNT
03R1	TEST1A	01-01-81	01-81	DPFD	03-00	10,000.000	2,500.000	2,500.000	600.000
03R1	TEST1B	01-28-81	01-81	DPFD	03-00	10,936.000	2,984.000	2,984.000	656.000
03R1	TEST1C	03-28-81	01-81	DPFD	03-00	10,000.000	2,500.000	2,500.000	600.000
03R1	TEST1D	06-01-81	01-81	DPFD	03-00	10,000.000	2,500.000	2,500.000	1,290.000
03R1	TEST1F	06-28-81	01-81	DPFD	03-00	2,500.005	2,500.005	2,500.005	1,320.751
03R1	TEST1G	10-31-81	01-81	DPFD	03-00	13,534.02	3,383.51	3,383.51	935.802
03R1	TEST1H	09-01-81	01-81	DPFD	03-00	2,125.19	531.300	531.300	255.027
03R1	TEST1J	09-28-81	01-81	DPFD	03-00	553.11	138.28	138.28	17.36
03R1	TEST1K	12-01-81	01-81	DPFD	03-00				
03R1	TEST1L	12-28-81	01-81	DPFD	03-00				
** CLASS CODE	03R1					61,340.31	15,335.09	15,335.09	4,886.39
05R1	TEST2A	01-01-81	07-81	STL	05-00	17,936.000	1,793.600	1,793.600	1,097.960
05R1	TEST2B	01-28-81	07-81	STL	05-00	10,000.000	1,000.000	1,000.000	1,136.000
05R1	TEST2C	03-01-81	07-81	STL	05-00				
05R1	TEST2D	03-28-81	07-81	STL	05-00				
05R1	TEST2F	06-01-81	07-81	STL	05-00	1,000.000	1,000.000		
05R1	TEST2G	10-04-81	07-81	STL	05-00	5,435.002	543.500	543.500	2,150.000
05R1	TEST2H	10-31-81	07-81	STL	05-00	1,435.02	1,435.02	1,435.02	
05R1	TEST2I	09-01-81	07-81	STL	05-00				1,550.602
05R1	TEST2K	09-28-81	07-81	STL	05-00	2,125.99	1,302.19	1,302.19	247.602
05R1	TEST2L	12-28-81	07-81	STL	05-00	553.11	155.31	155.31	118.092
** CLASS CODE	05R1					61,340.31	6,134.01	6,134.01	8,142.30
10R1	TEST15A	01-28-81	07-81	STL	35-00	10,000.000	142.85	142.85	1,600.000
10R1	TEST15B	03-01-81	07-81	STL	35-00	10,936.000	142.35	142.35	1,850.000
10R1	TEST15C	03-28-81	07-81	STL	35-00			1,000.000	
10R1	TEST15D	06-01-81	07-81	STL	35-00	10,000.000	142.85	142.85	2,150.000
10R1	TEST15G	10-04-81	07-81	STL	35-00	5,435.02	76.21	76.24	1,558.602
10R1	TEST15H	10-31-81	07-81	STL	35-00	13,534.02	193.34	193.34	277.602
10R1	TEST15I	09-01-81	07-81	STL	35-00	2,125.19	30.35	30.35	118.092
10R1	TEST15J	09-28-81	07-81	STL	35-00	553.11	7.90	18.80	
10R1	TEST15K	12-01-81	07-81	STL	35-00				
10R1	TEST15L	12-28-81	07-81	STL	35-00				
** CLASS CODE	10R1					61,340.31	876.24	876.24	8,142.30
** CORP	001					368,041.86	32,970.52	31,758.82	21,165.99

```
CORP 001   ABEL CORPORATION        INVESTMENT TAX CREDIT REPORT - FORM 3468      REL 5.02 PAGE 1
FA120-B                                 PERIOD END DATE  1-82                     REPT DATE 10-22-81
```

PROPERTY TYPE	LINE	LIFE YEARS	COST OR BASIS	PERCNT	QUAL INVESTMENT
NEW PROPERTY *	A	3 OR MORE BUT LESS THAN 5		33.3	
*	B	5 OR MORE BUT LESS THAN 7		66.6	
*	C	7 OR MORE		100.0	
**	D	COMMUTER HIGHWAY VEHICLE		100.0	
*	-	3 OR MORE BUT LESS THAN 99	60,000.00	60.0	36,000.00
*	-	5 OR MORE BUT LESS THAN 99	120,000.00	100.0	120,000.00
USED PROPERTY *	F	3 OR MORE BUT LESS THAN 5		33.3	
*	G	5 OR MORE BUT LESS THAN 7		66.6	
**	H	7 OR MORE		100.0	
*	I	COMMUTER HIGHWAY VEHICLE		100.0	
*	-	3 OR MORE BUT LESS THAN 99	28,031.70	60.0	16,819.01
*	-	5 OR MORE BUT LESS THAN 99	56,063.40	100.0	56,063.40

```
                LINE 2  - QUALIFIED INVESTMENT ...............      228,882.41
                LINE 3  - 10 PERCENT OF LINE 2 ..............       22,888.23
                LINE 4A - BASIC 1 PERCENT CREDIT ...........         1,453.57
                LINE 4B - MATCHING CREDIT ..................           443.95

                LINE 6  - TENTATIVE ITC - LINES 3-4B                24,785.75

                LINE 19 - ENERGY CREDIT - COST     102,188.04        8,856.30

             ** ITC AMTS INPUT BY USER  - COST                            .00

                TOTAL ITC - LINE 6,19, ITC AMTS INPUT BY USER      33,642.05
```

** COST OR BASIS NOT INCLUDED IN LINE 2

Figure 8.13. Tax reports. (Copyright 1979, McCormack and Dodge Corporation.)

REQUEST NO. 302
REPORT NO. 195
RUN DATE 02/12/79
RUN TIME 10.08.40

COMPANY 300 NAME   300

LEASE ACCOUNTING REPORT
JAN, 1977 THRU DEC, 1986

PAGE 1

ITEM NUMBER	UNIT NUM	AGREEMENT NUMBER	SCH NUM	LEASE RATE %	AMORTIZATION AMOUNT	INTEREST AMOUNT	MINIMUM PAYMENT	EXECUTORY COSTS	COMMITTED PAYMENTS
AGREEMENT TYPE LE									
LEASE TYPE C									
300000005		AGR-0005	05	7.0398	144,128.00	55,872.00	200,000.00		200,000.00
300000006		AGR-0006	06	7.8090	139,419.21	60,580.79	200,000.00		200,000.00
300000007		AGR-0007	07	7.0398	144,128.00	55,872.00	200,000.00		200,000.00
300000008		AGR-0008	08	7.6470	140,394.33	59,605.67	200,000.00		200,000.00
300000009		AGR-0009	09	7.7286	139,904.35	60,095.65	200,000.00		200,000.00
300000010		AGR-0010	10	9.4680	200.00	20.32	220.32		220.32
300000012		AGR-0012	12	9.4740	199.99	20.33	220.32		220.32
300000013		AGR-0013	13	9.4680	200.00	20.32	220.32		220.32
300000014		AGR-0014	14	2.0124	215.76	4.56	220.32		220.32
LEASE TYPE C				TOTALS	708,789.64	292,091.64	1,000,881.28		1,000,881.28
AGREEMENT TYPE LE				TOTALS	708,789.64	292,091.64	1,000,881.28		1,000,881.28
AGREEMENT TYPE LR									
LEASE TYPE F									
300000001		AGR-0001	01	7.7910	134,128.00	65,872.00	200,000.00		200,000.00
300000002		AGR-0002	02	8.4726	124,128.00	75,872.00	200,000.00		200,000.00
300000004		AGR-0004	04	7.7910	134,128.00	65,872.00	200,000.00		200,000.00
300000011		AGR-0011	11	9.4680	200.00	20.32	220.32		220.32
300000016		AGR-0016	16	4.0214	11,690.57	11,445.43	23,136.00		23,136.00
300000017	001	AGR-0017		11.3226	190,000.00	10,000.00	200,000.00		200,000.00
300000017	002	AGR-0017		12.5946	190,000.00	10,000.00	200,000.00		200,000.00
300000017	003	AGR-0017		11.2146	190,000.00	10,000.00	200,000.00		200,000.00
300000017	004	AGR-0017		9.5538	190,000.00	10,000.00	200,000.00		200,000.00
300000017	005	AGR-0017		13.9038	190,000.00	10,000.00	200,000.00		200,000.00
ITEM TOTALS					950,000.00	50,000.00	1,000,000.00		1,000,000.00
300000018		DESK CHK	01	11.9988	938.52	67.48	1,006.00		1,006.00
LEASE TYPE F				TOTALS	1,355,213.09	269,149.23	1,624,362.32		1,624,362.32
AGREEMENT TYPE LR				TOTALS	1,355,213.09	269,149.23	1,624,362.32		1,624,362.32
COMPANY 300 NAME 300				TOTALS	2,064,002.73	561,240.87	2,625,243.60		2,625,243.60

Figure 8.14. Lease accounting report, containing disclosure information required by the Financial Accounting Standards Board. (Courtesy of University Computing Company, Dallas, Texas.)

REQUEST NO. 102
REPORT NO. 465
RUN DATE 02/12/79
RUN TIME 10.08.40

COMPANY 100 NAME 100

YTD REPAIRS AND MAINTENANCE REPORT
JAN, 1977 THRU DEC, 1977

PAGE

ITEM NUMBER	UNIT NO	DESCRIPTION	COST CENTER	LOC/SECT	DEPT	PROP CATG	YTD MAINTENANCE		COMMENTS
							CALLS	CHARGES	
HEADQUARTERS	1010								
DEPARTMENT	A001								
100000014		HEAVY EQUIPMENT	1010	A001	A001	400	1	25.00	
100000016		HEAVY EQUIPMENT	1010	A001	A001	400	1	25.00	
100000017		HEAVY EQUIPMENT	1010	A001	A001	400	1	35.00	
100000019		HEAVY EQUIPMENT	1010	A001	A001	400	2	35.00	
100000021		HEAVY EQUIPMENT	1010	A001	A001	400	3	65.00	
100000023		HEAVY EQUIPMENT	1010	A001	A001	400	1	18.00	
100000028		HEAVY EQUIPMENT	1010	A001	A001	400	3	30.00	
100000031		HEAVY EQUIPMENT	1010	A001	A001	400	2	125.00	
100000038		HEAVY EQUIPMENT	1010	A001	A001	400	1	15.00	
100000042		SIDEWALK	1010	A001	A001	301	1	100.00	
100000043		HEAVY EQUIPMENT	1010	A001	A001	400	2	197.50	
100000044		COMPUTER EQUIPT	1010	A001	A001	401	2	1,006.00	
100000045		COPY MACHINE	1010	A001	A001	405	1	400.00	
100000046		AIR CONDITIONER	1010	A001	A001	409	2	250.00	
100000047		AUTO	1010	A001	A001	410	1	50.00	
100000048		TYPEWRITER	1010	A001	A001	501	1	62.50	
DEPARTMENT	A001	TOTALS					25	2,439.00	
HEADQUARTERS	1010	TOTALS					25	2,439.00	
COMPANY 100 NAME	100	TOTALS					25	2,439.00	

Figure 8.15. Repair and maintenance report. (Courtesy of University Computing Company, Dallas, Texas.)

```
        03/09/83    LEASING OPERATIONS INFORMATION SYSTEM        SCREEN: 01
ER:  W REG: 12 OFF: D44-0  TYPE: B 02 STAT: 0  OFF NAME CASALOMA  CO
                            LEASE DATA SCREEN   COMMON LEASE (SEE SCREEN 04)
OFFICE ADDRESS:                       LDLD ADDRESS:
2010    PARK CENTER DR 210            STE 200, 2621    PARK CTR BLVD
SOUTHFIELD            CO 80276        SOUTHFIELD            CO 80276
OFFICE TEL: 301 352 9100             LDLD TEL: 301 358 0033
MANAGER:    ANN JAMES                LDLD:   PARK CENTER ASSOCIATES

LEASE DATA -    TYPE: LESSOR    ORIG LEASE: 01 17 69    OCC SINCE: 02 01 69
TERM:  5             BEGIN: 02 01 79    END: 01 31 84   MTHLY TEN: NO
 OPTIONS:                 OPT RENT:                     OPT CUTOFF:
CAN PRIV:                                               CAN DATE:
                                                        RENT CU:
TYP BLDG: MO    FL LOC: 001    TYP SETUP:         ELEV:         PARKING: AO

 EXCESS SPACE:           CASE NO:
SUBLEASE DATA                   ANNUAL INCOME: $
  SUBLESSEE:
SUBLEASED            EXPIRATION:              AREA:          ESCAL CLAUSE:

EFFECTIVE DATE: 01 17 83 LAST UPDATED: 01/17/83 TIME: 10:49 AM OPER: EVERHARD
                                                            NEXT SCREEN:
```

Figure 8.16. Primary data screen for an online lease management system.
(Courtesy of Metropolitan Life Insurance Company, New York.)

Lease Management

Most large organizations lease some floor space. For many this is an inciden-
tal activity that does not require special computer support. But in large service organizations, the management of leases is a full-time activity for a large
staff. Such firms include banks, insurance companies, and brokerage houses,
which may have hundreds of leased spaces worldwide. The field sales arms of
large manufacturing companies may also have enough regional and local sales
offices to require computer support. Many organizations are also landlords,
leasing unused or investment properties to many tenants. In these situations,
a specialized lease management system can be a useful computer aid.

Where events (payments, changes, expirations, renewals) occur daily, it
may make sense to install an interactive, online system for up-to-the-minute
processing of transactions. The printouts shown in Figures 8.16 and 8.17
are two of several screens used in a leasing information system developed
for internal use by the Metropolitan Insurance Companies. The system is
called LOIS, for Leasing Operations Information System. It is designed to
process all lease-related transactions, and serves as the primary computer
aid to a staff of 40 people who negotiate and manage hundreds of leases.
LOIS produces more than 30 scheduled reports, some daily, others weekly,
monthly, and quarterly. These reports cover all aspects of leases, subleases, payments, and physical alterations.

The most common use of a lease management system is to produce a report like the one shown in Figure 8.18. This is a list of renewal options and
expirations. It serves as a monthly alert on those options and leases that are
about to expire. Such lists are vital when managing hundreds of leased facilities. Another useful report is shown in Figure 8.19, covering changes in
the amount of space rented and corresponding changes in annual rent payments. Both of these reports from the LOIS system are organized by territory, region, and office location.

```
        03/09/83      LEASING OPERATIONS INFORMATION. SYSTEM      SCREEN: 02
ER:  W REG: 12 OFF: D44-0  TYPE: B 02 STAT: 0  OFF NAME:CASALOMA  CO
                           RENTAL COSTS SCREEN
RENT DATA-  MTHLY: $    1452.21    ANNUAL: $   17426.50   ADJ ANN: $    19890.67

AREA DATA-      USABLE AREA:  2681      LEASE AREA:  2681      TOTAL AREA:  2681
        RATE/SQ FT-USABLE: $   7.42     LEASE:   $   7.42     TOTAL:   $    7.42

OTHER RENTAL COSTS (ANNUALIZED)
  HEAT:      NONE      ELECTRIC:   2145 F   AC-W-EL:   NONE     BULBS:      NONE
  JANITOR:   NONE      WINDOW:     NONE     RUBBISH:   NONE     EXTERM:     NONE
  WAT-LAV:   NONE      WAT-REF:    NONE     LAV-SUPP:  NONE     REPAIRS:    NONE
  PARKING:   NONE      REDECOR:    NONE     LAWN/SNOW: NONE     AC REP:     NONE
  ESCA-TAX:   319 A    ESCA-OP:    NONE     ESCA-ELEC: NONE     ESCA-OTH:   NONE
  MERCH FEE: NONE      COMM AREA:  NONE     MALL-AD:   NONE     ALTEREXP:      0

SERVICES PROVIDED BY LANDLORD
  HEAT:     YES      ELECTRIC: NO      AC-W-EL:   YES      BULBS:     YES
  JANITOR:  YES      WINDOW:   YES     RUBBISH:   YES      EXTERM:    YES
  WAT-LAV:  YES      WAT-REF:  YES     LAV-SUPP:  YES      REPAIRS:   YES
  PARKING:  YES      REDECOR:  YES     ELEVATOR:  YES      AC-WO-EL:  NO
  INSUR:    YES      OTHER:    YES

                                                              NEXT SCREEN:
```

Figure 8.17. Secondary data screen. (Courtesy of Metropolitan Life Insurance Company, New York.)

Vacating a space with time left on the lease is a common occurrence in changing organizations and markets. Reports like that shown in Figure 8.20 show the way in which a computer system can keep track of remaining obligations, as well as income from any subleases.

If the lease management system is well designed, it will use data base management and a report generator to produce a variety of nonstandard or unanticipated reports. Metropolitan uses a commercially available system called RAMIS for this purpose. Typical reports produced from the data base are shown in Figures 8.21 and 8.22. These summarize square footage assignments by organization unit, showing amounts and types of space and rental rates. Good report generator software also produces printer graphics like those shown in Figure 8.23 comparing rental rates by territory.

Systems such as LOIS can take several years to design and install. Dozens of man-years may be required. But the payoff in more effective leasing is of great value when several million square feet are leased at $10 to $20 per square foot or more.

Facilities Management Systems

The systems reviewed to this point have been rather specialized, each addressing one aspect of constructing and managing facilities. We will turn now to a general, multipurpose system with uses ranging from fixed asset management, to preventive maintenance, to lease management. The system we will examine is called INSITE, and it was developed at the Massachusetts Institute of Technology.

M.I.T. has over 8 million square feet of office, laboratories, and storage. More than 60,000 pieces of furniture and equipment are contained in roughly 12,000 rooms and spaces. Space occupancies and uses and equipment locations

LEASING OPERATIONS

REPORT OF RENEWAL OPTION DATA AND LEASE EXPIRATION DATA   AS OF: 5/31/82

P.I. TERRITORY: H
P.I. REGION: 70

OFFICE NUMBER & TYPE	OFFICE NAME	MO TO GO	OPTION CUT-OFF DATE	LEASE EXPIR. DATE	CURRENT ADJ ANNUAL RENT	ADJ ANNUAL OPTION RENT	CURRENT BASE ANNUAL RENT	BASE ANNUAL OPTION RENT	OPTION TERMS	USABLE AREA
D08-0 B 010	AZALEA MO	0	04/30/82	07/31/82	19629.00 5.90	19629.00 5.90	19629.00 5.90	19629.00 5.90	7 0 1	3325
C98-0 B 010	PEACH SPRINGS AR	0	08/31/82	11/30/82	9680.00 7.13	9680.00 7.13	9180.00 6.76	9180.00 6.76	3 0 1	1357
D06-6 T 010	COLDWELL MO	0	08/31/82	11/30/82	8865.07 7.48	8865.07 7.48	8190.24 6.90	8190.24 6.89	3 0 1	1187
R78 R 010	ST JOSEPH MO	1	09/30/82	12/31/82	17940.24 12.83	16509.48 11.80	12962.76 9.27	11532.00 8.24	2 0 1	1398
840-6 T 010	BLYTHE MO KIOSK	1	09/30/82	12/31/82	3780.00 6.00	4200.00 6.66	3780.00 6.00	4200.00 6.66	1 0 1	630
764-0 DK 010	SPRING BRANCH AK	4		12/31/82	6855.00 85.68	0.00 0.00	6675.00 83.44	0.00 0.00		80

Figure 8.18. Standard report of renewals and expirations, in sequence by option cutoff date. (Courtesy of Metropolitan Life Insurance Company, New York.)

PAGE 5

L E A S I N G   O P E R A T I O N S

A R E A / C O S T   T R A N S A C T I O N   C H A N G E S

YEAR-TO-DATE AS OF: 11/30/82

TERRITORY: SOUTHEASTERN
REGION: 61

OFFICE NBR & NAME	YEAR-TO-DATE CHANGES USABLE AREA FROM	TO	% CHG	USABLE NPSF FROM	TO	% CHG	TRANSACT	REASONS EFF-DATE	FOR CHANGES BASE ANNUAL RENT FROM	TO	USABLE AREA ADD(MED)	OTHER RENTAL COSTS (SYMBOL IND. CHANGES)
A21-0 B 01 0 ROSELAND GA BRANCH	3180	3180	0	9.01	9.01	0	CHANGE	9/02/82	28680.00	28680.00	0	H
A21-6 H 01 0 DAWSON GA DETACHED	800	800	0	7.83	8.59	9	CHANGE	10/12/82	5400.00	6000.00	0	E
858-0 B 01 0 COLUMBIA SC BRANCH	3911	2606	-33	8.17	11.04	35	CHANGE	10/13/82	18480.00	18480.00	(1305)	E
P92-0 B 01 0 CUMBERLAND TN BRANCH	1749	1749	0	13.39	13.39	0	CHANGE	9/02/82	23430.12	23430.12	0	
P92-0 BK 01 0 CUMBERLAND, TN KIOSK	145	145	0	155.31	155.31	0	CHANGE	9/02/82	22521.00	22521.00	0	
P95-0 BK 01 0 CUMBERLAND, TN KIOSK	145	145	0	155.31	155.47	0	CHANGE	9/22/82	22521.00	22543.44	0	
R91 K 01 0 ALABAMA, REGION	1225	1225	0	12.88	12.88	0	CHANGE	9/02/82	15750.00	15750.00	0	

OTHER RENTAL COSTS SYMBOLS:

A HEAT	E JANITOR	I MAT-LAV	M PARKING	U ESCA TAX	U MERCH FEE
B ELECTRIC	F WINDOW	J MAT-REF	N REDECOR	H ESCA UP	V COMMON AREA
C AC-W-EL	G RUBBISH	K LAV-SUPP	O LAWN/SNOW	S ESCA EL	H MALL AD
D BULB	H EXTERM	L REPAIRS	P AC REPAIR	T ESCA OTH	X ALT EXPENSE

Figure 8.19. Standard report of charges in square footage and costs. (Courtesy of Metropolitan Life Insurance Company, New York.)

LEASING OPERATIONS
AS OF: 11/01/82

REPORT OF VACATED/SUBLET QUARTERS CONTINUED LIABILITIES
BY DATE VACATED SEQUENCE

DATE VACATED	OFFICE NAME	LEASE EXPIR. DATE	CANCEL NOTICE DATE	USABLE AREA	BASE RENT ANNUAL --- RSF	ACTUAL BASE RENT PAYMENTS SINCE DATE VACATED	TOTAL BASE RENT LIABILITY REMAINING	ANNUAL SUBLEASE INCOME --- RSF	SUBLEASE AREA (USABLE)	SUBLEASE EXPIR. DATE
00/00/00	GOLDEN WEST CA	09/30/84		4682	27950.04 / 5.96	6987.51	-6987.51	0.00 / 0.00	0	
07/31/82	COFFEYVILLE KS	01/31/83		1587	9000.00 / 5.67	2250.00	+2250.00	0.00 / 0.00	0	
08/01/82	IRON MOUNTAIN MI	07/31/85		1813	26324.76 / 14.52	6182.33	+65512.47	0.00 / 0.00	0	
08/18/82	CIRCLE PARK IND	05/31/83		990	9329.04 / 9.42	2332.26	+5441.94	0.00 / 0.00	0	
09/01/82	LEE VA	06/30/84		3275	20632.56 / 6.30	5158.14	+32668.22	3624.96 / 10.99	3275	06/29/84
09/01/82	FLATBUSH NY	11/30/84		1044	7260.00 / 6.95	1815.00	+14520.00	0.00 / 0.00	0	
09/01/82	WESTWOOD CA	03/31/85		2033	17280.48 / 8.49	4320.12	+40321.12	0.00 / 0.00	0	
09/01/82	KEYSER, WV	09/30/83		895	3600.00 / 4.02	900.00	+3000.00	0.00 / 0.00	0	
09/01/82	CLAIREMONT MESA CA BR	02/29/84		2830	30164.40 / 10.65	7541.10	+37705.50	0.00 / 0.00	0	
09/01/82	GARDEN GROVE CA DV	08/31/85		840	4200.00 / 5.00	1050.00	+11550.00	0.00 / 0.00	0	
09/01/82	MIDWOOD NY BRANCH	01/31/84		775	4200.00 / 5.41	1050.00	+4900.00	0.00 / 0.00	0	
09/01/82	LAWTON OK BRANCH	06/30/84		3235	17856.60 / 5.51	4464.15	+28272.95	0.00 / 0.00	0	
09/01/82	PORT OF L A CA	05/31/85		2527	27380.04 / 10.83	6845.01	+68450.10	0.00 / 0.00	0	
09/01/82	SANTA ROSA CA BRANCH	12/31/84		4362	30000.00 / 6.87	7500.00	+62500.00	0.00 / 0.00	0	
09/01/82	SAN JACINTO TX BRANCH	01/31/86		5026	39593.88 / 7.87	9898.47	+125380.62	0.00 / 0.00	0	
09/01/82	ADRIAN MI	12/31/83		1141	8880.00 / 7.78	2220.00	+9620.00	3960.00 / 5.17	765	12/31/83
09/01/82	GUILFORD CT BV	10/31/85		4945	36790.80 / 7.44	9197.70	+107306.50	28710.00 / 5.80	4945	10/30/85
09/01/82	RHINELANDER WI	02/28/90		4383	27216.00 / 6.20	6804.00	+197316.00	0.00 / 0.00	0	

Figure 8.20. Standard report of continued liabilities. (Courtesy of Metropolitan Life Insurance Company, New York.)

RAMIS REPORT 2A - LISTING OF USABLE OFFICE SPACE BY TYPE AND REGION
WITHIN SOUTHEASTERN TERRITORY

REGION	TYPE	OFFICE IDENTITY	NAME OF OFFICE	ADDRESS OF OFFICE			BLDG TYPE	AREA	RATE PER SQ FT
22	B	A71-OB 010	JAMES RIVER VA BRANCH	3807 LEE HIGHWAY	RICHMOND	VA 23235	MO	1,505	14.04
		A81-OB 010	CHARLOTTESVILLE VA BR	1925 EAST MARKET STREET	CHARLOTTESVILLE	VA 22905	MO	2,188	7.42
		D11-OB 020	VIRGINIA BEACH VA	341 MAIN ST	VIRGINIA BEACH	VA 23452	MO	1,861	12.17
		D21-OB 010	TIDEWATER VA BRANCH	11214 LEE HIGHWAY	NORFOLK	VA 12636	MO	3,390	7.59
		D31-OB 010	PORTSMOUTH VA BRANCH	2825 SOUTH CRATER ROAD	PORTSMOUTH	VA 23705	MO	3,556	5.74
		D51-OB 010	COLLEGE PARK VA BRANC	2316 ATHERHOLT ROAD	VIRGINIA BEACH	VA 23464	MO	1,105	16.13
		P71-OB 010	MANASSAS VA BRANCH	6341-A COLUMBIA PIKE	FAIRFAX	VA 22030	MO	1,248	17.90
		O81-OB 010	ROANOKE VA BRANCH	8517 MIDLOTHIAN TURNPIKE	ROANOKE	VA 24015	MO	2,910	5.75
		110-OB 010	HARRISONBURG VA BRANC	8936 ARLINGTON BLVD	HARRISONBURG	VA 22801	MO	694	24.78
		340-OB 010	DANVILLE VA BRANCH	485 S. INDEPENDENCE BLVD	DANVILLE	VA 24543	MO	710	7.61
		425-OB 010	FAIRFAX VA BRANCH	18 KOGER EXEC CTR	FAIRFAX	VA 22030	MO	2,064	18.18
		539-OB 010	PETERSBURG VA BR OFF	430 CRAWFORD ST	PETERSBURG	VA 23805	MO	1,963	6.68
		860-OB 010	LYNCHBURG VA BR B	64-65 COLLEGE PARK SQUARE	LYNCHBURG	VA 24501	MO	2,196	7.09
		921-OB 010	ANNANDALE VA BRANCH	3015 NUTLEY RD AND LEE HGWY	BAILEY'S CROSSROADS	VA 22041	MO	788	17.08
	BK	P77-OBKO10	MANASSAS, VA KIOSK	27 WALNUT BOULEVARD	MANASSAS	VA 22110	MO	120	112.90
		533-OBKO10	PETERSBURG VA BK	8300 SUDLEY ROAD	PETERSBURG	VA 23803	MO	125	69.93
	BS	344-OBSO10	DANVILLE VA BS	SUITE 404, 8002 DISCOVERY DR.	MARTINSVILLE	VA 24112	MO	990	5.15
	D	A69-OD 010	JEFFERSON VA	703 NORTH COURTHOUSE RD	RICHMOND	VA 23230	SO	5,060	6.90
		D47-OD 010	PROVIDENCE SQUARE VA	5425 DUKE ST	VIRGINIA BEACH	VA 23464	MO	1,807	12.81
		P22-OD 010	CONGRESSIONAL VA	2013 CUNNINGHAM DRIVE	FALLS CHURCH	VA 22043	MO	4,035	9.62
		554-OD 01V	LEE VA	4928 WEST BROAD STREET	RICHMOND	VA 23235	MO	2,527	10.84
		934-OD 010	ALEXANDRIA VA	1023-1025 KEMPSVILLE ROAD	ARLINGTON	VA 22304	SO	4,710	8.37
		939-OD 010	PENINSULA VA	7700 LEESBURG PIKE	HAMPTON	VA 23666	MO	4,160	3.81
	R	R66 R 010	VIRGINIA	209 CLEVELAND AVE	RICHMOND	VA 23288	MO	1,296	11.03
	T	933-6T 010	FREDERICKSBURG VA	2146 WHISKEY ROAD	FREDERICKSBURG	VA 22401	MO	2,256	11.79
23	B	A21-OB 010	ROME GA BRANCH	2840 MT WILKINSON PARKWAY	ROME	GA 30161	SO	3,180	9.01
		B51-OB 010	COLUMBUS GA BRANCH	2300 HENDERSON MILL ROAD	COLUMBUS	GA 31904	MO	2,606	11.04
		E51-OB 010	MASTERS CITY, GA	2500 ATLANTA HIGHWAY	AUGUSTA	GA 30902	MO	3,796	6.56
		E61-OB 010	PALMETTO SC BRANCH	1058 CLAUSSEN ROAD	AIKEN	SC 29801	MO	2,010	7.32
		P91-OB 010	CUMBERLAND GA BRANCH	2300 HENDERSON MILL ROAD	ATLANTA	GA 30339	MO	1,749	13.39
		211-OB 010	NORTH ATLANTA GA BRAN	713 SHORTER AVE	ATLANTA	GA 30345	MO	3,219	7.26
		311-OB 010	ATHENS GA BRANCH	5650 WHITESVILLE ROAD	ATHENS	GA 30606	MO	2,865	5.56
		321-OB 010	AZALEA GA BRANCH	699 BROAD STREET, SUITE 800	AUGUSTA	GA 30907	MO	3,619	8.29
		591-OB 010	INTERSTATE NORTH GA B	1944 WILLIAM ST	ATLANTA	GA 30345	MO	750	7.46

REPORTED ON 11/17/82 12.11.51
DATE DATA LOADED - 10/29/82
E.P.C.

Figure 8.21. Ad hoc report of usable office space. Produced with a simple query in the RAMIS report generator. (Courtesy of Metropolitan Life Insurance Company, New York.)

RAMIS REPORT 2B - SUMMARY OF USABLE OFFICE SPACE BY TYPE OF OFFICE
BY REGION WITHIN TERRITORY

TERRITORY	REGION	TYPE	NUMBER OF OFFICES	TOTAL AREA	AVERAGE AREA	AVERAGE RPSF
SOUTHERN	20	B	14	26,178	1,869.86	9.01
		BK	2	245	122.50	11.04
		BS	1	990	990.00	6.56
		D	6	22,299	3,716.50	7.32
		R	1	1,296	1,296.00	13.39
		T	1	2,256	2,256.00	16.13
	21	B	10	25,600	2,560.00	17.90
		BK	2	225	142.50	5.75
		D	7	25,638	3,662.57	24.78
		H	1	800	800.00	7.61
		R	1	1,225	1,225.00	18.18
		T	6	5,162	860.33	6.68
	22	B	12	24,465	2,038.75	7.09
		BK	3	430	143.33	17.08
		D	3	10,653	3,551.00	8.82
		DK	2	385	192.50	12.90
		R	1	1,344	1,344.00	69.93
	23	B	16	37,281	2,330.06	8.71
		BK	3	470	156.67	5.15
		D	7	21,226	3,032.29	8.07
		R	1	1,380	1,380.00	6.90
		T	9	10,783	1,198.11	12.81
	24	B	18	44,334	2,463.00	9.62
		D	5	18,827	3,765.40	10.84
		R	1	1,307	1,307.00	8.37
		T	4	3,879	969.75	3.81
	25	B	12	37,438	3,119.83	11.41
		D	10	40,733	4,073.30	11.03
		DK	1	130	130.00	8.29
		H	2	1,601	800.50	11.79
		R	1	1,795	1,795.00	7.26
		T	3	3,361	1,120.33	5.56
	26	B	7	17,167	2,452.43	14.04
		BK	1	100	100.00	7.42
		D	4	16,865	4,216.25	12.17
		DK	2	250	125.00	7.59
		R	1	1,396	1,396.00	5.74

REPORTED ON 11/17/82 12.15.04
DATE DATA LOADED - 10/29/82
E.P.C.

Figure 8.22. Ad hoc summary of usable office space, showing average rental rate per square foot. (Courtesy of Metropolitan Life Insurance Company, New York.)

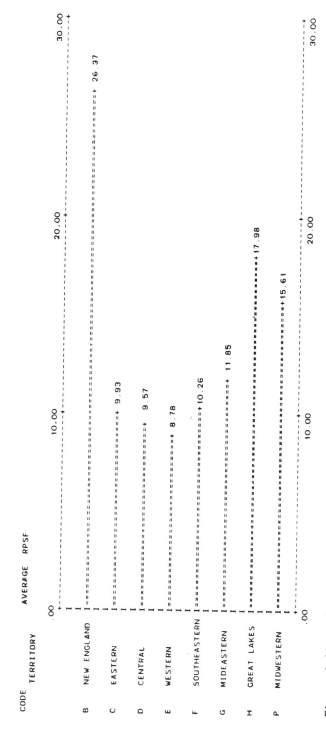

Figure 8.23. Printer graphics bar chart. Produced with the RAMIS report generator, using data from the online lease management system. (Courtesy of Metropolitan Life Insurance Company, New York.)

are constantly changing. There is a need to keep track of these changes,
since the costs of space and equipment can be recovered as an expense on
research projects. To improve control, M.I.T's Office of Facilities Management Systems developed the INSITE system in 1970. A fourth-generation
version was scheduled to be installed in 1984.

INSITE stands for Institutional Space Inventory Technique. It is primarily a space accounting system based on an inventory of individual rooms
or spaces. Each space can have a large number of attributes, including its
area, function (office, lab, etc.), occupancy (organizational unit), and furniture and equipment contents. Thus, pieces of equipment are associated
with the space in which they are housed. Further data on each item, such
as manufacturer, serial number, cost, and depreciation class can be entered
as well. In fact, the entire record set of a fixed asset system could be incorporated, including date of first use, replacement cost, and the like. The
data structure of the current INSITE 3 system is shown in Figure 8.24. Equipment is one of 17 record types that may be defined for each space record.
Within each record type, up to 128 attributes may be stored.

As an information system, INSITE has the following four capabilities:

1. *Data management*: such as that found in any data base management
    system.
2. *Problem-oriented language*: of English-like commands in a free for-
    mat. This permits facilities planners and managers, cost accountants,
    and other administrators to use the system directly, without the need
    for programmers.
3. *Report generation*: on virtually any mix of data elements in either
    outline or columnar format. Output can also be on magnetic tape,
    so that INSITE data can be transferred to other systems such as
    fixed asset management.
4. *Algebraic functions*: for complete computing capabilities of any quan-
    titative data elements within the data base. These functions are used
    to construct utilization ratios, measure deviations from specialized
    norms, and calculate depreciation.

Since 1973, the INSITE system has been made available to other institutions, businesses, and industry. Users are charged an annual fee, which
covers the use of software, training, and support by M.I.T's Office of Facilities Management Systems. A partial list of INSITE uses appears in Figure 8.25.

Unlike the specialized systems reviewed earlier in this chapter, INSITE
does not have pre-defined or standard reports. Each user constructs his
own, and produces them at the frequency desired. Two of the most common
uses and report formats are shown in Figures 8.26 and 8.27. The first report shows spaces occupied by the School of Science at M.I.T. The report
is organized by department and major function within department. Under
each function is a list of room numbers.

The second report shows equipment contents on a building-by-building
basis. The report is organized by department and funding source within department. Equipment is assigned to various fund sources for costing of research work. Under each funding source is a list of room numbers. Equipment tag numbers are listed within each room. At M.I.T., tag numbers are

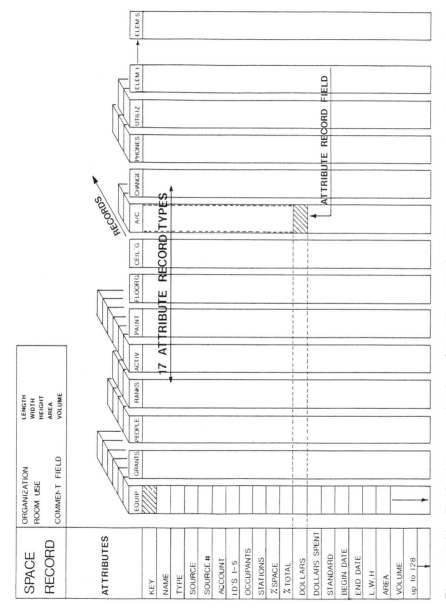

Figure 8.24. Basic data organization of the INSITE 3 facilities management system. (Courtesy of Massachusetts Institute of Technology, Cambridge.)

1. Maintain up-to-date inventory of facilities.

2. Provide accurate and timely inventory lists by user,
   by building, by floor, etc.

3. Automatic response for Federal, State or Local report requirements.

4. Information about space operations, allocation and planning.

5. Experiment with a variety of space allocations.

6. Review space allocation policy.

7. Match activities to appropriate spaces.

8. Track door key assignments and room access.

9. Inventory capital equipment by location.

10. Locate major pieces of machinery with an eye towards
    consolidating its use.

11. Analyze space change costs by sq. ft., room type, and space user.

12. Analyze telecommunications usage.

13. Analyze building use by occupant, room function, organizational
    use and program use.

14. Project future space needs based upon existing use.

15. Provide planning data to assist in "build" or "renovate" decisions.

16. Assist long-term and immediate planning decision-making.

17. Prorate building costs directly to building users.

18. Provide building maintenance budgeting/scheduling.

19. Inventory parking facilities and analyze parking assignments/users.

20. Analyze existing building maintenance by trade.

21. Project future building maintenance by trade.

22. Track exterior spaces by type, function and work requirements.

23. Track and analyze crime occurrences.

24. Relate space inventory data to personnel and financial data
    for analysis.

25. Study and produce overhead recovery/patient care cost allocation
    reports for submittal to government agencies.

Figure 8.25. Uses of INSITE. This list would apply to any multipurpose facilities management system. (Courtesy of Massachusetts Institute of Technology, Cambridge.)

converted to bar-coded labels and attached to each piece of furniture, equipment, and machinery. The room or space number is also converted to a barcoded label and attached to a door jamb or other surface at the entrance to the space. A property administrator tours every space on a periodic basis. Using a hand-held bar-code reader, the administrator scans the room number label, enters the room, and scans each label inside. The scans are recorded on a tape cassette which is used to update the INSITE data base, producing a variety of exception and change reports. This same bar-coding approach is being employed by an increasing number of facilities management groups.

In Figure 8.28, we see the use of INSITE as a pure equipment inventory. The listing, in tag number sequence, contains a room number as its only link to the space accounting portion of the system. The rest of the report covers descriptive information, assignments, acquisition dates, costs, and disposition information as might be found in a fixed asset management system. The value of a general purpose system like INSITE is that it integrates much of the fixed asset data into the space or room record, always showing *where* the asset is

INSITE 3 REPORT

MIT FACILITIES MANAGEMENT SYSTEM

SCHOOL SUMMARY REPORT

***** SCHOOL OF SCIENCE *****

PAGE 1

OFFICE OF FACILITIES MANAGEMENT SYSTEMS

SPACES BY MAJOR FUNCTION BY DEPARTMENT

REPORT DATE: 11/20/79

	AREAS	SPACES	AVERAGE AREA
CHEMISTRY			
ADMINISTRATION			
OFFICE			
2-101	140		
2-102	128		
2-105	162		
	----	---	
SUBTOTAL FOR OFFICE	430	3	143.33
RECEPTION			
2-101A	80		
2-105A	96		
	----	---	
SUBTOTAL FOR RECEPTION	104	2	52.00
SUBTOTAL FOR ADMINISTRATION	534	5	106.80
TEACHING			
CLASSROOM			
2-125	360		
2-126	290		
	----	---	
SUBTOTAL FOR CLASSROOM	650	2	325.00
SUBTOTAL FOR TEACHING	650	2	325.00
RESEARCH			
INSTRUMENTS			
2-133A	48		
2-135	30		
	----	---	
SUBTOTAL FOR INSTRUMENTS	78	2	39.00
LABORATORY			
2-133	330		
2-136	410		
2-137	440		
	----	---	
SUBTOTAL FOR LABORATORY	1180	3	393.33
LAB SUPPLY			
2-136A	84		
2-137A	110		
	----	---	
SUBTOTAL FOR LAB SUPPLY	194	2	97.00
SUBTOTAL FOR RESEARCH	2452	5	490.40
TOTALS FOR CHEMISTRY	3986	13	306.62

Figure 8.26. Space occupancy report. (Courtesy of Massachusetts Institute of Technology, Cambridge.)

```
                                    INSITE 3  REPORT                                    PAGE  1

OFFICE OF FACILITIES MANAGEMENT SYSTEMS   MIT FACILITIES MANAGEMENT SYSTEM        REPORT DATE:  11/20/79

EQUIPMENT BY FUND SOURCE BY DEPARTMENT    BUILDING EQUIPMENT SUMMARY   ***** BUILDING 2    *****
```

	NAME	MODEL	SER. NO.	COST	ACQ. DATE	ACCT. NO.
CHEMISTRY						
NAT INST HEALTH						
2-101						
000001703	CABINET,DIGITAL	H960-DA	X0760	3230.00	081778	845780
000001704	DEHUMIDIPIER	W18HAZ	84003	332.00	010573	734260
2-133						
000001701	AMPLIFIER,WIDEB	PRO 7825	190	603.00	051275	815740
000001702	BASE,ROTATING	WALKER 71	465	540.00	110971	721440
SUBTOTAL FOR NAT INST HEALTH				4705.00		
NAT SCI FOUND						
2-135						
000001705	ELECTROMETER	2450	1123	575.00	041572	730120
000001706	GAS LASER	SPECTRA	269	261.00	040570	713090
000001707	HOMOGENIZER,MIC	VIR TI545	12225	569.00	012777	230460
000001708	INTEGRATOR	227SG	23000	1788.00	121176	829950
2-136A						
000001709	JAW CRUSHER	2X6	1737	4380.00	310576	799540
SUBTOTAL FOR NAT SCI FOUND				7573.00		
SUBTOTAL FOR CHEMISTRY				12278.00		
TOTALS FOR BUILDING 2				12278.00		

Figure 8.27. Equipment locations report. (Courtesy of Massachusetts Institute of Technology, Cambridge.)

MIT OFMS   INSITE 3 EQUIPMENT INVENTORY   12/01/1981   PAGE 10

TAG NUMBER	1)ROOM 2)DEPT 3)PRIN INVEST	1)STD NOMEN 2)MANUFACT 3)MODEL	1)P O NUMBER 2)CLASS 3)SERIAL	1)ACCNT 2)SPONS 3)PREVS	1)GRANT 2)GOVT DCC 3)GOVT ID	1)ACQ DATE 2)INV DATE 3)DIS DATE	1)ACQ COST 2)TRADE-IN 3)DISPOS
0040963	11 - 300 MEDICAL DEPT BISHOFF,LAURENCE	TYPEWRITER, ELEC IBM 895	REQ119466 OFFICE EQUIP 6830988	16600 MIT		12/22/1980 05/07/1981	866. 0.
0041019	6A- 100 SPECTROSCOPY LAB FIELD,ROBERT	RECORDER 2M3	GFE CAPITAL EQUIP	86991 TIMIT NSF	7810178CHE SF122-8000081	09/19/1980 04/08/1981	614. 0.
0041020	6A- 100 SPECTROSCOPY LAB FIELD,ROBERT	OSCILLATOR, RADI	GFE CAPITAL EQUIP	86991 TIMIT NSF	7810178CHE SF122-8000710	09/19/1980 04/08/1981	520. 0.
0041021	6A- 100 SPECTROSCOPY LAB FIELD,ROBERT	AMPLIFIER DYMEC 2460A	GFE CAPITAL EQUIP 31100351	86991 TIMIT NSF	7810178CHE SF122-7904946	09/19/1980 04/08/1981	398. 0.
0041022	6A- 100 SPECTROSCOPY LAB FIELD,ROBERT	PREAMPLIFIER, LO TEKTRONIX RM122	GFE CAPITAL EQUIP 13456	86991 TIMIT NSF	7810178CHE SF122-8000713	09/19/1980 04/08/1981	207. 0.
0041023	6A- 100 SPECTROSCOPY LAB FIELD,ROBERT	PREAMPLIFIER, LO TEKTRONIX RM122	GFE CAPITAL EQUIP 13457	86991 TIMIT NSF	7810178CHE SF122-8000713	09/19/1980 04/08/1981	207. 0.
0041024	6A- 100 SPECTROSCOPY LAB FIELD,ROBERT	PREAMPLIFIER, LO TEKTRONIX RM122	GFE CAPITAL EQUIP 13436	86991 TIMIT NSF	7810178CHE SF122-8000713	09/19/1980 04/08/1981	207. 0.
0041025	6A- 100 SPECTROSCOPY LAB FIELD,ROBERT	AMPLIFIER DYMEC 2411A	GFE CAPITAL EQUIP 25700024	86991 TIMIT NSF	7810178CHE SF122-7904941	09/19/1980 04/08/1981	2250. 0.
0041027NPT	6A- 100 SPECTROSCOPY LAB FIELD,ROBERT	MODULATOR SPECTRA-PHYSICS 320	GFE CAPITAL EQUIP	86991 TIMIT NSF	7810178CHE SF122-8000081	09/19/1980 09/30/1980 09/30/1980	2335. CANNIBALIZED 0.
0041028	6A- 130 SPECTROSCOPY _AB FELD,MICHAEL	LASER, KRYPTON COHERENT 3000K	188862 LAB & SCI EQUIP 603	88776 NSF	7818580CHE	05/19/1980 04/09/1981	34000. 0.

Figure 8.28. Equipment inventory report in tag number sequence. (Courtesy of Massachusetts Institute of Technology, Cambridge.)

CU OFF PLAN & PROJ MGMT

INSITE 3 DM PROJECT INVENTORY

03/02/1982

BUILDING PROJECT COST SUMMARY

BUILDING	TOTAL	IMMEDIATE REPAIR	URGENT	PLANNED MAINTENANCE	ROUTINE MAINTENANCE	ENERGY	MODIFICATION /RESTORATION
ALUMNI RCRDS CTR	185000	43000	5000	82000	0	35000	20000
AVERY	995050	423200	37800	273950	51000	75500	133600
B F FIELD HOUSE	269000	27000	45000	120000	0	22000	55000
BUTLER LIBRARY	4051250	1163000	269100	1330950	514000	279700	494500
CASA HISPANICA	485250	195000	24750	163600	15000	26000	60900
CASA ITALIANA	514575	41250	33025	213200	41000	25800	160300
CHANDLER	1071500	125600	55300	565800	48000	115800	161000
COMPUTER CENTER	131540	52000	6000	56000	0	7000	10540
DODGE	1212380	28100	105980	670300	74000	98000	236000
DODGE P.F. CTR	473710	133000	36000	154000	0	60000	90710
EARL HALL	346980	30800	12530	224450	13000	39700	26500
EAST HALL	663950	200800	29550	285100	5000	99200	44300
ENGINEERING TERR	1124000	283000	33000	573000	0	94000	141000
FACULTY HOUSE	393700	96000	39000	98000	0	126000	34700
FAYERWEATHER	675590	55000	40540	211650	61000	114400	193000
FERRIS BOOTH	700000	174000	42000	212080	0	256000	16000
HAMILTON	1301225	160100	63125	722900	65000	111700	178400
HAVEMEYER	1896520	259670	86250	819100	98000	115300	518000
HOGAN	283900	18200	53600	62100	0	86500	63500
I.A. EXTENSION	59000	14000	0	35000	0	3000	7000
INT AFFAIRS	2010280	506000	89000	289000	0	1087000	39280
JOURNALISM	1156100	94800	68500	438500	65500	147700	341100
KENT	1628350	460050	49300	669000	84000	102200	263300
LAW	1497820	67000	88000	520000	0	762000	60820
LEWISOHN	579600	23800	29950	402350	44000	52000	27500
LOW LIBRARY	2221070	550800	75170	1252600	108000	59000	175500
MATHEMATICS	895600	361200	53200	166000	48000	74200	193800
MCVICKAR	542320	74000	69000	194000	0	123000	82320
MUDD	1744560	188000	123000	830000	0	579000	24560
PEGRAM	384540	6000	32000	175000	0	44000	127540
PHILOSOPHY	1127500	391100	52500	256400	50000	123500	254000
PRENTIS	2063300	457220	480600	479000	85000	162000	399500
PUPIN	2418000	530000	236000	436000	105000	252000	859000
SCHERMERHORN	1266250	62300	123200	392750	112000	198000	378000
SCHERMERHORN EXT	982450	29750	57800	517800	71000	144100	162000
ST. PAUL S	726200	433200	13000	85200	20000	119000	55800
UNIVERSITY HALL	326400	25400	41000	166200	0	4200	89600
URIS	1434390	493000	9000	473050	0	438000	21340
WATSON	448275	16772	20250	152500	26000	117550	115200
*** TOTALS	40287125	8293795	2728020	14768450	1803500	6379050	6314310

Figure 8.29. Preventive maintenance report using INSITE 3. Developed at Columbia University. (Courtesy of Massachusetts Institute of Technology, Cambridge.)

located, should an inspection be required. The alternative to this approach is to send out lists of assets that are supposed to be in the possession of a particular company unit, and have someone from that unit self-report if the assets are there or not. This approach rarely produces good results, yet most organizations are forced to use it because they do not yet have a facilities management system.

The use of INSITE to manage a preventive maintenance program is shown in Figure 8.29. Data is summarized by building, from an extensive inspection and survey of each space. One governmental user of INSITE employs the system primarily to manage leases. The reports shown in Figures 8.30 and 8.31 are very similar to those produced by the LOIS system.

A novel use of a facilities management system is shown in Figures 8.32 and 8.33. Here, the report generator has been linked to INSITE's room list to produce data collection and survey forms. The first is an evaluation form used to study the adequacy of existing spaces. The second is used to record the activities occurring in each room or space. Occupants are asked to indicate the percentage of time spent in each of 11 categories. Responses are used to update the relevant fields in the INSITE space record. With this capability the planner can quickly produce a complete set of forms for any of the area, building, or equipment-related surveys discussed in Chapter 2.

The current version of INSITE runs on IBM equipment (4300 series and larger) running OS or VS operating systems. It is possible to code a software link between a facilities management system such as INSITE and many of the CAD systems reviewed in the preceding chapter. As noted there, it is useful to be able to store non-graphic facilities data directly on a CAD system. An illustration was given in Figure 7.9. In theory, a planner could store an entire space accounting and equipment data base on his CAD system, record lengths and storage capabilities permitting. This would eliminate the need for a separate facilities management system. Several popular CAD systems include data base management to aid in such applications. But in practice, there are at least three significant problems involved with this approach:

1. *Data structure and storage*: A CAD system data base is drawing oriented. Typically, non-graphic data handling is provided to store information about graphic entities such as equipment symbols or floorplans. To have the entire non-graphic data base online for certain types of analyses may also require all drawings to be online as well—an impractical requirement where large facilities are involved. A large-scale facilities management system may require 100 Mb or more for non-graphic data. This is a problem on some CAD systems even without the associated graphic data.

2. *Conflicting uses and operations*: CAD systems are best used for detailed layout and design. As such, they are used by architects, interior designers, layout planners, plant engineers, and draftsmen. Large facilities management data bases may have 20 or 30 uses as we have seen in Figure 8.25. Few, if any, are related to detail design. Access to the non-graphic data base is typically through clerical personnel. Updates and large-scale analyses may require the entire core-memory (RAM) of the CAD system, degrading or even halting its design functions. The degree to which this occurs depends upon the manner in which the CAD vendor has configured the system's software, and which activities are given precedence in a multiple-user, multiple-task setting.

COMMONWEALTH OF MASSACHUSETTS   INSITE 3 LEASEHOLD INVENTORY   08/03/1982   PAGE 1

A SUMMARY OF LEASEHOLDS HELD BY STATE AGENCIES AT THE BEGINNING AND END OF FISCAL YEAR '82
THOUGH PRIMARILY OFFICE SPACE, THESE LEASEHOLDS INCLUDE OTHER TYPES OF SPACE AS WELL

	LEASES ACTIVE AS OF 06/30/81	AREA IN SQ FT AS OF 06/30/81	TOTAL ANN. COST AS OF 06/30/81	AVERAGE RATE/FT AS OF 06/30/81	LEASES ACTIVE AS OF 06/30/82	AREA IN SQ FT AS OF 06/30/82	TOTAL ANN. COST AS OF 06/30/82	AVERAGE RATE/FT AS OF 06/30/82
ADMIN + FINANCE	34	161906	1267717	7.83	28	71379	641423	8.99
ATTORNEY GENERAL	3	15803	121594	7.69	4	25803	214994	8.33
AUDITOR	6	9938	91235	9.18	6	9938	91235	9.18
CAMPN POLTL FIN	1	2478	14100	5.69	0	0	0	Z
COMMUN + DEVELMP	5	4828	43975	9.11	3	4120	39373	9.56
CONSUMER AFFAIRS	3	10558	62188	5.89	2	6279	45953	7.32
ED AFFAIRS	14	209301	1255605	6.00	11	164838	1493796	9.06
ELDER AFFAIRS	5	50837	397631	7.82	5	50837	397631	7.82
ENERGY RESOURCES	11	23269	191301	8.22	12	21251	163528	7.70
ENVMNTL AFFAIRS	46	179005	922072	5.15	44	170214	923388	5.42
GOVERNORS OFFICE	1	1747	26205	15.00	0	0	0	Z
HUMAN SERVICES	359	2365849	14015711	5.92	377	2488092	15850486	6.37
INTRSTATE COOP C	2	508	3194	6.29	2	508	3194	6.29
JUDICIAL	80	4162002	11751913	2.82	71	215514	1319022	6.12
MANPOWER AFFAIRS	149	1077903	5929727	5.50	142	1050113	5879267	5.60
PUBLIC SAFETY	57	329661	1603989	4.87	54	295285	1608910	5.45
SECTR OF STATE	5	14512	105777	7.29	5	14512	105777	7.29
TRANSPT + CONSTR	29	84067775	304413	0.00	24	83915515	295590	0.00
TREASURER	12	88354	813440	9.21	11	84829	760237	8.96
TOTALS:	830	92776234	38921787		801	88589025	29833804	

Figure 8.30. Lease management report using INSITE 3. Developed by the Commonwealth of Massachusetts. (Courtesy of Massachusetts Institute of Technology, Cambridge.)

COMMONWEALTH OF MASSACHUSETTS

INSITE 3 LEASEHOLD INVENTORY

LEASES EXPIRING IN CHRONOLOGICAL '82 SHOWN BY MONTH

	AGREEMENT NUMBER	EXPIRATION DATE	STREET ADDRESS		CITY OR COUNTY
BLIND COMMISSION	78 - 051	06/30/1982	COUNTY S	991	FALL RIVER
	82 - 100	06/30/1982	SECOND S	72	CAMBRIDGE
	82 - 129	06/30/1982	MADISON S	90	WORCESTER
COMMTY HEALTH PL	78 - 044	06/30/1982	WASHINGTON	600	BOSTON
	78 - 083	06/30/1982	BOYLSTON S	80	BOSTON
D E Q E	82 - 096	06/30/1982	STATE S	1414	SPRINGFIELD
DIV ST POLICE	82 - 042	06/30/1982	WIGGINS AIR		NORWOOD
FORESTS + PARKS	78 - 077	06/30/1982	MAIN S		SUNDERLAND
MASS HISTORC DIV	78 - 042	06/30/1982	WASHINGTO S	294	BOSTON
	80 - 140	06/30/1982	WASHINGTO S	294	BOSTON
MENTAL HEALTH		06/30/1982			
MNTL HLTH RGN 4B	82 - 056	06/30/1982	WATERTOWN S	429	NEWTON
PUB WKS DISTR 1	81 - 312	06/30/1982	MECHANIC S	3	GREAT BARRINGTON
PUB WKS DISTR 5	78 - 059	06/30/1982	RIVERSIDE D	1042	METHUEN
PUB WKS DISTR 6	78 - 064	06/30/1982	ROUTE 123		NORTON
PUB WKS DISTR 7	78 - 068	06/30/1982	ROUTE 139		PEMBROKE
PUBL WELFR RGN 3	78 - 017	06/30/1982	LOWELL S		PEABODY
PUBL WELFR RGN 4	82 - 101	06/30/1982	WASHINGTO S	320	BROOKLINE
PUBLIC HEALTH	78 - 045	06/30/1982	WASHINGTO S	600	BOSTON
	78 - 046	06/30/1982	WASHINGTON	600	BOSTON
	78 - 089	06/30/1982	WASHINGTO S	600	BOSTON
	78 - 090	06/30/1982	WASHINGTO S	600	BOSTON
	79 - 202	06/30/1982	WASHINGTO S	600	BOSTON
R M V	77 - 022	06/12/1982	ROUTE 132		HYANNIS
	82 - 063	06/30/1982	PRESIDENT A	1658	FALL RIVER
	82 - 066	06/30/1982	MARKET S	207	LAWRENCE
RECLAMTN BD	78 - 053	06/30/1982	COLUMBIA R	183	HANOVER
REVENUE DEPT	78 - 052	06/30/1982	PRESIDENT A	1670	FALL RIVER
	81 - 372	06/30/1982	PRESIDENT A	1658	FALL RIVER
	81 - 373	06/30/1982	MERRIMACK S	144	LOWELL
TRIAL COURT	82 - 019	06/30/1982	FOREST A	486	BROCKTON
	82 - 095	06/30/1982	MAIN S	501	SPENCER

Figure 8.31. Lease expiration report using INSITE 3. Developed by the Commonwealth of Massachusetts. (Courtesy of Massachusetts Institute of Technology, Cambridge.)

SPACE EVALUATION FORM

ROOM NO.	AREA	OCCUPANT RANK	PROXIMITY TO CORE	LIGHT AIR	ACTIVITY % TIME	CLUSTER PATTERNS	REMARKS
4 – 031	420						
4 – 033	570						
4 – 035	94						
4 – 035A	329						

Figure 8.32. Survey form for space evaluation. Produced by running off the INSITE room list onto a formatted but empty report. (Courtesy of Massachusetts Institute of Technology, Cambridge.)

3. *Data entry and start-up*: If the non-graphic data is reached through reference to graphic entities and drawings, these must all be entered in the system before the non-graphic data base can be used as a whole for summary level reports and analyses. This, in turn, requires the digitizing or drafting of all existing facilities--a process that can take years where several million square feet are involved. On the other hand, a non-graphic data base, by itself, can be loaded and used in a much shorter period.

One CAD vendor, Computervision, has recognized these limitations and conflicts and now offers a separate computer (the IBM 4300 series) for non-graphic data base management. Graphics operations remain on the CAD computer, and the two are linked as needed by software and communications lines. This may be the best approach where an extensive non-graphic data base is desired.

Corporate Data Base Design

Information systems developed during the 1970s use data structures known as hierarchies and networks (30). These are conceptually like an organizational

COLUMBIA UNIVERSITY INSITE 3 SPACE INVENTORY 01/15/1983 PAGE 1

DEPARTMENT: ALUMNI RECORDS

ROOM TYPE	AREA	SPON RSCH	TRN/ INST	DEPT ADMN	OTHR ACTS	GENL ADMN	SPON PROJ	O&M	STUD ADMN	LIBR	JOIN SPAC	UNA/ INAC
ALUMNI RCRDS CTR												
1 - B01 OFFICE SVC	294											
1 - B01A OFFICE SVC	7											
1 - B02 LOUNGE	310											
1 - B02A OFFICE SVC	10											
1 - B04A OFFICE SVC	12											
1 - 101 OFFICE	210											
1 - 101A EDP RM	68											
1 - 102 SECTY/RECP	158											
1 - 102A SECTY/RECP	112											
1 - 103 OFFICE SVC	57											
1 - 103A OFFICE SVC	4											
1 - 103B PRIVATE LAV	21											
1 - 201 OFFICE	100											
1 - 202 SECTY/RECP	277											
1 - 202A OFFICE SVC	72											
1 - 203 EDP RM	196											
1 - 203A OFFICE SVC	3											
1 - 204 PRIVATE LAV	57											
1 - 204A OFFICE SVC	12											
1 - 301 OFFICE	183											
1 - 301A OFFICE SVC	3											
1 - 301B OFFICE SVC	11											
1 - 302 OFFICE	124											

Figure 8.33. Survey form for space activity analysis. Produced at Columbia University. (Courtesy of Massachusetts Institute of Technology, Cambridge.)

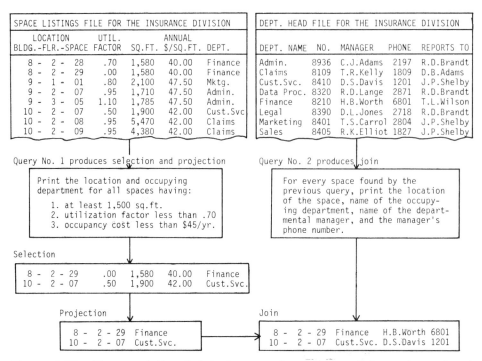

Figure 8.34. How a relational data base works. Sequential queries extract successive subsets of data from separate files, joining the extracted data as required. (Adapted from an example by Relational Database Systems, Inc., Sunnyvale, California.)

chart in the way they relate files to one another. The way data can be retrieved and manipulated is pre-defined, and some desirable operations may be precluded without re-defining the data structure. This is a major task to be avoided if at all possible. All users of information systems have experienced the frustration of having data on-file but needing a special program to use it in a newly desired way. Often the user does without, since the cost of the program or the time required to write it is excessive. In many of these instances, the user is encountering the limitations of a pre-defined data structure.

In recent years, a new approach to data base design has become popular for large-scale applications such as facilities management. The data structure used is called "relational." It permits large numbers of discrete records to be manipulated and their relationships defined interactively by the user of an information system. Programming, as such, is not required, although certain commands must be mastered. A relational data base is more flexible than the hierarchy or network, and it is the preferred choice whenever all the desired applications and extensions of the information system are not known in advance. Relational data bases are also appropriate where organizational change is frequent, since this often forces restructuring of a static, hierarchical design. A relational data base makes it easy to have several "views" of the data supporting different uses--equipment inventories, lease management, preventive maintenance--and still have a consistent overall structure.

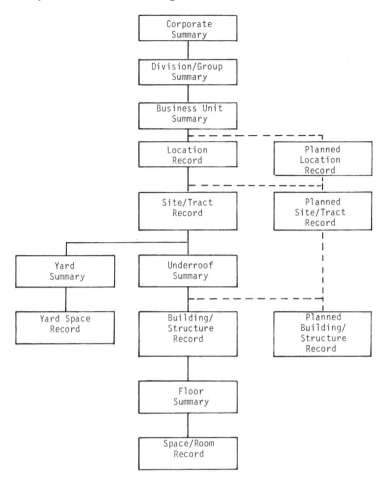

Figure 8.35. Profile (user view) of a corporate facilities data base. (From Ref. 19.)

A user's view of the way relational data bases generally function is shown in Figure 8.34. Resource Dynamics, Inc., of Stowe, Massachusetts, was one of the first companies to offer a relational data base for facilities planning and management. It is linked to CAD software also from RDI. Arrigoni Computer Graphics and Micro-Vector both use relational data base systems to integrate their facilities planning software. Other major CAD vendors use them for non-graphic data storage and manipulation. In nearly all cases, these vendors are using a relational DBMS licensed from another software house or system supplier. Arrigoni, for example, uses the UNIFY DBMS from UNIFY Corp. (formerly North American Technologies). This is a DBMS written for use with the UNIX operating system. Informix is another system written for UNIX by Relational Database Systems, Inc., of Sunnyvale, California. Many such products are available for a variety of operating systems.

A user's view of a corporate facilities data base is shown in Figure 8.35. It appears to the user as a hierarchy, with space/room and yard (outdoor) space records as its basic elements. These can be summarized by floor and yard location. Building/structure records appear to include all space/room records plus additional data that pertains to the building as a whole. In reality, there are pointers in the data base between the building record and all the spaces or rooms within it. Buildings and structures can be summarized as underroof space on a particular site or tract. Each site or tract has its own record, which consists of unique data about it plus pointers to the appropriate building/structure and yard space records. The highest level record is the location, which is a geographic entity rather than a physical one. The location record, for example, would contain data on the aggregate Detroit-area operations of the Chrysler Corporation. Information about specific facilities, however, would be contained only in the lower level site/tract, building/structure, and space/room records--each of which would contain pointers to the Detroit location record. The contents of a location record might include average facilities densities, output ratios, average wage rates, utility costs, taxes, rents, new construction costs, climatic data, and other matters of importance in long-range, strategic planning.

Above the location record are the organizational summaries. These are actually tables that relate business units to divisions or groups. The corporation is simply the set of all division or groups. There are no organizational units as such in the facilities profile, although these could be added if desired. But, as shown in Figure 8.35, these are merely meaningful ways of aggregating (relating) the underlying location and facilities records for purposes of planning and management.

Long-range planning decisions must consider facilities that are under construction or renovation for occupancy at a future date. Facilities that have been planned or approved but not yet begun must also be considered. For this reason the profile in Figure 8.35 includes records for planned entities. These stop at the building/structure "level," since space/room assignments are typically not made until occupancy draws near. At that time, the planned entity may be converted to real, and the appropriate space/room records added to the data base.

The data structures for our sample facilities profile are illustrated in Figure 8.36. The data base is a series of relations, each identified by a key, relating it to the larger physical and geographic entities of which it is a part. There are no relations for "underroof" or "floor" spaces, since these are aggregations of basic records. In practice, there are several ways to define the key identity of each relation. In the INSITE system, for example, each space/room number contains its site, building, and floor reference in a prefix. Thus room E19-451 is the 51st room on the 4th floor of building 19, located on the East Campus. Each room or space must have a unique identity or the system will malfunction.

The data structures show a graphic ID associated with each physical entity. This represents the place of a CAD system pointer or interface. Design graphic or CAD relations are illustrated in Figure 8.37. They include both vector and symbol relations. Vector data is used to create room, building, and site boundaries. Symbol relations cover the active placements of standard library symbols. These could be furniture, equipment, or architectural features. Thus, a floor plan is viewed as a collection of vector and symbol relations, each with its appropriate attributes in the form of origins, layers, angles, and line codes.

As suggested in Figure 8.38, the facilities profile can also contain data on equipment, furniture, inventories of finished goods or even work-in-process, and personnel--all by defining these relations to include the space ID.

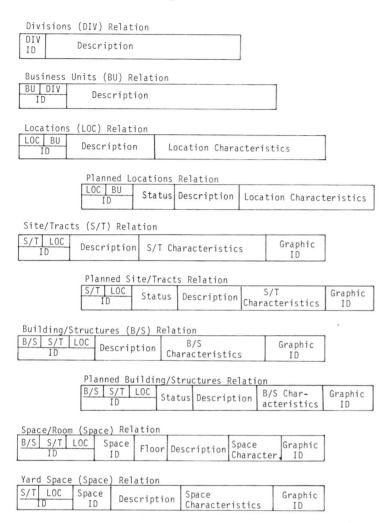

Figure 8.36. Data structures for a relational facilities data base. (From Ref. 19.)

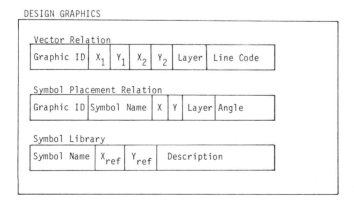

Figure 8.37. Data structures for design graphics (CAD) relations. These are an optional component or interface to the data structures in Figure 8.36.

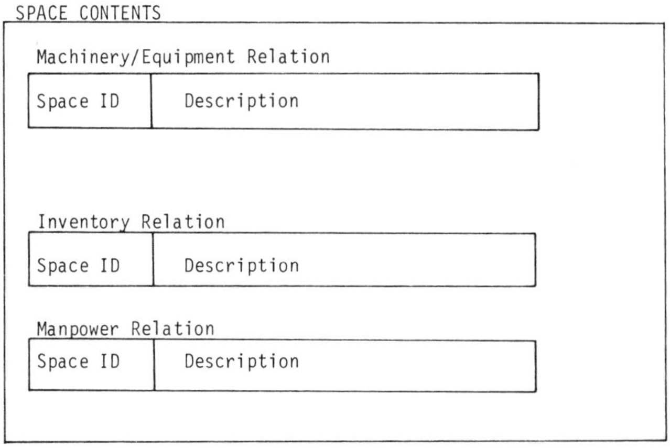

SPACE CONTENTS

Machinery/Equipment Relation

Space ID	Description

Inventory Relation

Space ID	Description

Manpower Relation

Space ID	Description

Figure 8.38. Data structures for space "contents"--equipment, inventory, and personnel. These are also optional components or interfaces to the data structures in Figure 8.36.

Developing a corporate facilities profile is a job for data processing professionals who have a good understanding of facilities planning and data base management systems. The process may take a year or more, passing through the following seven steps:

1. *Administrative procedures*: defining the jobs and organizational responsibilities of administrators and users
2. *System design*: covering both manual forms and information flow as well as computer-based
3. *Data definitions*: to insure unique space names and numbers, define dictionaries of space functions and assignments, and define the rules by which areas are measured (see Figure 8.39)
4. *Data collection*: beginning with the installation of a comprehensive room numbering system, followed by an audit of existing facilities, to establish room configurations and function, construction of up-to-date floor plans, area measurements, and finally room assignments
5. *Build data base*: beginning with the appointment and training of a data base manager and the key entry of all collected data
6. *Maintenance procedures*: including regular audits of facilities, capturing of data from other systems and paperwork, and production of verification reports
7. *Report generation*: the final step once the data base has been constructed and provisions made for keeping it up-to-date

This seven-step process is the one prescribed for INSITE users by M.I.T.'s Office of Facilities Management Systems. A companion process for developing an equipment data base includes the definition of standard nomenclature for all machinery and equipment, selection of bar-code labels, and an optical scanning device. Data collection procedures will typically include the capturing of

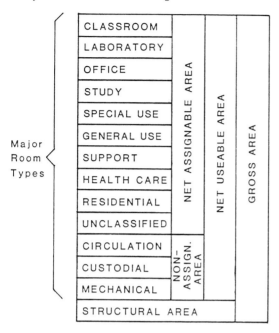

Figure 8.39. Building area definition summary. (From Federal Construction Report 50, 1964; Facilities Inventory Classification Manual, 1973; United States Government Printing Office, Department of Health, Education and Welfare Publication No. (OE) 74-11424.)

equipment purchase orders and vendor invoices, as well as the tagging of all newly received equipment.

The MIS applications covered in this chapter have addressed the provision and review phases of the facilities management cycle presented in the preface of this book. We have now come full circle and have surveyed the full range of computer aids available to the facilities planner. We have discussed them roughly in the order in which they would be encountered in practice--moving systematically through the phases of planning, followed by construction, and ongoing facilities management. We have also presented these aids roughly in order of the leverage they provide in achieving more effective facilities plans-- looking first at decision support, then CAD, and finally at MIS. In the chapters that remain, we will devote our attention to the important issues in successful implementation.

Part III

HOW TO ACHIEVE SUCCESS

9

Selecting and Developing Computer Aids

Basic Choices

The decision to acquire or develop computer aids leads to four basic choices. First, should existing hardware be used or should new hardware be acquired? The choice depends on the nature, capacity, and utilization cost of existing equipment. Certainly its use is favored if it has the capacity and can do the job at an acceptable cost.

Second, should software be acquired from outside sources or internally developed? Proven software is available for every significant facilities planning application. Internal development can be a costly, time-consuming proposition. It is often poorly documented and subject to many errors and deficiencies. For these reasons, outside sources are usually better. At the very least, an outside search should be made before internal development.

Third, should a turnkey system be acquired, or should separate hardware and software choices be integrated in-house? The turnkey choice assures that hardware and software are properly matched to the desired application. It also insures continuing maintenance and support. However, it may require duplication of replacement of existing hardware. For this reason, the choice may be an economic one, with the additional cost weighed against the additional security of the turnkey solution.

Fourth and finally, should computer support be acquired at all, or should a service bureau be used? In general, service bureaus or time-shared services are best under these conditions:

1. Experimental use only
2. Infrequent use
3. Non-urgent use
4. No suitable equipment available
5. No funds for acquisition
6. No qualified staff to operate or maintain

Naturally, the choices to be made depend upon the nature and complexity of the applications. Simple calculation routines, suitable for a personal computer, are easily developed in-house. Many facilities planning groups already have a personal computer. The task is made even easier if the planner can use general-purpose software such as an electronic spreadsheet, or a statistical package supplied by the computer manufacturer.

At the other extreme, continuous use of computer-aided design will almost always require an outside, turnkey solution. Likewise, development of several computer programs tied to a single, common data base will require either a turnkey solution or a blend of outside and internally developed software.

Getting Organized

Before any choices are made, the facilities manager or planner should organize a study team. The selection and development of computer aids, no matter how simple, is never a one-person job. The team should consist of 3-5 people. On major applications, the team must include the final-line management authority or approver of the funds. Without the commitment and early involvement of this person, there is a good chance of problems later on--delays, failure to get approvals, lack of follow-through during the start-up period.

The rest of the team should include: a key user of the output, a probable operator (if not the same), the supervisor responsible for day-to-day implementation and use, an open-minded data processing staffer (if available). The data processing staffer must be willing to objectively consider outside replacements for current hardware and software. If not, he will be an impediment to progress.

The role of consultants will be discussed more fully later. It should be noted here that an outside, knowledgeable consultant can be a useful team member, especially if internal expertise in computers is lacking. But, the consultant should not lead the team.

For best results, the team leader should be the person who will supervise or operate the system once installed, provided this person possesses the following:

1. Experience in the applications under study
2. Ability to write a clear specification of what is wanted
3. Ability to lead but not dominate a meeting and handle heated discussions of technical matters
4. Objectivity and desire for facts
5. Ability to debate issues on their merits, and not on the personalities of those putting forth the issues
6. Willingness to compromise
7. Businesslike appearance and demeanor in dealing with outsiders
8. Commitment to successful implementation--ideally his or her career advancement should in some way be tied to or influenced by the team's decisions and success

These are the minimum qualities required. Notice that knowledge of computers or programming is not listed. A team leader's lack of technical experience can always be overcome by the presence of the data processing staffer and the outside consultant--or any other knowledgeable team member, for that matter.

The Specification

Since so many computer aids are commercially available, there is a natural desire to reach for a catalog and start searching for the ones that might fit.
And, like any exercise in catalog shopping, there is a great deal of excite-
ment as apparently useful products are identified at reasonable or even low
prices--especially those that run on existing hardware.

But before any literature is even requested, the planner needs a written
specification of what is wanted. Without this document, a great deal of time
will be wasted investigating solutions that have little chance of satisfying the
planner's true requirements.

Focus first on what is wanted, then on what is available. Start with the
list of tasks, decisions, and computer assistance shown in Chapter 4. Which
one(s) are of immediate interest?

If you are starting from few or no computer aids, pick those which have
high leverage--low cost but high payoff in more effective plans. This will
generate interest and support from management for subsequent, more expensive aids.

Once you have identified the task(s) or decision(s) to be supported, prepare the written specification. It may be no more than one page for an internally developed personal computer application. It may be 20-50 pages with exhibits for a major turnkey or outside software investment. See Figures 9.1
and 9.2. Regardless of length, the following discussions must be present:

1. *Organization* (WHO?)
   a. What units or departments will house and use the computer aid?
   b. Who are the beneficiaries of its use (names, titles)?
   c. Who will be the operator or user of the computer aid (names, titles)?
   d. Who (names, titles) will have authority and control over use?
2. *Results* (WHAT?)
   a. What recurring problem, task, or decision is to be helped by the computer aid?
   b. What form will the output take (reports, tables, graphs, charts)?
   c. How specifically will the output be of use?
   d. What is the chief benefit of using the computer?
   e. How are results achieved today?
   f. What are the weaknesses of today's approach?
3. *Access* (WHERE? and WHEN?)
   a. At what physical locations will access to this aid be required?
   b. How many users will need simultaneous access?
   c. What other applications or aids will be running at the same time?
   d. What hours of the day will access be required?
   e. Is batch submission of input acceptable?
   f. What is the benefit of interactive input, if any?
   g. How fast is response required?
4. *Approach* (HOW?)
   a. What is the desired input data to the computer (attach examples)?
   b. What is the desired output from the computer (attach examples)?
   c. For analytical aids, what is the acceptable or desired procedure (attach logic flow chart of mathematical algorithm)?

```
┌─────────────────────────────────────────────────────────────┐
│                   SPACE PROJECTIONS SOFTWARE                  │
│                        Specification                          │
│                                                               │
│ 1.  The programs will be used by the Mining Equipment         │
│     Division -- Headquarters Facilities Staff.  Todd          │
│     Kelley will install and maintain.  All staff              │
│     members will be expected to learn and use the             │
│     programs.                                                 │
│                                                               │
│ 2.  The programs will be used to make long- and short-        │
│     range space projections.  They will eliminate the         │
│     need for manual compilation of head count surveys         │
│     and extensions.                                           │
│                                                               │
│ 3.  The programs must operate on the existing Apple IIe       │
│     with 96K RAM.                                             │
│                                                               │
│ 4.  Reports must fit on the existing Okidata Microline-80     │
│     printer.                                                  │
│                                                               │
│ 5.  The programs must have the following features:            │
│                                                               │
│     a.  Accept up to 50 departments;                          │
│                                                               │
│     b.  Accept up to 30 job titles or positions;              │
│                                                               │
│     c.  Store up to 20 space standards;                       │
│                                                               │
│     d.  Produce space projections for up to 5 future          │
│         periods.                                              │
│                                                               │
│     e.  Calculate gross square footage from a variable        │
│         net to gross factor.                                  │
│                                                               │
│     f.  Produce a summary report for all departments.         │
│                                                               │
│ 6.  The vendor must have an 800 number for user support.      │
│                                                               │
│ 7.  The vendor must supply a complete user manual             │
│     including sample cases.                                   │
│                                                               │
│ 8.  The programs should cost less than $1500.                 │
└─────────────────────────────────────────────────────────────┘
```

Figure 9.1. Example of a one-page software specification.

     d. How much information must be stored and manipulated to arrive at acceptable output (number of characters at each step and on file)?

     e. Who (names, titles) will enter the inputs and what are their skills?

     f. Who (names, titles) will interpret the outputs and what are their skills?

     g. Who (names, titles) will review and approve the internal logic or procedures of the software?

     h. What is the specific desired relationship (if any) between the computer and others already in use?

  5. *Budget* (AT WHAT COST?)

     a. How much are we willing to spend?

     b. What economic payoff (if any) do we expect?

     c. How will the payoff be verified (if this is possible, at all)?

  d. *Schedule* (WHEN NEEDED?)

     What is the target date for first use of the computer aid?

SPECIFICATION

PROPOSED CAD SYSTEM

Table of Contents

Figure 9.2. Sample table of contents for a CAD system specification. Note the links to two existing systems, and the provision of 3-D and space planning software. The configuration section is written to allow a minicomputer or standalone micros.

When preparing the specification, keep in mind that it must be clear and useful to several audiences:

1. *Management*: who will approve the funds
2. *The team itself*: as a basis for subsequent evaluation of solutions
3. *Consultants*: who may advise on the worth of various solutions
4. *Vendors*: of software, hardware, or turnkey systems
5. *Data processing staff*: to evaluate outside and internal solutions

The Selection Process

In general, the selection process for outside software or a turnkey system will follow the nine-step process outlined below.

1. *Get organized*: Appoint the team, make assignments; establish a budget and schedule; write the specification.
2. *Identify vendors*: Use such resources as directories, consultants, friends and associates, professional and trade associations. Request literature. Attend shows.
3. *Initial screening*: Identify 3-5 vendors with greatest potential based on: literature provided, first impressions, referrals, telephone contact, salesman's calls.
4. *Demonstration/test case*: Visits; hands-on test of the software or system using a realistic sample problem.
5. *Request for proposal (RFP)*: Written request adapted from initial specification, based on what has been learned since it was written. Send to 3 finalists.
6. *Final evaluation*: Based on demonstrations, response to RFP, and other factors.
7. *Negotiations and contracts*: Clarify what is to be delivered and when, warranties, liabilities, terms, conditions, dates, installation support, maintenance agreements, prices. (Keep the contract documents as simple and straightforward as possible.)
8. *Implementation planning*: Conversion from existing computer support, training, physical installation or connection.
9. *Installation*: Switch over, move in, connect, load data, run, audit.

If the intended purchase is small--say only a few hundred dollars--the selection process will be greatly abbreviated. A good deal of software is available by mail "as is." The vendors of such software may not have local or even regional sales people, conduct demonstrations, respond in detail to an RFP, or negotiate contracts. On the other hand, because they do not incur the expense of such activities, what they do have is available at a low price.

The Request for Proposals

Requests for proposals (RFP's) should be used only for major software or computer system investments. There is no need for an RFP, even on major acquisitions, if the preliminary screening yields a solution that meets all the needs contained in the specification. RFP's take a good deal of time to prepare, and still more to review the responses received. An RFP should be preceded and followed with a telephone call or a personal visit to each reviewing vendor. For these reasons, an RFP should be sent to no more than 5 firms, and possibly as few as 2 or 3.

The request for proposals should cover the following points:

1. Background on your organization and its approach to facilities planning
2. Specification for the computer aids of interest
3. Information desired about the vendor--how old, how big, what growth rate, what markets served, key personnel and their backgrounds
4. Information desired about the product (computer aid)--key features, hardware, software, training, support, maintenance, pricing, customization
5. Information on the schedule and deliverability of the product--when available, what dates for installation, training, de-bugging, etc.

The RFP should be kept as brief as possible, and should not require the vendors to respond in a rigid format. You will learn more by letting them tell their story in their own way. In fact, the responses to the RFP may cause you to revise the specification. Try to avoid an RFP that indicates the weights you place on various objectives and the scoring mechanism you intend to use. This may inhibit the vendor's response.

Test Cases

Unfortunately, you can never be sure how a computer system really works until you try it with one of your own problems--preferably one you have personally solved using sound manual methods. What you see at trade shows, on the vendor's premises, or in a sales call is what the vendor wants you to see because it stresses the strong points of his product. This does not necessarily mean that the product is weak. But until you sit down and try it, you are only seeing the product at its best.

If the investment in software, hardware, or turnkey system is more than a few hundred dollars, the surest way to evaluate performance is with a test case. Review recently completed projects or reports and find one that is representative of your organization's workload. The "80/20 rule" should apply in selecting a test case--that is, select a problem that comes up 80% of the time, not an extreme case.

Look for a project that was particularly well documented. Then, sit down with representatives from the 3-5 finalist vendors and review the project with them. Let them study it and explain to you how their computer aid would be used--with and without modification.

If you are reviewing software, check to see if your project documents contain all the necessary input data. If not, you may have to manufacture some missing pieces in order to try out the software. Once you are sure you have all the necessary inputs, schedule a hands-on demonstration session, using your test case. Allow plenty of time--up to half a day may be required. And make sure all key decision-makers will be present.

Expect some problems and inadequacies, but keep an open mind. Even the best computer aids can "have a bad day." Keep track of the time spent entering data. This is the most overlooked aspect of computer aids--especially at shows and sales presentations. Then, look carefully at the outputs obtained. Is all that you need provided in a useful format? Finally, have each member of the team write down his or her impressions at the close of the demonstration. Collect these and file them for use in the final evaluation.

If for some reason, the vendor is unwilling or unable to conduct a test case prior to purchase, the purchase agreement should contain a clause giving you the right to operate the system with your data before making a final payment. Some legitimate software houses, for example, simply do not have the sales force or resources to invest in test case demonstrations for every interested party. Yet they still should be willing to let you satisfy yourself within 30 or 60 days that the product will do your job.

Selection Criteria

Having identified vendors, screened them to a few finalists, attended demonstrations, and received proposals, it's now time to decide: which solution is best?

The best way to decide is with a cost/benefit approach. What is the cost or price of each alternative and what are its benefits? It becomes an easy decision if the lowest-cost solution also has the greatest potential benefits. Unfortunately, most choices are not so clear. It is necessary to weigh additional benefits and see if they justify additional costs.

When determining the cost of a computer aid, be sure to include all the sources of cost--some of which may be glossed over by the vendor. Consider these 9 cost factors:

1. *Hardware*: including all desired peripherals such as printers, plotters, tablets, terminals, drives, etc.
2. *Software*: both operating system, if required, and all applications programs
3. *Modifications*: if required, to hardware and software
4. *Training*: standard length plus any agreed-upon extras
5. *Documentation*: in the necessary number of copies
6. *Accessories*: such as cables, modems, workstation furniture, etc.
7. *Freight*: from specified FOB point(s)
8. *Maintenance*: covering hardware, software, and documentation
9. *Installation fees*: if any

Once you are satisfied that you understand the cost of each alternative, you should turn your attention to benefits. If not already done, make a list of the intangible factors or objectives that you are seeking from the alternatives at hand. In your author's experience, the following baker's dozen include the most important selection criteria (in no particular order):

1. *Ease of input*: Consider learning time, difficulty, typical time to enter required data. What is the appeal of the devices being used-- keyboard, touch screen, tablet, etc? Remember, what's easy for some is hard for others. Get consensus from the team, after tryouts with the test problem.

2. *Acceptable output*: Completeness, format. Is it easy to read and interpret? Quality and ability to reproduce. What is the appeal of the technology being used--printer, plotter, monitor, color vs. monochrome?

3. *Ease of set-up and operation*: Consider learning time, difficulty to start up and maintain. Simplicity of interfaces between hardware devices. Physical requirements for power, air-conditioning.

4. *Documentation*: Completeness, number of levels, clarity, internal reproduction rights, update frequency. If calculations are included in the software, check for complete exposition of all formulae.

5. *Training*: Quality of staff, topics covered, length. On major acquisitions ask to attend a training course or at least to review the training manuals. Reference checks should include a discussion of training.

6. *Apparent understanding of application*: Especially for software acquisitions. Do the vendor's salespeople and technical staffers speak your language? Do they have experience in facilities planning and management? Where? When? Do they want to change the way you do things? Why? Are they right? Be sure to keep an open mind and listen to what they are saying.

7. *Willingness to customize*: Again, especially important for software acquisitions. This factor is somewhat related to the previous one. Will the vendor customize to meet unique requirements? What about subsequent support?

8. *Liability*: Does the purchase agreement give you indemnity from patent or copyright infringements by the vendor? Watch for any unreasonable limitation of vendor liabilities.

9. *Service/maintenance*: Who provides it, the vendor or a contract service? Where are they located? What is the response time? The service hours? Consider both hardware and software.

10. *References*: Can be misleading. No vendor knowingly refers you to a dissatisfied customer. Furthermore, even your acquaintances or professional contacts may be unreliable sources--especially if they have made a costly mistake. No one likes to admit that he wasted his own or his employer's money. The bigger the investment, the more likely you are to get only favorable responses.

11. *Installation support*: In person? For how many days? By phone? Don't forget to clarify the charges for this support.

12. *Quality of vendor personnel*: Especially when looking at software acquisitions. Will new, valuable products be forthcoming? Do you know far more than they do about your applications? What is their background and work experience?

13. *Fit with facilities management approach*: How well do the product and the vendor fit with your approach to facilities management and planning? Do the products (especially true with software) and the people share your philosophies, if not your approaches?

Having identified the important factors, we need an organized way to evaluate our alternatives. This can be done with the weighted-factor approach

EVALUATING ALTERNATIVES

Plant/Area Corporate

Project Facilities Inventory   Date 8/10

Description of Alternatives: (enter a brief phrase identifying each alternative) . . . .   A. DBM, Inc.

B. Inventory Systems Co.   C. Bar-Scan Corp.

D.   E.

Weight set by   C. Lawson   Ratings by Property Staff   Tally by   JDM

FACTOR/CONSIDERATION	WT.	A	B	C	D	E	COMMENTS
1 Ease of use	8	E 24	A 32	I 16			
2 Scope and usefulness of output	10	E 30	O 10	A 40			
3 Quality of documentation	7	A 28	E 21	E 21			
4 Amount of training provided	6	O 6	O 6	E 18			
5 Quality of references	3	E 9	A 12	A 12			
6 Amount of installation support	2	O 2	O 2	E 6			
7 Service arrangement	5	I 10	U 0	I 10			
8 Acceptance by DP staff	3	I 10	X -3	O 3			
9							
10							
11							
12							
13							
14							
TOTALS		119	80	126			

NOTES

EVALUATING DESCRIPTION

A	Almost Perfect	4pts	O	Ordinary Results	1
E	Especially Good	3	U	Unimportant Results	0
I	Important Results	2	X	Not Acceptable	1̄

RICHARD MUTHER & ASSOC. INC. - 173

Figure 9.3. The weighted-factor approach to evaluating alternatives.

illustrated in Figure 9.3. Here three computer aids for a facilities inventory are being evaluated.

The relevant factors are listed at the right. The relative importance of each factor is indicated by a weight from 10 to 1, with 10 as most important. The alternatives are listed across the top. The alternatives are rated, one

- EFFICIENCY

  . Lower planning expense

  . Decrease in planning cycle time

  . Ready access to relevant information

  . Standardization of planning approach
    - Common planning language
    - Pooled data base and management

- EFFECTIVENESS

  . Shorter reaction times

  . Reduced effort to perform more sophisticated analysis

  . Development of a greater number of alternatives

  . Longer lasting plans

  . Potential for increased contingency planning

- "HARD" SAVINGS

  . Reassignment of staff

  . Attrition

  . Hiring freeze or avoidance

  . Budget reduction

- "SOFT" SAVINGS

  . Time for new or neglected tasks

  . Cost avoidance without budget reductions

Figure 9.4. Key issues in justifying computer support. (From Ref. 19.)

factor at a time, using a vowel-letter rating scale. After all ratings have been made, the vowel-letters are converted to points and multiplied by the weight to get a score. The factor scores are totaled, with a 12%-15% spread indicating a clear winner. For best results, the ratings should represent the consensus of the study team.

Cost Justification

We have discussed costs and cost/benefit analysis. But in many situations an additional step of cost justification will be required. It is not enough to say that a computer aid is needed, or that it offers great intangible benefits. Some economic return on the investment must be demonstrated.

Unfortunately, cost justification often becomes a very creative exercise when applied to something as intangible as computer support. There are exceptional cases--notably investments in computer-aided design, where productivity gains are a key issue. Other kinds of aids, however, typically yield soft savings or benefits. The justification of CAD systems will be addressed fully in the next chapter. We will dwell for a moment here on other types of computer systems and support. The key issues are summarized in Figure 9.4.

First, recognize the difference between efficiency and effectiveness. In the context of computer-aided facilities planning, this amounts to the difference

between *faster planning* and *better plans.* Savings in both areas can be sig-
nifcant and can easily be shown to justify most computer investments.

Under the banner of faster planning come the savings from shorter planning projects and potentially from the reduced calendar time between identified needs and occupancy of the new or renovated facility. Shorter planning
projects may allow reduction in planning staff, or at least the postponement
of staff additions. But these savings are trivial compared to the potential of
faster occupancy. Saving a month or even a week on large projects can be
worth a good deal of money if construction or interim financing is involved.
If the project is an industrial expansion, for products that are in demand,
the ability to deliver even 2 or 3 weeks sooner may yield a handsome payoff
in sales or market share that might otherwise be lost.

Harder to quantify, but of even greater payoff, are the savings achieved
by better plans. A few percentage points in operating efficiency--if they can
be attributed to the facilities plan--will pay for a lot of computer support.
One of the hallmarks of a good plan is its longevity, i.e., its avoidance or
delay of a rearrangement or expansion. Savings here can be enormous and
may be attributable to the use of computers in generating the plan.

Periodic measurement improves performance in just about every arena of
life. The same is true of facilities and space. When computer aids are used
to monitor and control space and fixed assets, utilization of these assets can
be expected to improve. Again, the savings can be significant.

The facilities manager or planner must recognize, however, that most of
these payoffs are "soft," i.e., potential savings that are difficult to audit.
But that does not mean they are "pie in the sky." Those who would use the
computer to achieve these savings will simply have to do some internal selling
to get the support they need.

The Internal Development Process

In spite of the many commercially available aids, some planners will choose the
internal development process. Perhaps they have a simple application in mind,
an available computer, and a knowledgeable programmer on the staff. With the
added ingredient of time, it is possible to develop quite a few useful applications. Or, there may be a good programming group in the data processing department whose "charges" are reasonable and whose past performance has been
good.

If either approach is to be taken, it should be with a reasonable amount of
planning and organization. The essential steps in the process are:

1. Write a specification (no matter how brief).
2. Prepare a project plan, including schedule and budget.
3. Develop a test case.
4. Write the code.
5. Test and improve the code.
6. Final demonstration using the test case.
7. Document all code.
8. Write a user's manual (no matter how brief).
9. Install and use.
10. Modify and refine (including updates to documentation and user's
    manual.

Steps 1, 2, 6, 7, and 8 are often skipped, to the user's ultimate chagrin. The most common failings of internal efforts are documentation and user's manuals. If they are ever done at all, they are rarely current with the latest version of the code. Some experts recommend writing the user's manual *before* developing the software, to insure that it gets done and to provide a more definitive specification.

It is becoming very common, with the spread of the personal computer, to find a facilities group in which one or two people have written a collection of programs. They often reside on diskettes in a desk drawer with no code listing to be found and no hint of a user's manual. The "software writers" teach others in the group how to use the programs. For convenience, these new users may copy the code, which then is subject to lack of update when the original version is changed by the "experts." This is a vulnerable state of affairs, subject to errors and wasted time.

If programs are developed in-house, the department or group manager must exert some discipline and control over the effort. Cataloging, libraries, and sign-outs are in order. But most of all, insist on documentation.

Internal development is required for some situations and applications. These include the following:

1. *Large, non-graphic data bases for facilities management and planning*: Using data base management systems, planners and their data processing staffs can create very large, specially tailored information and reporting systems comparable to the INSITE system discussed in Chapter 8. The key issues in such an effort are listed in Figure 9.5.

2. *Links between CAD systems and non-CAD data bases*: As suggested in Figure 9.5, it may be worth the effort to link the CAD system's drawing of a space to the non-graphic records about it. This keeps the CAD system free of unnecessary data, but still permits the planner to merge the two types of information when desired.

3. *Microcomputer-to-mainframe links*: These can be physically accomplished with a variety of commercially available hardware and software products. But internal applications programming is still required to accomplish the desired tasks. A common application is the maintenance of a local facility data base on a microcomputer, some portion of which is copied and merged with other locations' data in a central headquarters data base. A related application is the ability to pass project files and data from a micro in one facility to a micro or mainframe in another.

4. *Links between CAD systems and mainframes used for computer-aided engineering (CAE)*: Structural analysis and other computationally intensive routines may require a mainframe computer. The CAD system or a microcomputer may serve as the input and control device for executing the mainframe routines. Internal programming is usually required to link the various components of such a system. An example is shown in Figure 9.6.

5. *Large, transaction-oriented, facilities management systems*: The LOIS lease management system discussed in Chapter 8 is a good example. Work order systems for maintenance and plant engineering, and inventory systems for supplies are other examples, although these are also available commercially.

1. <u>Purposes</u> -- Charge-back/overhead recovery; governmental/
    contractual reporting; inventory/asset control; maintenance
    planning and budgeting; future space planning; other (list).
    Rank order the purposes of the data base.

2. <u>Users</u> -- Who are they? Where are they? What tasks will the
    data base support? How? How often? Are there other solutions
    to the need for information? Rank order the users and their
    needs.

3. <u>Single vs. Multiple Data Bases</u> -- Can one data base serve
    the needs of all users? Would separate, perhaps partially
    shared data bases be more appropriate -- for space, equipment,
    utilities, maintenance?

4. <u>Graphics</u> -- should there be a separate system for the display
    of spaces and buildings? Or should this be integrated with
    the non-graphic data base? Why? What kind of graphics are
    required? Key plan vs. as-built, working drawings?

5. <u>Size</u> -- How many spaces (indoor and outdoor)? How many
    buildings and structures? How many leases? How big is
    each record? How many pieces of equipment? How many floor
    plans? Overlays?

6. <u>Batch vs. On-Line</u> -- What printed reports are required from
    the system? (Make a list) To whom are these distributed?
    With what frequency? How are these reports used? Would an
    on-line display take the place of a report? Which ones?
    What non-standard inquiries will be made of the system?
    (List at least 10 examples) How often will these occur?

7. <u>Available tools</u> -- Is a CAD system in use or planned for
    facilities design? Has a data base management package been
    licensed for company-wide use? Which one(s)? What assistance
    will be provided from the data processing group? What is
    the policy on purchased software vs. in-house development?
    What systems are available to implement the facilities
    data base?

Figure 9.5. Key issues in facilities data base design. Internal data base development must first resolve these issues.

The desire to integrate DSS, CAD, and MIS applications is sometimes
cause for an internal development effort. If you plan offices and labs, the
internal approach is less likely since you can already choose from a number
of commercially available, integrated systems. These take space projections
and relationships as input, and give back clusters, multistory assignments
and, in a few cases, block layouts, as we have seen in earlier chapters.
Once block plans are complete, the same system is used for detailing, with
full CAD capabilities. The MIS component may be weak, but is at least capable of maintaining graphic, drawing-based space and furniture inventories. Decision Graphics, DFI/Systems, Resource Dynamics, Auto-trol, Calcomp, and Sigma Design have all offered varying degrees of integrated systems for some time.

Unfortunately, planners of factories and warehouses cannot yet find commercial systems with the same degree of integration or functional standardization. While the office-oriented systems can be used for plant layout, they

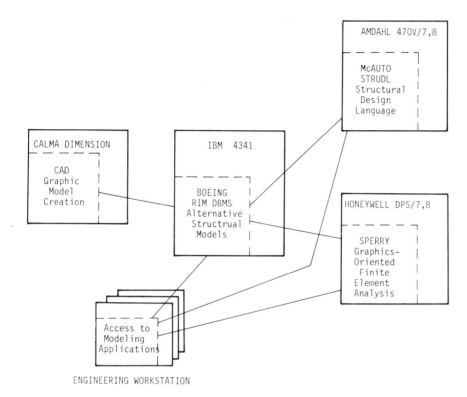

Figure 9.6. Internally developed links are often required to integrate various aspects of computer-aided engineering. This example is based on the systems in use at Southern Company Services, Inc., Atlanta, Georgia, for power plant design.

are missing critical functions in group technology and materials handling analysis. At present, these must be performed with separate systems or software.

There is nothing inherently wrong with discrete systems for DSS, CAD, and MIS. In fact, a good case can be made for separate, specialized aids which do not compete for a single computing resource, or a single-color display, or degrade response time when used simultaneously. Even in office space planning, not every facilities group needs an integrated solution. The two chief conditions that favor this solution are:

1. A single facilities group has comprehensive, internal responsibility from concept through design to installation and ongoing management. All tasks are performed in-house, at one location.
2. The output of one planning step is *directly* usable as the input to the next. The same original data is used repeatedly for different planning tasks, and is maintained during the ongoing management of the facility.

For those who believe an integrated system is realistic and justified, careful planning is required. With today's networking and telecommunications

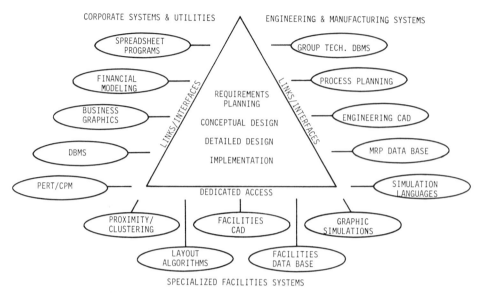

Figure 9.7. Computer-integrated facilities planning. Integrated computer aids for industrial facilities planning.

products it is becoming easier and less expensive to physically link various pieces of hardware. But difficulty remains, and programming is required to get access or share data and applications software, once the hardware has been connected. As suggested in Figure 9.7, industrial facilities planning needs to draw from a variety of other "non-facilities" systems. It also needs its own dedicated capabilities of the kind we have seen throughout Part II of this book.

The notion of "computer-integrated" facilities planning as presented in Figure 9.7 is largely conceptual. It represents an ideal spectrum of support for computer-aided facilities planning not found in use today. However, examples can be found, such as the systems used at Deere & Company, which suggest that integrated industrial planning aids are not too far in the future. Facilities planning at Deere & Company is aided by five partially integrated systems:

1. *JD/GTS (John Deere Group Technology System)*: is a collection of group technology data bases, one for each product unit (tractors, combines, etc.). These hold non-graphic part descriptions and related data on production rates and routings.
2. *FROM-TO ANALYSIS (actually part of the JD/GTS)*: extracts a variety of material flow measures from the routing files of the parts data base. Output can be in tons, pieces, number of parts, or even containers per day, week, or month. Output can be entered directly as the volume array for CRAFT.
3. *CRAFT*: the original layout improvement routine, accepts volumes manually or automatically from JD/GTS (via FROM-TO). The planner enters the distance array and move costs manually.

4. *GPSS (General Purpose Simulation System)*: one of the first and still most powerful simulation languages is used to model and size conveyors and other automated systems. Once constructed, the models of newly installed systems are kept intact, for subsequent analysis of changes and improvements.
5. *GMS (Geometric Modeling System)*: using the CADDS 4 system from Computervision. This system contains graphic part descriptions, including those for pallets and containers. The system is used for 2-D and 3-D graphic simulations of clearances, movement of parts, and equipment. It is also used for detailed layout, once CRAFT has verified the block plan.

The systems at Deere & Company have been evolving since 1976. The JD/GTS was developed in-house by engineer-programmers, after initial experimentation with a commercial group technology system. The CRAFT code was installed without modification, as was GPSS. Roughly 70% of the GMS CAD commands and menus had to be developed by Deere's engineers. Only 30% came from the basic, vendor-supplied software. All systems but GMS reside on IBM mainframes in the corporate data center. They are accessible from any manufacturing plant. Each plant with a CADDS 4 system can use the GMS menus and commands. Files are shared between plants by exchanging magnetic tapes. Future goals include an interface between the JD/GTS and GMS, and a system for process planning. Clearly, these accomplishments required foresight and many man-years of sustained commitment--not to mention know-how in both planning and computer systems development. Commercial products have played an important role, but key developments and all of the system integration were accomplished in-house.

The Role of Outside Consultants

Qualified consultants can play a useful role in selecting and developing computer aids. A qualified consultant is one who has:

1. Experience in or a good understanding of facilities planning
2. Knowledge of or experience with available software for facilities planning
3. Knowledge of hardware and directions of computer technology
4 Independent, unbiased perspective, i.e., does not sell or hold an active interest in the sale of software or computer systems for facilities planning

Finding a consultant who meets all of these criteria is almost impossible. Many top facilities consultants and designers offer their own facilities planning software for sale or lease, either directly or through others. Or, they have a computer, and they will offer you a service bureau arrangement. With this type of personal financial involvement, their assistance to you may be somewhat biased.

There are many well-informed consultants in the fields of data processing and computer graphics. Unfortunately, few of them have any significant knowledge of facilities planning. This is not a problem if they are used in a limited way to advise on the purely technical aspects of the decision.

A consultant should be brought in early and have a place on the study team. Functions performed include:

1. Sounding board for ideas and approaches
2. Source of leads on available solutions
3. Advice on technical or hardware matters
4. Advice on implementation
5. Resource for internal development efforts

Working effectively with a consultant requires time and money. This is justified for major computer investments--especially those which may require you to modify your approach to facilities planning. On modest investments-- say personal computers, or limited software packages--a consultant is probably not needed--certainly not on a continuing basis.

When choosing a consultant, check references carefully for a general tone of objectivity and knowledge of the computer aids in question. A list of consultants active at the time of this writing is included in Appendix B.

10

CAD Selection and Installation

Basic Choices

Of all the computer aids described in this book, CAD systems are potentially the most expensive and the most difficult to evaluate. As noted in Chapter 7, some CAD systems offer such non-graphic routines as: space projections, lease management, project management, and furniture inventories. These systems can provide the core of a computer-based approach to facilities planning and management. This possibility, as well as the potential expense of a CAD system, requires an orderly, thorough approach to selection and installation (52).

The selection process and the suggestions made in the previous chapter will apply to computer-aided design. They will not be repeated here. The purpose of this chapter is to address those considerations that are unique to CAD.

The facilities planner can choose from four basic approaches to CAD. These are:

1. Use an existing CAD system.
2. Purchase software and peripherals for an existing mainframe or mini-computer.
3. Use a service bureau.
4. Purchase a turnkey CAD system.

Many facilities planners may find a CAD system already in use by their company's engineering or manufacturing groups. The questions are: Will their software work for facilities planning? Can enough time and data storage be provided? How will the planner be charged for such use? In many cases, existing CAD systems will be hopelessly specialized in functions such as design of printed circuits or mechanical parts. And, the facilities department may be restricted to off-hours, or to an inconvenient location.

It may be possible for the planner to use an existing mainframe or mini-computer. If so, purchases can be reduced to software, displays, and peripherals. This approach, in fact, is becoming easier every year, with several proven architectural software packages now available. The critical requirements with this approach are the knowledge, experience, and commitment of the planner's data processing group. There may be interferences between the CAD software and other applications on the host computer. The planner and the data processing group will need a practical, long-range plan for the CAD applications--one that identifies the number of terminals, activity levels, and amount of data storage that will be required. Otherwise, the whole effort may lead to a premature dead end.

Some facilities planners may avoid a CAD purchase by using a service bureau. As mentioned in the last chapter, service bureaus are a good way to experiment, or to get work done that is infrequent and less than urgent in nature. To get good results, however, will require someone to spend time supervising the work that is done. The use of a service bureau is no different than turning work over to an outside contract draftsman. This one simply uses a computer. Occasional problems of interpretation, style, and quality will occur. Experienced users of service bureaus suggest starting with simple, schematic drawings, and keeping in frequent contact with the operators doing the work.

Purchasing a system from a turnkey vendor is by far the most popular approach to computer-aided design. Installing a properly selected, dedicated system insures that it will not interfere with other groups or systems and that it can be easily expanded when the need arises. On the other hand, the initial cost is naturally a lot higher than the preceding three approaches.

Which approach to CAD is best? The answer depends upon the planner's specifications, budget, and timetable for action. Since the turnkey approach is the most common and most expensive, it will receive most of our attention.

Turnkey CAD Vendors

A few years ago, there were only a handful of turnkey CAD vendors. Today there are over 40. Most offer systems that could be used by the facilities planner. When approaching such a well-populated marketplace, the planner will benefit from some overview and orientation to the types of vendors involved. (Reasonable detail on specific products can be obtained through the directories listed in Appendix A.) CAD vendors can be grouped as follows:

1. Large computer vendors and service companies
2. Traditional CAD vendors
3. CAD divisions of non-computer firms
4. Specialty CAD vendors
5. Drafting equipment companies

Several of the "big names" in computers offer systems for computer-aided design. Companies in this category include: IBM, Control Data, Honeywell, Tektronix, McAuto, and Boeing Computer Services. Most software is licensed. The service companies typically sell a package of licensed software running on popular, name-brand minicomputers such as the VAX from Digital Equipment Corporation, or the NOVA from Data General. Large firms represent a "safe" choice for the planner because of their size and reputation. There is less resistance from data processing staff, who might not be familiar with CAD. However, biggest does not necessarily mean best. The products in this category should be compared with others on a feature-by-feature basis.

Traditional CAD vendors include the firms that pioneered the technology. Each currently offers some products for architects and facilities planners, although commitments to these applications vary a great deal. The firms in this category best known for facilities applications include: Autotrol, Computervision, Intergraph, and Calcomp.

In their early years, these CAD vendors offered fairly limited adaptations of underlying, general-purpose capabilities. The facilities planner received essentially the same software offered for engineering, circuit design, and mapping. Today this is no longer true. Vendors who emphasize architectural and facilities applications offer software that is more tailored and, in some cases, entirely different from that used in other applications.

Traditional CAD vendors provide medium- to large-scale systems running on popular minicomputers. Their secure place in the CAD industry gives them credibility with corporate data processing groups. However, their products are being challenged by those of the large firms just discussed-- now that CAD is becoming largely a matter of software. The traditional CAD vendors are also being challenged by the newer, specialty vendors and their lower-cost, microcomputer products. Autotrol was one of the first to respond, offering a low-cost system using the Apollo microcomputer. Computervision has announced its intention to use a similar device from Sun Microsystems. Other vendors of minicomputer-based systems can be expected to follow shortly.

CAD divisions of non-computer firms comprise something of a catch-all category. Calma (division of General Electric) and Applicon (division of Schlumberger) are both early-day pioneers of CAD that have been acquired by large manufacturing companies. Both have systems in use for architectural and facilities-oriented applications. Also in this category are: Bausch and Lomb, which acquired Houston Instruments, a long-time vendor of CAD peripherals; and Herman Miller, a furniture company which now offers CAD systems through its CORE division.

Specialty CAD vendors have taken a clear aim at architectural and facilities applications. These vendors, while small, have put all their development talent to work on architectural and facilities-related problems. The results have been impressive, both in drafting and in such related tasks as block layout, inventories and space utilization reports. Unfortunately, the small size of the firms involved is a source of discomfort for some data processing personnel. Furthermore, several of these systems are based on new microcomputers which are still foreign to most data processing groups. The planner who is interested must typically convince data processing to back an innovative solution. Among the vendors in this group are: Decision Graphics, DFI/Systems, Formative Technologies, Sigma Design, and Resource Dynamics.

A final vendor category consists of drafting equipment companies that also offer CAD systems. As might be expected, these use general-purpose software which runs on brand name computers. The vendors include: Bruning, Keuffel and Esser, Ozalid, and J. S. Staedtler. In effect, these firms are distributors of CAD systems who act as turnkey vendors.

With so many firms to choose from, the planner's first move should be to narrow the field, possibly by eliminating one or more of the categories

above. The next move must be to develop a specification which can be used
to evaluate the remaining vendors and systems.

CAD Specifications

While there is no such thing as a standard CAD specification, a number of issues should be resolved before any purchase (45). These should be addressed
in the written specification.

1. *Basic drafting capabilities*: The most important features were identified earlier in Chapter 7. Particular attention should be paid to the
   creation of floor plans and elevations and to the creation and manipulation of symbols. Essential capabilities must be separated from those
   which are simply nice to have. How important are multiple type fonts,
   for example? What about creation of three-dimensional views? The
   planner should assign relative weights to all desired features so that
   those of lesser importance do not overly influence a decision.
2. *Non-drafting capabilities*: Most planners expect far too much from
   what are primarily drafting systems. Bills of material or take-offs
   are reasonable. This capability can also serve for inventories. Are
   other facilities planning capabilities such as space projections and
   vertical stacking essential? Desirable? Can they be provided more
   cost-effectively on a separate low-cost computer? Or on an existing
   minicomputer? What about project management (PERT or CPM)? Or
   leasing and property management? Many planners are enamored with
   the idea of a single "facilities computer" which will perform all applications. This is the same approach taken 20 years ago by data processing personnel. It has been rejected in recent years in favor of
   specialized, smaller computers. There is a lesson here for facilities
   planners and managers.
3. *Volume of activity and output*: It is essential to know the number of
   projects, drawings per project by size, and number of potential overlays per drawing. The number of existing drawings is important only
   if they are going to be loaded into the computer. Figure 10.1 is an
   example of volume data for a CAD specification.
4. *Storage capacity*: A common desire is to have all drawings available
   online at all times, i.e., without resorting to an archive of tapes or
   disks. This leads to unnecessary expense. This is a bit like en-
   larging one's desk to avoid a filing cabinet. In practice, only a few
   drawings are needed on any given day. The time it takes to load
   them from a well-organized archive may not be significant compared
   to the cost of a large, online data base. A related issue is the desire to reach any drawing from any terminal at any time. In practice, facilities departments do not work this way. Projects and design disciplines are assigned to various individuals. There may be
   a "handoff" at some point in the life of the project, but spontaneous
   copying of others' drawings is not common. To have this capability
   may cost more than it is worth.
5. *Configurations and expansions*: Which individuals need terminals,
   and for what purposes? Where are they located? Who could share?

ANNUAL PROJECT AND DRAWING VOLUME

Project Size	Number of Projects	Drawings per Project	Total Drawings
PEAK YEAR			
Small	220	1	220
Medium	220	5	1100
Large	220	10	2200
TOTALS	660		3520
AVG. YEAR			
Small	150	1	150
Medium	150	5	750
Large	150	10	1500
TOTALS	450		2400

There are approximately 8,000 existing drawings. Four sizes are used as follows:

Drawing Size	Number of Drawings	Percent of Total
21" x 30"*	400	5%
24" x 36"	2000	25%
26" x 38"**	3200	40%
30" x 48"	2400	30%
TOTAL	8000	

* cut from 24" x 36"
** cut from 30" x 48"

Base building plans are retained indefinitely; others as long as necessary.

Figure 10.1. Volume data for a CAD specification.

Who could not? Why not? Who will need terminals in the future? At what locations must plotted output be provided? What about inputs?

6. *Interfaces to existing systems*: These are often more trouble than they are worth. Beware of spending large sums to avoid minor duplications of data and key entry tasks. Again, watch out for the "single system" concept. It may not make sense for every application.

7. *Programming capabilities*: Some planners and managers like to do their own programming. Others do not or cannot. Some CAD systems are not programmable. They come with a set of commands and functions that can be executed at will by the operator; however, creation of new functions requires vendor involvement. By the time a specification is written, the planner should have some concrete idea of how a programming capability would be used.

There is no one right answer to the issues and questions posed above. The CAD specification may evolve with time, through two or three versions, as the planner expands his understanding of what the market has to offer.

Above all else, the specification needs to be results-oriented. It should state how many drawings need to be reached at any moment. How many active projects need to be supported online? How many items need to be tracked in inventory? How many characters of information must be maintained for each? Do not rely on vendors to answer these questions for you.

Shows, Demonstrations, and Benchmarks

Trade shows are a fast, effective way to become familiar with CAD systems. Most vendors attend the major industry shows. This gives you a chance to see their systems, ask questions, and form opinions about each vendor's personnel. You also have a chance to collect current literature and obtain current ballpark price information. If you represent a large firm or a potentially large order, you may get special treatment from the larger vendors--typically a private demonstration in a nearby location, or in a partitioned enclosure at the vendor's booth.

Bear in mind that large, expensive booths and small armies of people are designed to impress you with the vendor's stature and stability. Be sure to look beyond this to the products themselves. One reason smaller firms are able to offer lower-cost systems is that they are not yet paying for large booths and hospitality suites. The most innovative products may be found in smaller, less-frequented booths.

It is important to see a "live" demonstration, as opposed to one that is "canned." The least informative is one in which the display is automatically controlled by preprogrammed software. This is virtually the same as watching a videotape. Such demonstrations reveal absolutely nothing about ease-of-use and the time required to accomplish a particular drafting task. At best, the viewer learns about the display capabilities and the intended applications of the system.

A better, though still "canned" demonstration is the "microphone talk," in which a speaker follows a pre-arranged script while a trained operator executes the required commands. Such demonstrations do give some idea of speed. Of necessity, the talk is generally rather fast and formal with little or no opportunity to interrupt. And the script describes only what the system does well. Try to come back for a more informal demonstration, perhaps in the off-hours when the vendor's personnel are not too busy.

The ideal demonstration is "live," allowing for response to questions and digressions from the script. This is difficult and risky for the vendor as the digression may reveal a limitation in the system. Vendors who will conduct such a demonstration for you should earn a degree of respect. Do not jump to quick conclusions about the limitations that are revealed. All systems have them. Your job is to find and compare them very carefully. Keep in mind, too, that trade show computer equipment is subject to extreme handling and operating conditions. The fact that a vendor's equipment was not working properly at a trade show has little bearing on its reliability in actual use.

For maximum benefit, planners should attend a show *after* writing their initial specification. This way, the specification can serve as a checklist and a means of focusing on those system features that are truly important. At best, a show will tell which vendors should be considered and which should not. The task of selection can begin at a trade show. It cannot end there.

Benchmarks are the best way to evaluate the capabilities of a CAD system. These consist of controlled but realistic test cases submitted to each vendor for execution on its CAD system. Benchmarks require a great deal of a vendor's time. Typically, vendors will not respond unless they are among the top three or four candidates for a likely sale. Planners should not expect or demand benchmarks from more than this number of firms.

Typical benchmark problems might include:

1. Digitizing and rearranging 5,000-10,000 square feet of plant or office, including creation of necessary symbols. The number of symbols should be limited to between 25 and 50 of various types.
2. Drafting from a dimensioned sketch to produce preliminary floor plans for a 5,000-square foot building addition. This may include digitizing the relevant portions of the existing building.
3. Drafting a common construction detail from a dimensioned sketch and placing it onto a larger drawing.

If the vendor offers non-CAD software for facilities planning, project management, or other applications, these too should be examined through the benchmark process.

The planner must allow the vendor plenty of time to study the problem and formulate an approach that takes advantage of his system's strengths. However, the planner needs to be present with the clock running from the moment when data is entered into the computer. Keep in mind that the vendor's operator is experienced. Do not expect to reach the same speed and proficiency for at least one year. Watch for ease of use, the amount of know-how required, and any problems or "bugs" that occur. Overall benchmark performance should then be rated along with other selection factors in reaching a final decision.

Cost Justification

As we have seen in earlier chapters, limited personal CAD systems can be bought today for as little as $10,000. Software is oriented toward general-purpose, two-dimensional drafting.

At the opposite end of the spectrum, a 4- to 6-station system with large color displays, E-sized plotter, and large digitizer may cost between $400,000 and $500,000. Software will be architecturally oriented, two- and three-dimensional, with non-drafting applications also provided.

Even $50,000 can be hard to obtain for many facilities planning groups. Justifications must be compelling and easily understood by management. They must also be realistic if the facilities planner's reputation for good estimates is to be upheld.

Improved productivity is the central issue in all CAD justifications (49, 50). In most applications, it is faster than the best manual methods, including overlay, cut-and-paste, and pin register drafting. As a result, more

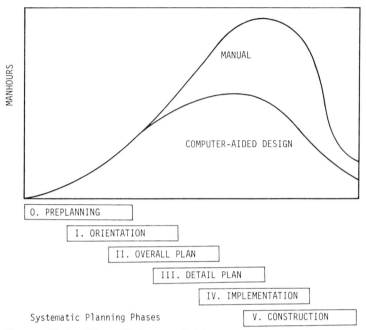

Figure 10.2. The efficiency of CAD.

work can be accomplished in less time. In the context of our earlier discussions, the efficiency of CAD occurs primarily in detail planning (Phase III) of a facilities project. See Figure 10.2. There is a link between efficiency gains in the latter phases of a project and the creation of more effective facilities plans. This link is illustrated in Figure 10.3. If one assumes that calendar time is moved from the later to the earlier phases (without shortening the total project time), then it should be possible to examine more alternatives, thus arriving at potentially better plans and decisions. In practice, however, it takes a great deal of commitment and discipline to achieve this reallocation of planning time.

A compelling economic argument can be made for using CAD to shorten the overall planning cycle. In the case of a major project, such as a new plant or office, taking weeks out of the schedule can lead to substantial gains in financing and to additional revenues from earlier start-up. This type of payoff on one large project will probably pay for the system.

Some companies have used the efficiency of CAD to bring design and drafting work in-house, lessening the use of outside professional firms. The economic benefits of this approach are not always clear, as the installation of the CAD system adds a permanent fixed expense to the cost of doing business. This may not be offset unless there is the prospect of continuing fees to outside professionals or contract draftsmen.

Another approach to justification involves the elimination of backlogs. This is essentially a cost-avoidance approach that assumes a backlog of truly important work. Most of us operate routinely with enormous backlogs, so the planner should exercise some judgment in the application of this approach. The procedural steps are outlined in Figure 10.4.

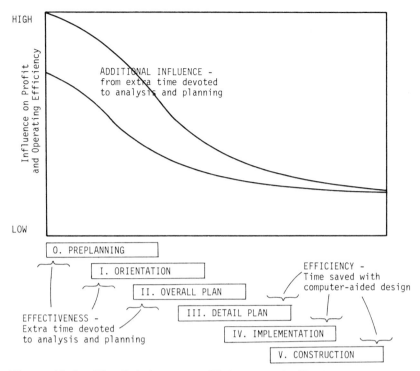

Figure 10.3. The link between efficiency and effectiveness.

To see if a dedicated CAD system is justified for facilities planning, the following procedure may help.

1. List all potential areas of CAD assistance (facilities planning, engineering, and management).

2. Estimate (in man-hours) the current backlog of work in the areas just defined. (This assumes a backlog of truly important work.)

3. Estimate the cost of eliminating the backlog by hiring additional personnel. (Include all costs: hiring, benefits, space, etc., in addition to salary.)

4. Estimate the cost of eliminating the backlog by using CAD. (Include all costs: start-up, data entry, debug, learning curve, maintenance of hardware and software.)

5. Compute the savings and the payback.

Figure 10.4. Justification of CAD for facilities planning activities. (Adapted from W-K-M Valve Division, ACF Industries, *Manufacturing Engineering*, November 1980.)

The most common approach to justification is the reduction or control of staff through reassignment, layoff, attrition, or freeze. Such claims are easy to make and very hard to live by. In your author's experience, it is preferable to advance the foregoing arguments first, using this one last.

Regardless of approach, it is necessary to quantify the efficiency or productivity gains to be expected from the system. Many improvement ratios have been reported by users, vendors, and observers. Some of these are reproduced in Figure 10.5. The problem with published ratios is their susceptibility to system application and operator skills.

Every CAD system does some tasks well, some poorly, and others not at all. Productivity gains can only occur when the system is applied to those tasks that it does well. Consequently, ratios claimed for one system on one type of drawing may not apply equally to other systems on the same type of drawing.

In general, CAD systems yield productivity gains on the following specific tasks:

1. Placement of repetitive symbols
2. Simple line work
3. Cross-hatching
4. Changing scales
5. Counting objects in a drawing and performing related calculations
6. Measuring lengths and areas
7. Revising existing drawings
8. Storing and retrieving drawings
9. Copying drawings

Productivity gains are only achieved by those who operate the CAD system. By itself the system achieves nothing. It follows that a trained operator will do better than a novice. The typical CAD learning curve is illustrated in Figure 10.6. The curve applies to people--not to "the system." A year-old system may exhibit 3-6 month productivity ratios if there is a lot of turnover among operators and trainees.

Your author is aware of several CAD systems that yielded overall productivity ratios of 0.5 to 1.0 or worse. Unfortunately, no one wants such failures made public. In each case, the root causes were misapplication, inadequate training, and turnover. Published productivity ratios of 2:1, 3:1, 4:1, or better are targets to be achieved by sound system selection and management. They are not automatic benefits for those who buy a CAD system.

A realistic approach to CAD justification requires the use of conservative productivity ratios--probably no greater than 2:1 or 3:1 overall. At these levels, labor savings alone may not justify the purchase. This forces some value to be placed on the very real intangible benefits involved.

The savings of a CAD system are offset by considerable expenses, some of which are frequently overlooked. The main expense categories include:

1. *Hardware*: Including all cables and accessories.
2. *Software*: Including all applications and any licensee fees.

SPIF Phase	Planning Activities	CAD Productivity Ratio per Drawing Type			Overall Drafting Ratio	Drafting Content	Overall Phase Productivity
0	Preplanning	Miscellaneous charts, graphs, maps and plans			1:1	10%	1:1
I	Orientation				3:1	20%	1:1
	.Engineering analyses	2:1					
	.Graphs	2:1					
	.Topographic maps	3:1					
	.Plant site plan	3:1					
	.Cross-hatching	4:1					
		Schematics Not Dimensioned Not-to-Scale	Typicals Piping, Isos Dimensioned Not-to-Scale	Working Drawings Dimensioned To-scale			
II	Overall Plan				3:1	33%	1.7:1
	.Engineering analyses	2:1	2:1	--			
	.Graphs	--	2:1	--			
	.Process flow diagrams	4:1	--	--			
	.Electrical single line	4:1	4:1	5:1			
	.Structural plans	--	3:1	4:1			
	.Block layouts	--	--	3:1			
	.Architectural drawings	2:1	3:1	3:1			
III	Detail Plan				5:1	67%	3.7:1
	.Office layout	--	--	8:1			
	.Equipment layout	--	--	5:1			
	.Piping arrangement	2:1	4:1	4:1			
	.Electrical arrangement	1.5:1	3:1	3:1			
	.HVAC	2:1	4:1	5:1			
	.Mechanical	2:1	4:1	4:1			
	.Structural/Plumbing	2:1	4:1	5:1			
	.Material handling	1.5:1	--	3:1			
	.Installation drawings	4:1	8:1	8:1			
	.Architectural drawings	2:1	3:1	3:1			
	.Take-offs; listings	7:1	10:1	12:1			
IV	Implementation				4:1	10%	1.3:1
	.CPM/PERT charts	4:1					
V	Construction				6:1	10%	1.5:1
	.CPM/PERT charts	4:1					
	.As builts	8:1					

Figure 10.5. Productivity ratios, developed largely from unpublished research by CAD vendors and from interviews with users of CAD systems. Estimates of drafting content in each phase are those of your author. These ratios assume fully trained, experienced operators. No provision is made for the time required to build symbol libraries, enter existing drawings, or develop specialized commands.

3. *Shipping*: Including any special in-transit insurance that may be required.
4. *Installation*: Electrical, air-conditioning, fire protection, furniture, and any required enclosure. Also any vendor fees, if installation support is priced separately.
5. *Training*: Fees, if any, and travel expenses for sessions held at the vendor's facilities.
6. *Maintenance*: On both hardware and software.
7. *Supplies*: Such as plotter paper, ink, pens; also diskettes and tapes.

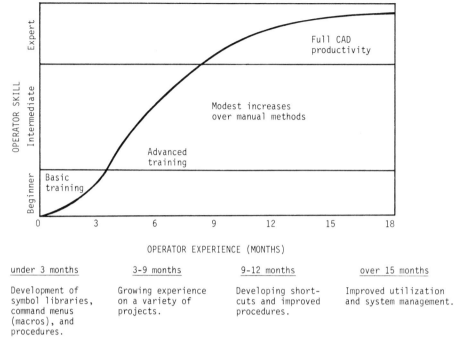

OPERATOR EXPERIENCE (MONTHS)

under 3 months	3-9 months	9-12 months	over 15 months
Development of symbol libraries, command menus (macros), and procedures.	Growing experience on a variety of projects.	Developing short-cuts and improved procedures.	Improved utilization and system management.

Figure 10.6. Learning curve for CAD operators. (From R. E. Blauth and E. J. Preston, "Computing the Payback for CAD," *Machine Design*, August 20, 1980.

8. *Utilities*: Power, light, phone, and air-conditioning.
9. *Insurance*: On the system and possibly on the stored data.
10. *Personnel*: Yes, a CAD system may require additional personnel, especially a supervisor for large installations.
11. *System selection*: Do not overlook the expense of attending shows, visiting vendors, and retaining consultants. These may constitute a significant percentage of total expense.
12. *Contract data entry*: Some facilities planners and managers see the installation of CAD as an occasion to survey, update, and digitize existing facilities drawings. This is often done on a contract basis.

Some knowledgeable observers estimate that the 5-year cost of a CAD system including data entry and learning curve is easily 3-5 times the price of the hardware. The numerous cost categories above suggest that this may well be the case.

We have, until now, avoided the issue of multiple shifts. Sooner or later in every CAD justification, someone proposes its use on a second shift to maximize return on investment. This may be when the computer begins to run the planner instead of the other way around. Some studies have shown that a percentage of the population enjoys working nights. Planners should poll their own staffs before assuming this condition applies. If willing second-shift draftsmen are available, *and* they are self-supervising, *and* no system support staff is required at night, a second shift might be practical. In your author's judgment, it may be wiser to buy a lower-cost CAD system and use it in one shift instead of an expensive one that can only be justified with two-shift use.

Negotiations

Most facilities planners will rely on their purchasing and data processing departments to negotiate the purchase of a CAD system. The planner should, however, be aware of the major issues and negotiating points between his organization and the vendor. These include:

1. *Acceptance testing*: To insure that all hardware and software performs as specified with your data. Testing must be completed by a given date, so organize your initial efforts accordingly. Know what happens next if the system fails the test. Be sure to clearly define acceptance.

2. *Scope of sales agreement*: Also called a merger or integration clause. Many standard vendor terms and conditions make no references to the proposals, sales literature, or correspondence which may have been used to describe the system's capabilities. Make sure these are referenced in the sales agreement, to support acceptance testing.

3. *Warranties*: Should be extended wherever possible. They should apply to working days, excluding holidays and weekends. Be sure that hardware and software are covered. Clarify starting dates-- from time of installation or date of acceptance?

4. *Liabilities*: Vendors will not assume liability for any delays, errors, or damage caused by their hardware or software. This is not unreasonable, as the system may be misused by untrained operators.

5. *Indemnification*: Is necessary in the event that a vendor has infringed on someone else's patents or copyrights.

6. *Payment terms and title*: Be sure final payment is withheld until acceptance tests are complete. Understand when title passes to your firm--date of shipment? acceptance? final payment? Attempt to make the vendor responsible for risks during transportation. Be sure the shipment is insured.

7. *Maintenance*: A variety of plans may be offered. Be sure to understand what you are getting. This will be discussed further in the next chapter.

8. *Software updates*: Be sure to understand and clarify the difference between fixes and enhancements. The former should be free. The latter will be priced but may be included in a full maintenance agreement.

Do not be afraid to ask for concessions during negotiations. The CAD industry is a very competitive one. Discounts are occasionally given, as well as concessions on terms, installation, training, and even software updates.

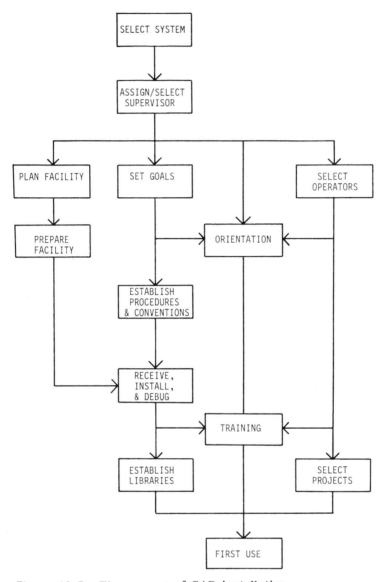

Figure 10.7. The process of CAD installation.

Installation Planning

Getting a CAD system installed is more of an organizational task than it is physical (47). The general process is outlined in Figure 10.7. Activity begins when the system is selected--at least several months before delivery. The first step is the designation of a supervisor. As recommended in the previous chapter, this should be the same person who conducted the search for the system. It need not be a computer expert. And, if the selected system is one of the newer microcomputer-based products, the supervisor's job may not be a full-time responsibility. In some cases, the supervisor may also be the lead operator of the system. In addition to the attributes listed in the previous chapter, the CAD supervisor should possess:

1. Drafting experience
2. Ability to supervise draftsmen--especially in scheduling their time
3. Ability to train peers and subordinates, and to develop procedures manuals
4. Ability to understand the basics of CAD hardware and operation
5. Ability to communicate effectively with vendor personnel

The manager of the facilities group must also become knowledgeable about the selected CAD system. Otherwise, he or she will be incapable of supporting and evaluating the system supervisor. As a practical matter, this means that the manager must go through the same initial system training as the supervisor.

The supervisor has three important responsibilities in the early stage of implementation planning. These are:

1. Setting goals for use of the system
2. Planning and preparing the facility
3. Selecting the CAD operators

Goals are the target dates by which certain capabilities will have been achieved. These should include:

1. Pre-installation training
2. Development of procedures and conventions
3. Readying the facility
4. Delivery, de-bugging, and acceptance testing
5. Post-installation training
6. Development of symbol libraries and the drawing data base
7. Selection of initial projects

The final goal should be the attainment of target productivity ratios and benefits claimed in the cost justification.

The degree of facilities preparation depends on the type and size of the selected CAD system. Those using larger minicomputers will require special facilities. Those using microcomputers will require little more than floor space and lighting adjustments. In general, the following conditions should be observed:

1. *Power supply*: typically 115 volts AC, 50/60 Hertz current with a ± 5% tolerance. Circuits should not be shared with large motors or

other equipment that may produce voltage variations in excess of 5%. Impulses must also be avoided.

2. *Temperature and humidity*: depends upon the type of CPU. The majority of large minicomputers should be enclosed in rooms that range from 60°F to 75°F with relative humidity between 35% and 70%. Microcomputers will tolerate a normal office environment. (This may represent a significant cost avoidance.) Plotters and digitizers can tolerate temperature swings between 68°F and 78°F and humidity between 25% and 75%. Environmental changes do affect the accuracy of plotters and digitizers due to expansion and contraction of surfaces and the media being used. Graphics displays can be used in a normal office environment.

3. *Air-conditioning*: may be necessary to keep temperatures within limits. Remember, each piece of peripheral equipment generates heat, along with the CPU and displays. In close proximity, they will raise ambient temperatures. Air filtering may also be desirable to reduce dust and cigarette smoke, especially near the CPU and disk drive.

4. *Noise*: can be a problem for operators and nearby personnel. Plotters and printers should be partitioned or secluded. If the system is minicomputer-based, enclosure may be necessary to suppress fan noise.

5. *Lighting*: is troublesome wherever computer terminals are used. Lights should be placed to avoid any direct reflection on the display. They should be controlled by dimmers for flexibility. Work tables and surfaces should be directly lighted for ease of viewing prints and sketches.

Most vendors have well-prepared installation guides that address these issues. The planner should be sure to involve his data processing staff as well as the vendor if in doubt about the adequacy of the available facilities.

The selection of operators is crucial to the success of the CAD system. The attributes of a good operator are summarized in Figure 10.8. In smaller facilities groups, it may be reasonable to expect all staff members to become users of the CAD system. But one should not assume that every staff member will become an effective user. Aptitude and attitude are important. Problems in this regard cannot be overcome by training. It often happens that the vendor's personnel will spot a weak operator during initial training. The supervisor and the group's manager should be prepared to reassign such individuals in a graceful way. Figure 10.9 reproduces an excellent questionnaire which was used by one firm to screen potential CAD operators. It addresses the most pertinent issues in a very direct way.

The CAD vendor will provide initial training for a small group of operators. This will be just enough to get started. A great deal of follow-on training will be required. This must be the responsibility of the supervisor, although much of it may still be supplied for a fee by the vendor's personnel. Guidelines for effective in-house, on-the-job training are summarized in Figure 10.10.

Most facilities planners and managers underestimate the amount of discipline, standardization, and organization required to make effective use of CAD. At a minimum, standard operating procedures will be required in the following areas:

The proper selection of the first CAD operator is a critical element for a rapid installation. It is management's responsibility to select from their design staff the best suited individuals to introduce this new technology.

The following are some observations on what operator characteristics are important in the selection of the <u>first</u> operator in a CAD installation:

1. <u>A General Knowledge of the Design Process</u> - Implementing CAD within any design group requires full knowledge of the workshop conventions. Working with vendor staff, the operator will have to be able to recognize the subtle difference between today's conventions and the workings of the new CAD system.

2. <u>Experience in Drafting</u> - An eye for fine details and aesthetics is important. This comes with experience in drafting.

3. <u>Good Communication Skills</u> - The first operator trained is likely to be involved in assisting or training other users. An operator who has the ability to communicate will not only learn quickly, but will be able to convey the concepts of computer design to other workers, management, marketing staff and customers. They will be effective in demonstrating the uses of the system to clients.

4. <u>Exposure to Computers</u> - Although not required, it is desirable for the first operator to have some exposure to computers. If they have used a home computer, they may have a higher interest level and may be more motivated in mastering the system in a shorter period of time. They will also be less intimidated by the new technology.

5. <u>Flexibility and Adaptability</u> - The ability to quickly adapt to new tools and give up old established routines is a quality that will lead to faster learning and reduced user resistance.

6. <u>Organized and Patient</u> - A well-organized and patient person is better able to learn the large number of new conventions faster. They will also make better instructors as more designers learn this new tool.

7. <u>A Proper Training Environment</u> - It is important to remove as many of the daily responsibilities and time commitments while the operators are in training. The operators should be allowed to devote full-time attention to this new tool. Being required to demonstrate CAD to clients before they have gained proficiency may impede the learning process. Thus, it is managment's responsibility to insure that the early periods of training are conducted as to promote rapid installation.

Figure 10.8. Attributes of a good CAD operator. (Courtesy of Arrigoni Computer Graphics, Los Gatos, California.)

1. *Symbols*: All systems come with a variety of available symbols. These should be reviewed in advance, and their application should be understood by all. Creation of new symbols should be governed by some type of review.
2. *Templates*: Of machinery and equipment will need to be created. A standard approach should be observed. It should indicate the

Questionnaire

(Designed and administered by Cummins Engine Company, Inc., Columbus, Indiana, Kenneth K. Durham.)

We are embarking on a pilot project to determine the usefulness of a computer-aided design and drafting (CADD) and computer-aided manufacturing (CAM) system.

We will need senior drafters and designers to participate in the pilot project. The success or failure of this development project will depend heavily on the attitude and abilities of the people who will be using the equipment. The final selection of the members for the development team will be made by: _____

The following questionnaire was prepared to give you an idea of the requirements of this project as we see them now, and to provide us with information to help select a team:

- Why are you interested in the CADD system?
- In order to use the system successfully, the CAD/CAM system operator will have to be enthusiastic, adaptable, creative, open to new ideas, and capable of working with a minimum of supervision.
  - Do you feel you fit this description? _____ Why?
- One of the objectives of developing a CADD system is to improve our productivity.
  - Are you willing to work under this pressure? _____
  - Are you willing to have your productivity measured on this system? _____
  - Are you willing to accept your removal from the development team for productivity-related reasons? _____
- Introduction to this system will require several weeks of training, both here and at the supplier's sites. Typically, each user will spend one week at the supplier's factory site, one week here, and possibly several 2-5 day periods at the supplier's district office. The first users of the CADD system would be expected to help in the training of later users and conducting system demonstrations, should the need arise.
  - Are you willing to travel for as much as a week at a time for training purposes? _____
  - Are you willing to accept your removal from the development team, during training, based on the recommendation of the supplier? _____
  - Are you willing to train others? _____
  - Are you willing to demonstrate the system to visitors? _____
  - Are you willing to travel occasionally to attend related meetings, if asked? _____
- Since this is a pilot system, we have little experience in how the users should approach their work. Much of the organization of the project and the procedures will be undefined at the start, and will have to be developed by the development team.
  - Are you willing to work with a group to help solve problems, resolve conflicts and develop CADD standards and procedures? _____
- Initially, the working conditions may also be ill-defined. We are looking for a place which will house the system. We are also considering several alternative working schedules. With an eye toward maximizing the utilization of the system...
  - Are you willing to be located away from your department? _____
  - Are you willing to work shifts other than the first shift? _____
  Preferred shift (1, 2, or 3)? _____
  2nd preference? _____
  - Within any shift, are you willing to be flexible about your starting and ending times? _____
- As mentioned previously, the success of this system will depend heavily on the dedication of the members of the implementation team. Part of this dedication includes a commitment to staying with the project to minimize turnover. Note that participation in this project will not restrict your normal promotional prospects within your department.
  - Are you willing to stay with this project for at least two years or until the termination of the project, whichever occurs first? _____

Figure 10.9. Questionnaire for screening CAD operators. (From *The CAD/CAM Handbook*. Bedford, Mass.: Computervision Corporation, 1980.)

1. Full Time Operation–	A minimum of two weeks with five to six hours a day directly following the training session. This assumes that other work responsibilities have been removed.
2. Low Stress Environment–	The operator should be free to work on the system without interruptions. A test project should be selected, rather than an important or urgent job. The operator should not be required to give demonstrations during the learning process.
3. Structured Task Program–	A work program should be devised that will systematically bring the operators up to full proficiencies and familiarize them with the potential uses of the system.
4. Shop Conventions Defined–	Drafting and filing conventions should be worked out in advance, but it is assumed that some time will be spent in developing new conventions.
5. Encouraged to Call for Help–	The operator should be encouraged to call the vendor on a routine basis in the beginning, for questions on the functions of the system, shop conventions and applications support. A problem log should be maintained to assist vendor staff in identifying the source of the problem.

Figure 10.10. Guidelines for an environment conducive to training. (Courtesy of Arrigoni Computer Graphics, Los Gatos, California.)

information to be shown and the layer in which it is to be contained. Use of color, line weights, and textures should also be addressed.

3. *Line weights*: Should be standardized in various contexts and related to pen numbers when plotting.

4. *Textures*: Should also be standardized in various contexts such as plant layout, site planning, and building components in working drawings. (The International Materials Management Society has adopted a shading or texturing code for layout planning which lends itself well to use on CAD systems. This code appears in Muther and Hales, *Systematic Planning of Industrial Facilities, Volume I*, together with a code for site planning (8).)

5. *Color*: Should be controlled and standardized in all contexts. Color communicates, but its value is quickly lost if it is applied at random. The layout and site planning codes just described also come in color and are cross-referenced to textures.

6. *Lettering*: Both height and font should be prescribed in various contexts. This is necessary to achieve uniform appearance.

7. *Overlays*: Are a critical area for standardization. Operators must be constrained to put various sets of information into common layers. Without this discipline much time is lost and confusion created. One of the productivity benefits of CAD will be quickly lost.
8. *Archiving*: Will be necessary sooner or later as data storage is consumed. Successful archiving requires discipline and control over the naming and numbering of drawings. Remember, the computer must find them on tape or disk. This is not the same as rummaging through a flat file.

The best way to address these issues is with a CAD procedures manual begun in advance of installation and expanded as experience is gained with the system. This should be a primary task of the supervisor. Starting without standard procedures is a certain prescription for starting over.

The first live projects attempted on the system should be selected with care. For motivational reasons the first project needs to be a success. Start with small, simple projects that do not tax an operator's knowledge of the system. The drawings should contain numerous repeated elements, and the project should not be urgent. The first uses of a CAD system will often take longer than manual methods. In time, as familiarity increases and more standard symbols and details are loaded, speed will increase.

The supervisor should be sensitive to scheduling, especially in the initial start-up period. Operators may experience eye fatigue and nervousness. Some operators may prove to be better than others at certain applications. Work should be assigned accordingly. With manual methods, the mind of the draftsman, designer, or planner is more free to wander. His pencil does not prompt him for the next input as soon as a line is drawn. On the other hand, the computer does not tire. It always awaits the next command. The more creative the task, the more the CAD system may drain the energy of the user. It will take several months before an operator can set a proper pace and vary it depending upon the task to be performed.

Common Misconceptions

Less than 10% of facilities planners today (1983) have had any direct experience using CAD. Yet most of us have read and heard a great deal about it in professional journals and groups. Not surprisingly, some misconceptions have emerged. The most common include:

1. *CAD systems are used intensively in the design stage of a facility.* Not really. CAD systems are primarily used for drafting working drawings and detailed layouts.
2. *Low-cost systems are acceptable for working drawings and detailed layouts.* Keep in mind that these are the most taxing applications for a CAD system. If the low cost has been achieved at the expense of memory, data storage, and response, such systems may prove inadequate for detailed planning tasks. On the other hand, they may be great for the early conceptual planning phases.

3. *The smaller facilities department cannot support or justify CAD.* On the contrary, we have seen several approaches which are low in cost but still provide the usual high payoff of CAD. Use of an existing mainframe or mini, or obtaining time on an existing CAD system are potentially good approaches. And the low-cost turnkey systems should be given serious consideration.

4. *CAD systems are easy to use.* Some are easier than others, but they are all a great deal more difficult than other, common uses of computers. No system works by itself. CAD systems are no exception. A great deal of time and effort is required to become proficient.

5. *CAD systems require a full-time computer expert to operate them.* Large systems using central mini-computers may require a full-time supervisor. However, there are plenty of intelligent workstation systems that can be mastered and maintained by a well-motivated draftsman.

6. *CAD systems are the best way to begin a computer-aided approach to facilities planning and management.* For most of us, CAD should probably represent the final stage of computer support, not the first. Decision support systems using personal and microcomputers offer a much lower cost and lower risk way to introduce computerization.

As with any form of automation, CAD systems are occasionally misapplied. Some turn out to be failures. Hopefully, by eliminating misconceptions, we can increase our chances for a sound, successful application of computer-aided design.

11

Managing Computer Resources

Common Failures

Far too many computer systems fail to return the money invested in them. When this happens, the problem is usually found in management--not in hardware or software. So, in this closing chapter, we will dwell on effective management in the hope that facilities planners might avoid the more common failures associated with computers.

What are the most common failures? Experience shows the following ten to be very "popular":

1. *Lack of management commitment*: leading to frustration, premature complaints from management, and often to withdrawal of support.
2. *Choosing the wrong application*: installing a particular computer aid because it is a fad, or "Company 'X' did it" or it "looks like a good place to start" only to discover that results are more elusive than planned.
3. *Failure to write a specification*: reacting instead to product descriptions prepared by the vendor. Discovering too late that the computer aid does not really accomplish the intended tasks.
4. *Overestimating benefits*: assuming unrealistic productivity ratios (4:1, 8:1, etc.), also that people can be eliminated or new hiring avoided.
5. *Underestimating costs*: focusing on quoted, out-of-pocket expenses for hardware and software--overlooking the internal "costs" associated with search, selection, installation, training, and maintenance (not to mention any internal software costs).
6. *Expecting results too soon*: assuming the learning curve is what the vendor says it will be (or overlooking the learning curve altogether).
7. *Inadequate personnel*: assigning inexperienced, poorly trained and poorly motivated people to key positions.
8. *Failure to document software*: especially internally developed code. Also, settling for whatever the vendor supplies no matter how inadequate. The result is poor utilization of the computer.

```
 1.  Get management involved.

 2.  Choose a key application.

 3.  Write a detailed specification.

 4.  Make a thorough cost/benefit analysis.

 5.  Pay attention to maintenance.

 6.  Assign qualified people.

 7.  Pay attention to training.

 8.  Document all software.

 9.  Always audit results.

10.  Plan for growth and improvements.
```

Figure 11.1. Prescription for successful computer aids.

9. *Failure to audit results*: repeating the same mistakes over and over, as additional aids are acquired.
10. *Outgrowing the system*: a problem when it occurs in less than 3 to 5 years. Usually related to a poor specification.

Reversing these failures, and expressing the issues in a positive way, we can make the prescription for successful computer aids shown in Figure 11.1.

Getting Management Involved

Major investments in computer aids need the endorsement of a key line manager--general manager, plant manager, or administrative or divisional executive. These are the beneficiaries of facilities planning and management--the true "end-users" of our computer support. Typically they control the purse strings for computer aids. Before allocating funds, they will usually ask themselves these key questions:

1. Do we understand the proposed application?
2. Do we agree that it is important?
3. Do we trust the facilities staff to achieve the intended result? Are they committed?
4. Do the cost estimates appear thorough?
5. Do the benefits claimed look reasonable?
6. Are they quantifiable? Will we be able to see them? (How?)
7. Are there some key decision points (go/no-go) before we have to "buy the farm?"

The facilities planner or manager should work for a strong "yes" to every one of these questions. Most projects take longer than expected to show results. And sometimes additional funds are required for special software, extra hardware, extra training, etc. When this happens, line management's patience may wear thin, especially if the initial approval was made with reservations. On major projects--more than a few thousand dollars--a phased approach is best, with final commitment of funds made only after a testing or evaluation period.

Choosing a Key Application

Acquiring, installing, and de-bugging computer aids is a time-consuming, risky process. Most facilities planners are overworked to begin with. For these reasons, planners should pursue only those applications that are truly vital to their organization.

In earlier chapters we concluded that first priority should be given to decision support technology applied to the early phases of a project--preplanning, orientation, overall plan.

Let us be more general here and state that key applications should do at least three of the following things:

1. Improve control of facilities (land, buildings, and equipment) by providing information not currently available
2. Improve the accuracy of information already available
3. Aid line management in making frequent, routine decisions
4. Reduce clerical errors and free the planner for more productive work
5. Handle increased workloads

Be very suspicious of applications whose primary purpose is to eliminate people or avoid additional hiring. These generally fail to deliver. The facilities planner should focus instead on making better plans. The results will be of far greater benefit to his organization.

Writing a Specification

We have covered this subject in earlier chapters. It is so important that a reminder is in order. If the facilities planner is unable to write a clear specification, he is unlikely to achieve useful results. Only by writing the specification can existing problems or issues be properly defined. If the specification is vague, both vendors and internal programming staffs will inflate their estimated costs. Still worse, however, is the risk that they may promise results that cannot be delivered.

Making a Cost/Benefit Analysis

This subject, too, has been covered in earlier chapters. Still, some final thoughts are in order, expressing management's perspectives. First, the

facilities planner should set his sights on a very high return--at least 3 to 4
times the total funds invested over a 5 year period. This way, the computer
aid will be a winner--even if the costs are higher and benefits lower than expected.

Second, the planner should always show three sets of cost and return figures: high, low, and most probable.

Third, returns or results should be as measurable as possible with target
dates for achievement, specific people held accountable, and a plan for audit-
ing performance.

On major investments such as a turnkey system, look beyond hardware
costs. Look at software, conversion, preparation, and training. Also look
at operation and maintenance over the life of the system--at least 5 years.
And don't overlook the cost of the initial study and specification. Taken together, do not be surprised if total, 5-year costs are 4-5 times the cost of
the hardware.

Don't count on benefits coming too soon. Expect almost none in the first
6-12 months. On internal development projects, 50%-75% of the total cost may
be incurred before any results are seen. This is especially true when developing large data bases or converting to CAD.

Paying Attention to Maintenance

Maintenance contracts should be taken on all computer aids--both hardware
and software. The monthly cost will generally fall between 0.5% and 2% of
purchase price.

An outlay of 0.5%-1.0% per month typically purchases an extended war-
ranty. Service is provided at the vendor's location, with the defective device sent back and forth by low-priority mail. For 1% to 1.5%, service can
usually be found locally on a carry-in basis. Full, on-site service is hard
to find for less than 1.5%-2.0% per month.

Maintenance expenses can be reduced by limiting hardware coverage to
mechanical devices--printers, plotters, tape or disk drives--on the theory
that these are most likely to give trouble. In your author's experience, full
maintenance, especially on software, is a worthwhile investment. If the benefits of the computer are not great enough to justify the cost of full maintenance, the whole investment is probably a waste. Further, if the computer
aid is heavily or frequently used (as it should be), one should question
whether mail-in or carry-in service will be adequate (51).

Assigning Qualified Personnel

Without people, hardware and software accomplish nothing. People are the
key ingredient in successful computer aids. If the development or investment is significant, management should choose the team carefully. People
who succeed with computer aids are patient self-starters, willing to work
extra hours to master and de-bug software. But they must have more than
desire. They need prior experience either with computers, the applications,
or, ideally, with both. Projects that are staffed and led by first-time computer

users can succeed, but those involved should expect some problems and delays along the way.

Too often, management expects relatively unskilled or inexperienced staff to become immediately more effective by using the computer. Clerical staff are expected to become draftsmen by the use of CAD. Or a draftsman is expected to become an analyst with the use of a decision support system. In general, computer aids will not turn a junior-level staffer into one of senior caliber. Only time and experience will.

Paying Attention to Training

When computers or software are first purchased, there will usually be some initial training provided by the vendor. If this is the only formal training ever provided to the facilities planners, then problems are almost certain to arise. Those trained initially will inevitably move on to other locations, assignments, or employers. Those who come after will get only hit-or-miss training by the existing users. Fewer and fewer staffers will really know how to use the available computer aids. In time, one or two people will have emerged as the "operators" and the rest of the staff will be dependent on them. To avoid this turn of events, managers should take these actions:

1. Conduct an extensive initial training program for as many staffers as possible. Consider paying vendors for extra training.
2. Assign initial users who are good communicators, capable of training others.
3. Have the initial users develop a tailored training program around "real-life" company projects. Use current facilities data, and realistic case problems.
4. Insist that new staffers become proficient with the computer, with each person completing the test case problem.

Most planners will say they are too busy to develop and conduct this type of training. By the same token, most planners are not getting full staff-wide use from their computer aids.

Document All Software

Documentation is closely related to training. Both are essential to full, continued use of computer aids. Look around you. How many people are doing the same job they were 2-3 years ago? Now look at your existing computer aids. Do they function as they did when first installed? How do you know? Look at your computer-generated reports. Can everyone who uses them explain what each entry means? How it is obtained? Who provided the source data? How correct are their answers? How correct are yours?

Most computer systems are learned by informal, word-of-mouth explanations. These are usually given by people who did not write the software and probably got their information second- or thirdhand. In fact, it is a rare staff group that has more than one or two people who have read the system documentation. Internally developed computer aids often have none. And the most poorly documented program may be one of the facilities planner's own--developed in haste on his personal computer.

Managers of facilities groups should not let this situation occur. Those who are given permission to develop or acquire computer aids should be required to produce documentation. It should be intelligible to first-time users, even if this means writing a summary of vendor-supplied documentation.

Always Audit Results

Managers should make it known, in advance, that each significant investment in computer aids will be audited, to see if the investment was warranted. On major projects, those conducting the audit should be independent of the team that developed or installed the computer system in question.

By announcing the audit at the outset, those involved are aware that results will matter. And it is a forthright way to avoid any feeling of a witch-hunt if the investment turned out to be a mistake.

Evaluations should be made against the project's original objectives, whether or not they remain current. It is unfair to judge against a new situation which may not have been foreseen. The purpose of an audit is to learn from success and from failure, so that we may avoid mistakes in the future.

Plan for Growth and Improvement

Few, if any, computer aids remain as they were initially installed. It would be disappointing if they did. With use, the planner will discover improvements, extensions, and enhancements to both software and hardware. While it may solve a particular current problem, each new aid in fact offers a beginning, a new approach, with many more ideas and uses to follow. Like the process of facilities planning itself, there is always more to learn.

Part IV

HOW TO LEARN MORE

Appendix A

Information Sources

Computer-aided facilities planning is an active, growing field. Several hundred vendors offer software, hardware, and turnkey systems of interest to facilities planners. With new vendors and new products appearing all the time, how can the facilities planner keep up to date?

There are five ways to stay informed. In order of increasing cost, they are: (1) periodicals, (2) newsletters, (3) directories, (4) conferences, and (5) seminars.

Periodicals are the least expensive (occasionally free) way to keep up to date. Feature articles will often include case histories and more general "how-to" discussions of computer use. Periodicals are also a good way to keep up with trends in hardware. Most publish an annual directory issue listing vendors, their products, and base prices. Read the advertisements as well. Many ads for hardware and software are educational in nature and can help you master terminology, trade names, and trends. Ads will also alert you to directories, conferences, and seminars that may be of value.

Newsletters generally cost between $50 and $150 per year. Their advantage over periodicals is their timeliness, depth, and exclusive focus on specific computer applications. Newsletters offer "insider" information and assessments of specific vendors and products. Some of it is gossip, but most is very informative. Often, newsletter editors have business relationships with periodicals and conference sponsors. As a general rule, they print their best material first in their own newsletter. It may appear later in periodicals or form the basis of a presentation at seminars and conferences.

A variety of directories can be obtained for prices of $25 all the way up to $200 or more. Directories can be a useful starting point for a serious product search. However, because of the lead time to produce a directory, the product descriptions may be out of date before the directory reaches your shelf. Because directories are constructed from other directories and mailing lists, there will always be some vendors who are overlooked. Moreover, not every vendor chooses to be in every directory. Therefore, the planner should not reject a potential vendor simply because it was not listed.

Conferences and trade shows are a useful but expensive way to see a variety of products first hand. With travel expenses, it typically costs $500 to $1000 to attend a distant conference or show. Still, if the planner expects to spend $25,000 or more on a computer system, the expense may be worthwhile. Conferences are also a good place to meet and assess various industry consultants. There will typically be two or three invited to speak in conjunction with the show.

Seminars, like directories, are a valuable starting point for a major product search. They are a useful place to learn from vendors, other planners, and consultants. Seminars provide a distraction-free, concentrated dose of information on products, costs, trends and implementation.

Vendor-sponsored seminars are free, except for travel expenses, and provide an in-depth look at the vendor's product. More importantly perhaps, the planner has a chance to evaluate the vendor's personnel, to see if they really understand his problems and requirements.

Consultant-sponsored seminars are expensive--$100 to $200 per day plus travel expenses. Consultant's seminars are usually of high quality but at a general level. They are best as a forum for learning about specific techniques or about implementation success factors. Watch out for consultants who evaluate vendors' products in their seminar. Such evaluations are mere generalizations. They cannot possibly be definitive, and they are not made with reference to the planner's specifications. Some consultants have a financial interest in or close friendship with certain vendors. Unfortunately, it is difficult for the planner to recognize this and discount any bias the consultant may have.

Association- and university-sponsored seminars may be packages conducted by professors or consultants under contract to the sponsor. Or they may provide a forum in which facilities planners (not consultants or professors) speak about their experiences with computers. While the speakers may vary in quality, such meetings often provide the most objective information available.

The lists below are not definitive. They have been constructed from first-hand knowledge and experience during three years of steady research. Each entry leads to still other information sources, so the reader will not be wanting for long by using these limited lists.

Newsletters

A-E-C Automation Newsletter: E. Forrest, P. E., editor and publisher; 7209 Wisteria Way, Carlsbad, CA 92008. Monthly. Covers CAE and CAD. Hardware-oriented.

A/E Systems Report: George Borkovich, editor; MRH Associates, Inc., publisher; P.O. Box 11316, Newington, CT 06111. Monthly. Primarily CAD hardware, software, and implementation.

Anderson Report: 4505 E. Industrial #2J, Simi Valley, CA 93063. Monthly. Primarily graphics computers--people, companies, and product announcements.

CCAN--Construction Computer Applications News: Construction Industry Press, 1105-F Spring Street, Silver Spring, MD 20910. Monthly. Covers CAE and MIS for construction contractors and managers.

Computer-Aided Design Report: 711 Van Nuys Street, San Diego, CA 92109. Monthly. Covers CAE and CAD. In-depth assessment of specific vendor offerings.

ECAN--Engineering Computer Applications News: 5 Denver Tech Center, P.O. Box 3109, Englewood, CO 80155. Monthly. Covers CAE, CAD, MIS; software, hardware, and implementation.

F-M Automation Newsletter: Automation Group, Inc., 9501 West Devon, Suite 203, P.O. Box 507, Rosemont, IL 60018. Monthly. Use of computers and computer graphics in managing facilities.

S. Klein Newsletter on Computer Graphics: Technology & Business Communications, Inc., 730 Boston Post Road, P.O. Box 89, Sudbury, MA 01776. Biweekly. Covers full range of computer graphics applications. Occasional articles on facilities planning.

Periodicals

Building Design & Construction: Cahners Publications, 1750 S.W. Skyline Blvd., Box 6, Portland, OR 97221. Monthly. Regular coverage of CAD for architects and engineers.

Business Computer Systems: Cahners Publishing Company, Div. of Reed Holdings, Inc., 221 Columbus Avenue, Boston, MA 02116. Monthly. Products and trends in personal and microcomputer systems.

Computers for Design and Construction: Meta Data Publishing Corp., 310 East 44th Street, New York, NY 10164. Bimonthly. CAE and CAD; software, hardware, and case histories.

Computer Graphics World: PennWell Publications, 1714 Stockton Street, San Francisco, CA 94133. Occasional coverage of architectural CAD applications.

Facilities Design and Management: Gralla Publications, 1515 Broadway, New York, NY 10036. Monthly. Regular coverage of computer aids for planning administrative and institutional facilities.

Industrial Engineering: Institute of Industrial Engineers, 25 Technology Park / Atlanta, Norcross, GA 30092. Monthly. Occasional coverage of computer aids for planning industrial facilities.

Directories

CCAD--Construction Computer Applications Directory: Construction Industry Press, 1105-F Spring Street, Silver Spring, MD 20910.

CAD/CAM, CAE--Survey, Review and Buyer's Guide: Daratech, Inc., 16 Myrtle Avenue, P.O. Box 410, Cambridge, MA 02138.

Datapro Directory of Software: Datapro Research Corp., 1805 Underwood Boulevard, Delram, NJ 08075.

Major Software Sources for Consulting Engineers: American Consulting Engineers Council, 1015 15th Street, N.W., Suite 802, Washington, D.C. 20005.

McGraw-Hill A/E Computer Systems Update: McGraw-Hill Book Company, 1221 Avenue of the Americas, New York, NY 10020.

Conferences (listed by sponsor)

A/E Systems: Conference on Automation and Reprographics in Professional Design Firms; annually each spring; location varies. Contact A/E Systems Report, P.O. Box 11316, Newington, CT 06111.

Gralla Conferences: Computer-Aided Space Design and Management Conference, annually each fall in New York. Contact Gralla Conferences, 1515 Broadway, New York, NY 10036.

Industrial Development and Research Council, Inc.: Spring and Fall conferences; locations vary. Contact IDRC, 1954 Airport Road, N.E., Atlanta, GA 30341.

International Facilities Management Association: Annual Conference and Trade Show; each fall; location varies. Contact IFMA, 3971 South Research Park Drive, Ann Arbor, MI 48104.

World Computer Graphics Association: Computers/Graphics in the Building Process; annually; conference location varies. Contact WCGA, 2033 M Street, N.W., Suite 250, Washington, D.C. 20036.

Seminars

Check the calendar sections of periodicals and newsletters for current seminars of interest. The following organizations give occasional seminars on computer-aided facilities planning:

A/E Systems Report: P.O. Box 11316, Newington, CT 06111.

Facilities Management Institute: (Division of Herman Miller) 3971 South Research Park Drive, Ann Arbor, MI 48104.

Gralla Conferences: 1515 Broadway, New York, NY 10036.

Institute of Industrial Engineers: 25 Technology Park/Atlanta, Norcross, GA 30092

Appendix B

Consultants

Good consultants typically belong to an association that monitors standards of conduct and performance. One such group is the Independent Computer Consultants Association, P.O. Box 27412, St. Louis, MO 63141. This group requires its members to reveal any and all interests in software or hardware products whether offered by the consulting firm or another vendor in whom the consultant holds an interest. (This is not an uncommon situation in the field of software development and computer consulting.) Members must also decline assignments which they cannot reasonably expect to complete with professional competence.

Another similar organization is the Institute of Management Consultants, Inc., 19 West 44th Street, New York, NY 10036. This group certifies individual consultants and attests to their competence in their chosen specialties. The IMC prohibits its members from accepting fees, commissions, or other considerations from vendors whose services or products they might recommend during the course of an assignment. Members are also pledged to decline an assignment of such limited scope that they could not serve the client effectively. This touches on an important issue involving consultants in general: their ability and value as advisors on facilities planning and management, quite aside from the issue of computer aids. There are plenty of facilities groups that could benefit more from reorganization or better methods than from computer aids. A good consultant will discover and point out such opportunities, even if it cuts the initial assignment short.

No consultant should be retained without first providing a written description of the work to be performed and a firm estimate of the total fees to be incurred. Advice on computer-aided facilities planning is not high-level management consulting. It is technical consulting to middle management. As such, daily fees (1983) should range between $500 and $800 per day, not including expenses.

Choosing a consultant can be almost as difficult as choosing a computer. Experience, satisfied clients, and objectivity are the most important selection criteria. The firms listed below are knowledgeable on project management systems, CAD, and microcomputers for architectural and facilities applications. Principals of these firms are authorities in the field. They publish and lecture frequently at conferences and seminars. Their listing here, however, does not constitute an endorsement of their consulting abilities. For this, the reader should conduct a personal interview and check any references which are offered.

Design & Systems Research, Inc.
Dan Raker, President
55 Upland Road
Cambridge, MA 02140

H. Lee Hales, Inc.
Management Consultant
5907 Lodge Creek
Houston, TX 77066

G. Anthony DesRosier
1217 Wagon Wheel Road
Hopkins, MN 55343

Robert J. Krawczyk
5219 N. Lockwood Avenue
Chicago, IL 60630

Graphic Systems, Inc.
Eric Teicholz, President
180 Franklin Street
Cambridge, MA 02139

O'Connor Consulting, Inc.
Timothy O'Connor, P.E.
28630 Southfield Road, Suite 234
Lathrup Village, MI 48076

Engineering Computer Applications, Inc.
Kenton H. Johnson, P.E.
P.O. Box 3109
Englewood, CO 80111

Orr Associates, Inc.
Joel Orr, President
21 Chambers Road
Danbury, CT 06810

Facility Management Institute
3971 South Research Park Drive
Ann Arbor, MI 48104. (A division
of the Herman Miller Corporation)

Practice Management Associates, Ltd.
David Rinderer
126 Harvard Street
Brookline, MA 02146

Guidelines
Fred Stitt
P.O. Box 456
Orinda, CA 94563

Michael P. Sherman, Ph.D.
2202 East Vermont
Urbana, IL 61801

The following organizations have a great deal of expertise in computer-aided facilities planning. While they provide software and turnkey systems, this does not preclude the interested planner from retaining them as sources of up-to-date advice.

BASICOMP, Inc., 1345 East Main, Suite 209, Mesa, AZ 85203. This is a software house that specializes in interior design and office space planning applications. The staff members are experienced and capable of consulting in these areas.

Massachusetts Institute of Technology, Office of Facilities Management Systems, Room E19-451, 77 Massachusetts Avenue, Cambridge, MA 02139. This is an administrative arm of M.I.T. that supports outside users of the M.I.T. INSITE system. Staff members may be consulted on large-scale information systems, bar-coding, and data bases for facilities management.

Mitchell, Reeder & Hammer--The Computer-Aided Design Group, 2407 Main Street, Santa Monica, CA 90405. This consulting firm specializes in advice to architects and facilities planners where computers are concerned. The firm also sells software for facilities planning and management.

Moore Productivity Associates, 1607 Greenwood Drive, Blacksburg, VA 24060. This firm provides software for layout and facilities planning. The president, James M. Moore, is an authority on computer aids.

PLANIMETRON, 129 Nyac Avenue, Pelham, NY 10803. Consulting, software, and microcomputer systems for many aspects of facilities planning and management.

References

The references below are organized into three lists, one for each part of this book.

Part I--How Facilities are Planned

1. Apple, James M. *Plant Layout and Material Handling*, New York, NY, John Wiley and Sons, 1977.

2. Francis, R. L., and J. A. White. *Facility Layout and Location*, Englewood Cliffs, NJ, Prentice-Hall, 1974.

3. Gallagher, C. C., and W. A. Knight. *Group Technology*, London, England, Butterworth, 1973.

4. Lewis, Bernard T., ed. *Management Handbook for Plant Engineers*. New York, NY, McGraw-Hill, 1977.

5. Lewis, Bernard T., and J. P. Marron, eds. *Facilities and Plant Engineering Handbook*. New York, NY, McGraw-Hill, 1973.

6. Muther, Richard. *Systematic Layout Planning (SLP)*, 2nd ed. Boston, MA, Cahners Books, 1973.

7. Muther, Richard, and Knut Haganaes. *Systematic Handling Analysis (SHA)*, Kansas City, MO, Management and Industrial Research Publications, 1969.

8. Muther, Richard, and Lee Hales. *Systematic Planning of Industrial Facilities (SPIF)*, Vols. I and II. Kansas City, MO, Management and Industrial Research Publications, 1980.

9. Saphier, Michael. *Planning the New Office*. New York, NY, McGraw-Hill, 1978.

Part II--How Computers Can Help

10. Buffa, E. S., G. C. Armour, and T. E. Vollman. "Allocating Facilities with CRAFT." *Harvard Business Review*, Vol. 42, No. 2, Mar.-Apr. 1964.

11. Burger, Amadeus M. "Computerized Plant Modeling--Practical Experiences." Construction System Associates, Inc., Marietta, GA, April 1983.

12. Carrie, A. S. "Computer-Aided Layout Planning--The Way Ahead." *IE News*, Facilities Planning and Design Division, Institute of Industrial Engineers, Winter, 1982.

13. "Computer-Aided Everything." *Engineering News-Record*, Dec. 3, 1981, reprint.

14. "Computers and the Building Team." *Building Design and Construction*, 1982, collection of reprints.

15. Cytryn, Allan, and William H. Parsons. "Planning ADES, a system for Computer-Assisted Space Planning." *Interior Design*, May 1977.

16. Dosset, Royal. "Choosing Floppy Disk Microcomputers: A Checklist of Features for IE Applications." *Industrial Engineering*, Oct. 1981.

17. Dosset, Royal. "Personal Computers" What They Can Do for IEs and How to Choose One." *Industrial Engineering*, Oct. 1982.

18. Hales, Lee. "Computer-Aided Facilities Planning--A Brief Survey of Current Practice." *Proceedings of the Numerical Control Society*. Annual Technical Conference, Los Angeles, CA, 1979.

19. Hales, H. Lee, and Harvey C. Jones, Jr. "Facilities Decision Support." Master's thesis, Massachusetts Institute of Technology, 1980.

20. Horwitt, Elizabeth. "When More May Not Be Better." *Business Computer Systems*, Feb. 1983.

21. Hosni, Yasser A., and Timothy S. Atkins. "Facilities Planning Using Microcomputers." *Proceedings of the 1983 Annual Industrial Engineering Conference*, Institute of Industrial Engineers, Norcross, GA.

22. "Integrated system combines design drafting." *Building Design & Construction*, July 1983.

23. Jacobs, Robert F., John W. Bradford, and Larry P. Ritzman. "Computerized Layout: An Integrated Approach to Special Planning and Communications Requirements." *Industrial Engineering*, July 1980.

24. Khator, Suresh, and Colin Moodie, "A Microcomputer Program for Plant Layout." *Industrial Engineering*, Vol. 15, No. 3, March 1983.

25. Lee, Kaiman. "Computer Programs for Architects and Layout Planners." *Proceedings of the American Institute of Industrial Engineers*, National Conference, Boston, MA, 1971.

26. Lee, R. C., and J. M. Moore. "CORELAP-Computerized Relationships Layout Planning." *Journal of Industrial Engineering*, Vol. 18, No. 3, March 1967.

27. Liggett, Robin S., and William J. Mitchell. "Optimal Space Planning in Practice." *Computer-Aided Design*, Vol. 13, No. 5, Sept. 1981.

28. MacDonald, Joseph A., ed., *Computers for Design & Construction, Handbook 82/83* (Vol. 1, No. 1). New York, NY, MetaData Publishing, 1983.

29. Machover, Carl, and Robert E. Blauth, eds. *The CAD/CAM Handbook*. Bedford, MA, Computervision Corp., 1980.

30. Martin, James. *Principles of Data-Base Management*. Englewood Cliffs, NJ, Prentice-Hall, 1976.

31. Mitchell, William J. *Computer-Aided Architectural Design*. New York, NY, Mason/Charter Publishers, 1977.

32. Moore, James M. "Computer-Aided Facilities Design: An International Survey." *International Journal of Production Research*. Vol. 12, No. 1, Jan. 1974.

33. Moore, James M. "Computer-Aided Facilities Design: Help, Hoax or Hex." *Proceedings of the American Institute of Industrial Engineers*. Systems Engineering Conference, 1976.

34. Moore, James M. "Computer Program Evaluates Plant Layout Alternatives." *Industrial Engineering*, Vol. 3, No. 8, Aug. 1971.

35. Moore, James M. "Summary of Questionnaire on the Use of Computers in Facilities Design." *AIIE Newsletter*, Facilities Planning and Design Division, American Institute of Industrial Engineers, Sept. 1976.

36. Moore, James M. "Who Uses the Computer for Layout Planning?" *Proceedings of the 4th International Conference on Production Research*. London, England, Taylor & Francis, 1978.

37. Moore, James M., and Yasser A. Honsi. "Can CAD/CAM?" *IE News*, Facilities Planning and Design Division, Institute of Industrial Engineers, Summer 1983.

38. Muther, Richard, and Kenneth McPherson. "Four Approaches to Computerized and Layout Planning." *Industrial Engineering*. Feb. 1970.

39. Ritzman, L. P. "The Efficiency of Computer Algorithms for Plant Layout." *Management Science*, Vol. 18, No. 5, Jan. 1972.

40. Ryan, Daniel L. *Computer-Aided Architectural Graphics*. New York, NY, Marcel Dekker, 1983.

41. Sato, Keiichi, and Charles L. Owen. "A Prestructuring Model for System Arrangement Problems." *Proceedings of the 1980 Design Automation Conference*, Association of Computing Machinery, New York, NY.

42. Seehof, J. M., and W. O. Evans. "Automated Layout Design Program." *Journal of Industrial Engineering*. Vol. 18, No. 12, Dec. 1967.

43. Tompkins, James A. "Computer-Aided Plant Layout." *Modern Materials Handling*, seven-issue series. May-Nov. 1978.

44. Tompkins, J. A., and J. M. Moore. *Computer-Aided Layout: A User's Guide*. American Institute of Industrial Engineers, Norcross, GA, 1978.

Part III--How to Achieve Success

45. DesRosier, G. Anthony. *Implementing an Interactive Computer Graphics System: Some Initial Thoughts*. Thorndale, PA, GSB Associates, 1981.

46. Keen, Peter G. W., and Michael S. Scott Morton. *Decision Support Systems: An Organizational Perspective*. Reading, MA, Addison-Wesley, 1973.

47. "Managing and implementing a computer system," *Building Design & Construction*, July 1983.

48. "Microcomputers provide entry to computerization," *Building Design & Construction*, July 1983.

49. Nilsson, John. "Analysis of Cost Benefits of Computer-Aided Design Systems" (unpublished paper). Southboro, MA, Decision Graphics, Inc., 1980.

50. Nilsson, John. "Cost Benefit Analysis in Computer Graphic Systems" (unpublished paper). Southboro, MA, Decision Graphics, Inc., 1980.

51. Raker, Daniel S. "Can Your Vendor Support Your Low-Cost CAD System?" *Plan and Print*, Vol. N-55, No. 10, pp. 8-11.

52. "Selecting a computer system." *Building Design & Construction*, July 1983.

53. Stasiowski, Frank A. *How to Select Data Processing Systems*. Dedham, MA, Practice Management Associates, 1981.

Index